Physics of Gravitating Systems II

Nonlinear Collective Processes.
Astrophysical Applications

A. M. Fridman
V. L. Polyachenko

Physics of
Gravitating Systems   II

Nonlinear Collective Processes: Nonlinear
Waves, Solitons, Collisionless Shocks,
Turbulence. Astrophysical Applications

Translated by
A. B. Aries and Igor N. Poliakoff

With 59 Illustrations

Springer-Verlag
New York Berlin Heidelberg Tokyo

A. M. Fridman
V. L. Polyachenko
Astrosovet
Ul. Pyatnitskaya 48
109017 Moscow Sh-17
U.S.S.R.

Translators
A. B. Aries
4 Park Avenue
New York, NY 10016
U.S.A.

Igor N. Poliakoff
3 Linden Avenue
Spring Valley, NY 10977
U.S.A.

Library of Congress Cataloging in Publication Data
Fridman, A. M. (Alekseĭ Maksimovich)
  Physics of gravitating systems.
  Includes bibliography and index.
  Contents: 1. Equilibrium and stability
— 2. Nonlinear collective processes. Astrophysical applications.
  1. Gravitation. 2. Equilibrium. 3. Astrophysics.
I. Poliâchenko, V. L. (Valeriĭ L'vovich) II. Title.
QC178.F74 1984   523.01   83-20248

This is a revised and expanded English edition of: *Ravnovesie i ustoĭchivost' gravitiruiushchikh sistem*. Moscow, Nauka, 1976.

Typeset by Composition House, Ltd., Salisbury, England.

9 8 7 6 5 4 3 2 1

ISBN 978-3-642-87835-0   ISBN 978-3-642-87833-6 (eBook)
DOI 10.1007/978-3-642-87833-6

Contents (Volume II)

Contents (Volume I)

Non-Jeans Instabilities of Gravitating Systems

In the previous chapters, we have faced mainly instabilities of a "Jeans" nature (cf. Introduction) or instabilities similar to those occurring in rapidly rotating systems of incompressible liquid.

This chapter deals with some *non-Jeans* instabilities of gravitating systems investigated so far. The various mechanisms of excitation of similar instabilities are well studied in plasma physics and in the mechanics of continua.

First of all, there are the beam instabilities, to which we devote §1. In §2 we study the gradient instabilities. Section 3 deals with the theory of "hydrodynamical" instabilities (Kelvin–Helmholtz instabilities and flutelike instability) with a growth rate much greater than the Jeans one. In the last section (§4), the general approach to the problem of kinetic instabilities in the collisionless gravitating systems is considered, and also, briefly, the question of the original "cone" instability at the central regions of systems with baled out stars of small angular moments (for example, due to a fall onto a "black hole"). Most frequently, consideration is given to the framework of the simplest models, such as the uniform cylinder with an infinite generatrix or a uniform flat layer.

§ 1 Beam Instability of a Gravitating Medium [88]

1.1 Theorem of a Number of Instabilities of the Heterogeneous System with Homogeneous Flows [64a[ad]]

Let us consider first of all the simplest case of the system, consisting of an arbitrary number n of moving homogeneous components. Recall that the analogous problem for the components at rest was solved in the Intro-duction, where we showed that the instability (Jeans) may occur only on one branch of oscillations while all remaining branches are the branches of "combined sound." ·

The picture described above changes qualitatively in presence of relative motions of components with velocities which exceed corresponding sound velocities. Then nonincreasing ("sound") oscillations occur on the wave-lengths smaller than Jeansonian ones (see Table I, Introduction, case 1). In the opposite case (point 4 in Table I), when the wavelength of the perturba-tion exceeds the Jeansonian wavelengths, the combined sound oscillations are absent. All the roots may be complex: then n different instabilities are developed in the system.

If the undisturbed velocity of the cold component is v_{0c}, and of the hot component is v_{0h}, then the disturbed densities of these components are

$$\rho_c = \rho_{0c} \frac{k^2 \Phi}{(\omega - kv_{0c})^2 - k^2 c_{sc}^2},$$

$$\rho_h = \rho_{0h} \frac{k^2 \Phi}{(\omega - kv_{0h})^2 - k^2 c_{sh}^2},$$

and the corresponding dispersion equation is

$$\frac{\omega_{0c}^2}{(\omega - kv_{0c})^2 - k^2 c_{sc}^2} + \frac{\omega_{0h}^2}{(\omega - kv_{0h})^2 - k^2 c_{sh}^2} = -1.$$

Similarly, for n components we have

$$\sum_{i=1}^{n} \frac{\omega_{0i}^2}{(\omega - kv_{0i})^2 - k^2 c_{si}^2} = -1.$$

Roots of this equation determine in general form the solution of the problem.

Let us consider first of all the simplest example, when the densities and pressures of the cold and hot components are identical: $\omega_{0c}^2 = \omega_{0h}^2 = \omega_0^2/2$; $c_{sc}^2 = c_{sh}^2 = c_s^2$. In the inertial coordinate system where $v_{0c} = -v_{0h} \equiv v_0$ we have the following dispersion equation:

$$\frac{\omega_0^2}{(\omega - kv_0)^2 - k^2 c_s^2} + \frac{\omega_0^2}{(\omega + kv_0)^2 - k^2 c_s^2} = -2.$$

Solution of this equation is

$$\omega^2 = k^2(v_0^2 + c_s^2) - \frac{\omega_0^2}{2} \pm \sqrt{4k^4 v_0^2 c_s^2 - 2\omega_0^2 k^2 v_0^2 + \frac{\omega_0^4}{4}}.$$

Let us choose three limit cases: (a) $\omega_0^2 \gg k^2 v_0^2$, $k^2 c_s^2$; (b) $\omega_0^2 \ll k^2 v_0^2$, $k^2 c_s^2$; (c) $k^2 v_0^2 \gg \omega_0^2 \gg k^2 c_s^2$. In case (a) we obtain two roots: $\omega^2 = -\omega_0^2$, $\omega^2 = -k^2(v_0^2 - c_s^2)$. The first root describes the Jeans instability; the second root, the beam instability provided $|v_0| > c_s$. As we see, the necessary condition of the beam instability for the heterogeneous gravitating medium coincides with the analogous plasma condition [31].

In case (b) both roots are positive, which corresponds to the oscillatory regime.

In case (c) we have

$$\omega^2 = k^2(v_0^2 + c_s^2) \pm i\sqrt{2}\omega_0 k v_0.$$

Here, two roots describe increasing solutions, and two other roots damping solutions.

We can represent the dispersion equation obtained above for the case of two beams with identical densities and velocities in the form

$$f(\omega) = -2.$$

The function $f(\omega)$ is depicted in Figs. 86(a) and 86(b) for the cases $v_0^2 \ll c_s^2$ and $v_0^2 \gg c_s^2$, respectively.

As it follows from the Fig. 86(a), in the case $v_0^2 \ll c_s^2$, only the Jeans instability is possible [provided that $\omega_0^2 > k^2 c_s^2$ (dotted line): two roots on the real axis ω (ω_2 and ω_s) are then absent]. For $\omega_0^2 < k^2 c_s^2$ we have four real roots.

In the case $v_0^2 \gg c_s^2$, as is seen from the Fig. 86(b), for $\omega_0^2 < 2k^2 c_s^2$ all the roots are real, and for $\omega_0^2 > 2k^2 c_s^2$ all the roots are imaginary. In the last case the beam instability occurs in the system (apart from Jeans instability).

The general dispersion equation for n moving beams may be graphically investigated similarly to the above investigation for the components at rest. Let us try to formulate the theorem of the number of unstable roots in the general case.

First of all we enumerate the components of the heterogeneous system in order of decreasing values of $(v_0 + c_s)_i$:

$$(v_0 + c_s)_1 > (v_0 + c_s)_2 > \cdots > (v_0 + c_s)_n.$$

Each of these numbers defines the flow. We shall speak of two arbitrary flows i and j ($i > j$) to be *connected* if

$$\Delta v_{ij} \equiv v_i - v_j < c_i + c_j.$$

We agree to mark such connected flows by curved lines: for example,

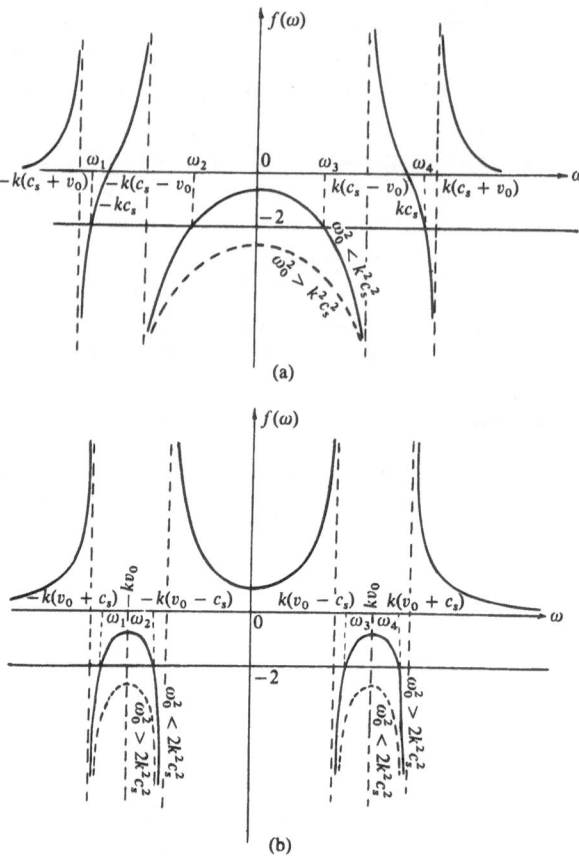

Figure 86. Determination of a number of unstable roots for two homogeneous bounded (a) and nonbounded (b) flows.

We cross out the curved lines which are completely covered by other curved lines (1–2, 2–3, i–j in the given example). The remaining curved lines determine *the independent elements*. To the independent elements we attribute also the *unconnected* flows (flow 5 in this example).

Theorem (of a number of instabilities of the heterogeneous system with moving homogeneous flows): *The number of the unstable roots of the heterogeneous system with moving homogeneous flows equals the number of independent elements.*

The theorem may be proved by the method of mathematical induction.
Let us perform further investigation of the beam instability by using, as the basis of the self-consistent model of the collisionless gravitating system, a model of a rotating cylinder cold in the plane of rotation (x, y). This model has already been considered in Chapter II for "nonbeam" distribution functions of the particles in longitudinal velocities (Maxwell and Jackson). In addition, we show in Section 4.3 that the beam instability is obtained in the model of a flat gravitating layer.

1.2 Expression for the Growth Rate of the Kinetic Beam Instability in the Case of a Beam of Small Density (for an Arbitrary Distribution Function)

As follows from the results of Chapter II, the Jeans instability of a uniformly rotating cylinder (with circular orbits of the particles) with the Maxwellian distribution function in longitudinal velocities, develops with an exponentially small growth rate if the thermal dispersion is sufficiently large: $v_T > v_0$. Therefore, the rearrangement of the initial spatial distribution of particles (the formation of "sausages") will occur very slowly. Under these conditions, one may speak of the quasistationary state of the Maxwellian subsystem and consider the excitation of nondamping oscillations of this subsystem by a group of fast particles.

Further, we assume that the particle distribution in the longitudinal velocities $f_0(v_z)$ can be split into two parts:

$$f_0(v_z) = f^{(0)}(v_z) + f^{(1)}(v_z), \tag{1}$$

where $f^{(0)}(v_z)$ is the Maxwellian function with the dispersion v_T while $f^{(1)}(v_z)$ is a certain function nonzero for $v_z \gg v_T$. An example of such a function is given in Fig. 87.

In other words, we assume that the system of gravitating particles under consideration consists of two subsystems: a slow one and a rapid one. Assume that the slow subsystem has a larger density than the fast one, so that

$$\frac{\int_{-\infty}^{\infty} f^{(1)} \, dv_z}{\int_{-\infty}^{\infty} f^{(0)} \, dv_z} \equiv \alpha \ll 1. \tag{2}$$

By making use of the smallness of the parameter α, the solution of the dispersion equation may be found by the method of subsequent approximations by defining in the next order the growth rates of these oscillations due to the interaction of the latter with the particles of the fast component. Such a settlement of the problem is characteristic for the theory of interaction of a beam of charged particles with a dense plasma.

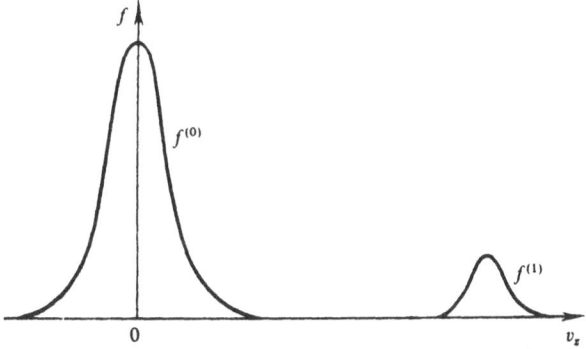

Figure 87. An illustration of the beam distribution function.

Using Eq. (24) and formulae (10) and (11), §2, Chapter II, we have the following dispersion equation:

$$1 + \frac{\omega_0^2}{k^2} \int_{-\infty}^{\infty} \left[\frac{k_\perp^2}{(\omega - k_z v_z)^2 - 4\Omega_0^2} + \frac{k_z^2}{(\omega - k_z v_z)^2} \right] f_0(v_z) \, dv_z = 0, \quad (3)$$

where $k^2 = k_\perp^2 + k_z^2 \simeq k_\perp^2$, since from the inequalities $k_z R \ll 1$ and $k_\perp R \gtrsim 1$ it follows that $k_z^2 \ll k_\perp^2$. For the sake of concreteness we represent formula (1) in the form

$$f_0(v_z) = (1 - \alpha) \frac{n_0}{\pi v_T^2} \exp\left(-\frac{v_z^2}{v_T^2} \right) + \alpha \frac{n_0}{\pi u_T^2} \exp\left[-\frac{(v_z - V)^2}{u_T^2} \right]. \quad (4)$$

Here v_T and u_T are the particle thermal velocities of the cylinder and of the beam, respectively; V is the velocity of the beam.

According to the results of Chapter II (§2), the long-wave ($k_z R \ll 1$), short-scale ($k_\perp R \gtrsim 1$) perturbations of high frequency ($\omega^2 \gg k_z^2 v_T^2$) practically do not damp (the decrement of damping is exponentially small). This means that if the cylinder has a small portion of the beam particles, the dispersion equation is composed of two parts. The first (main) part determines the nondamping (rotational) oscillatory branch of the cylinder (see §2, Chapter II), and the small addition connected with the presence of the beam is interesting to us in its imaginary part only.

Substituting (4) into (3), we obtain, taking into account the above, the following dispersion equation:

$$1 + \frac{\omega_0^2(1 - \alpha)}{\omega^2 - 4\Omega^2} + \frac{\alpha \omega_0^2}{(\pi u_T^2)^{1/2}} \left\{ \frac{k_z^2}{k^2} \int_{-\infty}^{\infty} (\omega - k_z v_z)^{-2} \exp\left[-\frac{(v_z - V)^2}{u_T^2} \right] dv_z \right.$$

$$\left. + \int_{-\infty}^{\infty} [(\omega - k_z v_z)^2 - 4\Omega_0^2]^{-2} \exp\left[-\frac{(v_z - V)^2}{u_T^2} \right] dv_z \right\} = 0. \quad (5)$$

Due to inequality (2), we can solve Eq. (5) by the method of successive approximations. Let us assume that

$$\omega = \omega_0 + i\gamma, \qquad |\gamma| \ll \omega_0. \quad (6)$$

Setting the imaginary part of Eq. (5) equal to zero, we find γ:

$$\gamma = \alpha \omega_0 \left\{ \frac{\omega_0^2}{k^2 u_T^2} \cdot \frac{k_z V - \omega_0}{k_z u_T} \cdot W\left(\frac{\omega_0 - k_z V}{k_z u_T} \right) \right.$$

$$\left. + \frac{\Omega_0}{4 k_z u_T} \left[W\left(\frac{3.4\Omega_0 - k_z V}{k_z u_T} \right) - W\left(-\frac{0.6\Omega_0 + k_z V}{k_z u_T} \right) \right] \right\}. \quad (7)$$

From formula (39), §2, Chapter II, it is clear that in (5) we neglected the exponentially small terms (provided that $\Omega_0 \gg k_z v_T$) describing a damping of the rotational oscillatory branch. Now we are interested in the possibility of the excitation of this branch by the beam with a small density, which is the case when increments of beam instability exceed decrements of damping.

This is obviously true if at least one of the following conditions,[1]

$$|k_z V - \omega_0| \ll k_z u_T, \qquad V > \frac{\omega_0}{k_z}, \tag{8}$$

$$|3.4\Omega_0 - k_z V| \ll k_z u_T, \tag{9}$$

is fulfilled. One can see that in the case (8) it is also necessary that the beam velocity be larger than the phase velocity of oscillations with the frequency ω_0. The increments of the beam instability are equal to

$$\gamma \simeq \sqrt{\pi} \alpha \omega_0 \frac{\omega_0^2}{k^2 u_T^2} \cdot \frac{k_z V - \omega_0}{k_z u_T}, \tag{10}$$

if the conditions in (8) are satisfied, and

$$\gamma \simeq \frac{\sqrt{\pi}}{4} \alpha \omega_0 \frac{\Omega_0}{k_z u_T} \tag{11}$$

in the case (9).

The increment (10) due to Cherenkov resonance $\omega = k_z v_z$, and the increment (11) due to rotational resonance: $\omega + 2\Omega_0 = k_z v_z$.

1.3 Beam with a Step Function Distribution

The beam instability is associated above all with the velocity asymmetry of fast particle distribution $f(v_z) \neq f(-v_z)$, rather than with the presence of a second maximum on the total distribution function. In order to make sure that this is the case consider the asymmetric distribution of particles of the beam in the form of a step

$$f = \frac{\alpha}{4v_1} \begin{cases} 1, & 0 < v_z < v_1 \\ 0, & v_z < 0, v_z > v_1. \end{cases}$$

In this case, the single contribution into the increment yields the resonance of the type $\omega + 2\Omega_0 = k_z v_z$, so that

$$\mathrm{Im}\,\omega = \frac{\pi}{8\sqrt{2}} \alpha \frac{\omega_0^2}{|k_z| v_1}.$$

This expression is valid for $|k_z| v_1 > 4\Omega_0$, from which follows the estimate

$$\mathrm{Im}\,\omega \lesssim \alpha \omega_0,$$

coincident with (10) at $v \simeq v_{T1}$.

---

[1] We assume here that $k_z > 0$.

1.4 Hydrodynamical Beam Instability. Excitation of the Rotational Branch

If the directed velocity of the beam is high in comparison with the scatter of particles of the beam in longitudinal velocities, $v \gg v_{T1}$, then, apart from the kinetic instability as considered, in the two-beam gravitating medium, there may develop a hydrodynamical beam instability. Let us show that by assuming that the thermal scatter of the beam is, nevertheless, finite: $v_{T1}^2 \gg \alpha v_0^2$, so that the development of the Jeans instability in the beam is impossible (within an exponential accuracy).

We follow from Eq. (3) with the function f of the form of (4) and assume that $|\omega + 2\Omega_0 - k_z v| \gg |k_z| v_{T1}$. Then, from (3), follows

$$1 + \frac{(1 - \alpha)\omega_0^2}{\omega^2 - 4\Omega_0^2} - \frac{\alpha \omega_0^2}{4\Omega_0(\omega - k_z v + 2\Omega_0)} = 0.$$

Assuming that $k_z v = \omega_0 + 2\Omega_0$, we find that oscillations with Re $\omega = \omega_0$ are excited with the growth rate

$$\text{Im } \omega = \frac{\alpha^{1/2}}{2^{5/4}} \omega_0.$$

This instability is similar to the cyclotron excitation via a monoenergetic beam of charged particles of cyclotron oscillations of plasma in the magnetic field.

If one compares the gravitational kinetic instabilities with those of the plasma [86], then one notices that, in case of gravitation, the region of instability is much narrower than that in the plasma. For example, the beam instability in the plasma with the distribution function

$$f(v_z) = \frac{n_0 \Delta}{\pi^2} \left[(1 - \alpha) \frac{1}{v_z^2 + \Delta^2} + \frac{\alpha}{(v_z - u)^2 + \Delta^2} \right],$$

as follows from [240], takes place for all velocities $u > \Delta$ within a broad interval of the wave numbers, while here instability takes place only for a definite relation between the velocity and the wave number, in the vicinity of the value $y = u^2 k^2/\Omega^2 = \frac{2}{3}$. It is probable that the conditions of equilibrium in the gravitating medium impose tighter links on the parameters of the system, which makes the region of the existence of unstable equilibrium solutions narrow also.

For the beams in the plane of rotation of the system, a similar inference is made below.

1.5 Stabilizing Effect of the Interaction of Gravitating Cylinders and Disks

It is also possible to analyze a heterogeneous system consisting of two cylinders rotating with respect to each other, of the kind of (2), §1, Chapter II, with the densities $\alpha_1 \rho_0$ and $\alpha_2 \rho_0$ ($\alpha_1 + \alpha_2 = 1$) [111, 113]. Such a system turns out to be stable with respect to perturbations $k_z = 0$. For the mode $n = 4, m = 2, \gamma_1 = 0, \gamma_2 = 1$ (rotation of a cold cylinder in a hot one at rest), the spectrum has the form $\omega^2 - 14\omega + 8\alpha_2 = 0$. The condition of the stability

$$16\alpha_2^2 - (\tfrac{14}{3})^3 < 0$$

is always valid since $\alpha_2 < 1$.

From the exact spectra of small perturbations of gravitating disks (cf. Section 4.4, Chapter V) it follows that the maximum growth rates of instability take place at $\gamma = 1$. This case corresponds to the cold disk model, all the particles of which rotate in circular orbits in the same direction. If a cold disk is composed of two subsystems rotating in opposite directions, then the growth rates of instability decrease. Thus, the effect of the "beam nature" can exert a stabilizing influence. However, in the case of the density increasing toward the edge, the effect of "beam nature" is destabilizing [20] (cf. next subsection).

1.6 Instability of Rotating Inhomogeneous Cylinders with Oppositely Directed Beams of Equal Density [20]

To begin with, consider a uniform dust cylinder consisting of two mutually penetrating cold, in the plane of rotation, beams with velocities $\pm v_{\varphi 0}$ and identical density $\rho_0/2$. In the stationary state

$$v_{r0} = 0, \qquad \frac{1}{r}\frac{d}{dr}\left(r\frac{d\Phi_0}{dr}\right) = 4\pi G\rho_0, \qquad \frac{v_{\varphi 0}^2}{r} = \frac{d\Phi_0}{dr}. \qquad (12)$$

The dispersion relation for the case of oscillation of the flute type ($k_z = 0$) can be obtained in the following way. Write first of all, for each beam, the system of linearized equations of hydrodynamics:

$$-i\left(\omega \mp m\frac{v_{\varphi 0}}{r}\right)v_{r_1}^{\pm} \mp 2\frac{v_{\varphi 0}}{r}v_{\varphi_1}^{\pm} = -\frac{d\Phi_1}{dr}, \qquad (13)$$

$$\pm\left(\frac{v_{\varphi 0}}{r} + \frac{dv_{\varphi 0}}{dr}\right)v_{\varphi_1}^{\pm} - i\left(\omega \mp m\frac{v_{\varphi 0}}{r}\right)v_{\varphi_1}^{\pm} = -i\frac{m}{r}\Phi_1, \qquad (14)$$

$$\left(\frac{\rho_0/2}{r} + \frac{d\rho_0/2}{dr}\right)v_{r_1}^{\pm} + i\frac{\rho_0}{2}\frac{m}{r}v_{\varphi_1}^{\pm} - i\left(\omega \mp m\frac{v_{\varphi 0}}{r}\right)\rho_1^{\pm} + \frac{\rho_0}{2}\frac{dv_{r_1}^{\pm}}{dr} = 0. \quad (15)$$

Hence, for $v_{\varphi 0}/r = \Omega_0 = \text{const}$, we find ($\rho_1^{\pm} \equiv \rho_1^{(1,2)}$)

$$\frac{\rho_1^{(1)}}{\rho_0/2}[4\Omega_0^2 - (\omega - m\Omega_0)^2] = \Delta\Phi_1, \qquad (16)$$

$$\frac{\rho_1^{(2)}}{\rho_0/2}[4\Omega_0^2 - (\omega + m\Omega_0)^2] = \Delta\Phi_1. \qquad (17)$$

Comparing (16) and (17) with the Poisson equations

$$\Delta \Phi_1 = \frac{1}{r}\frac{d}{dr}\left(r\frac{d\Phi_1}{dr}\right) - \frac{m^2}{r^2}\Phi_1 = 4\pi G(\rho_1^{(1)} + \rho_1^{(2)}), \qquad (18)$$

we find

$$\frac{1}{4\Omega^2 - (\omega - m\Omega_0)^2} + \frac{1}{4\Omega_0^2 - (\omega + m\Omega_0)^2} = \frac{1}{\Omega_0^2}, \qquad (19)$$

where the condition of equilibrium of (12) is employed; for ω^2 we obtain the expression

$$\omega^2 = \tfrac{1}{2}\omega_0^2(m^2 + 3 \pm 2\sqrt{3m^2 + \tfrac{1}{4}}). \qquad (20)$$

The value ω^2 is minimal for $m = 2$, when $\omega^2 = 0$, i.e., indifferent[2] equilibrium takes place; $\omega^2 > 0$ for $m \neq 2$.

When the cylinder is nonuniform (and again consists of two identical rotating flows), from the system of equations in (13)–(15), one can obtain the equation [(1), §7, Chapter II, in a new form]:

$$2\frac{\Delta \Phi_1}{\omega_0^2} = K_+ + K_-, \qquad (21)$$

where

$$K_+ = -\frac{1}{y^2}\left[\Delta\Phi_1 + A_1\Phi_1' + \frac{2m\Omega_0}{rx}\left(A_1 + \frac{\Omega_0'}{\Omega_0}\right)\Phi_1\right],$$

$$y^2 = x^2 - 2\alpha\Omega_0, \qquad \alpha = 2\Omega_0 + r\Omega_0', \qquad x = m\Omega_0 - \omega,$$

$$K_0 = \frac{1}{\rho_0}\frac{d\rho_0}{dr}, \qquad A_1 = K_0 + 2\frac{\Omega_0(\alpha + \alpha'\Omega_0/\Omega_0' - mx)}{x^2 - 2\alpha\Omega_0}\frac{\Omega_0'}{\Omega_0}, \qquad (22)$$

and the quantity K_- ensues from K_+ by substituting Ω for $-\Omega_0$ in all expressions of (22).

At $m/kr \ll 1$, we find

$$\frac{1}{2\alpha\Omega_0 - (\omega - m\Omega_0)^2} + \frac{1}{2\alpha\Omega_0 - (\omega + m\Omega_0)^2} = \frac{2}{\omega_0^2}. \qquad (23)$$

One can easily make sure that Eq. (23) transforms to (19) for $\rho_0' = \Omega_0' = 0$. Making use of (22), we get from (23) the dispersion equation in the form

$$\omega^4 - 2\omega^2[(4 + m^2)\Omega_0^2 - \tfrac{1}{2}\omega_0^2 + 2r\Omega_0\Omega_0']$$
$$+ [(4 - m^2)\Omega_0^2 + 2\Omega_0\Omega_0'r]^2$$
$$- \omega_0^2[(4 - m^2)\Omega_0^2 + 2\Omega_0\Omega_0'r] = 0. \qquad (24)$$

The discriminant of Eq. (24), biquadratic with respect to ω, is positive; therefore, the instability described by this equation may be only aperiodical, i.e., the growth rate $\gamma = i\omega$. The above investigation of the stability of a

---

[2] If the densities of two beams are different, the dispersion equation becomes more complex; however, it may be shown that it has real roots only.

uniform cylinder with oppositely directed beams shows that the maximum growth rate in the case of a nonuniform cylinder should be sought for at $m = 2$. Indeed, under the condition of smallness of the density gradient, instability takes place only when $m = 2$:

$$\omega^2 = -r\Omega_0\Omega_0' \frac{\omega_0^2}{8\Omega_0^2 - \frac{1}{2}\omega_0^2}; \tag{25}$$

$8\Omega^2 > \omega_0^2/2$ for small ρ_0' and Ω_0'. As is seen from (25), the necessary condition of the beam instability in this case is the condition of the growth of the angular velocity of rotation from the center of the cylinder.

§ 2 Gradient Instabilities of a Gravitating Medium [28, 114]

In plasma physics, we know of a vast class of the so-called gradient (drift) instabilities due to the spatial plasma inhomogeneity, which frequently play a decisive role. Gradient instabilities involve, in particular, the instability due to temperature inhomogeneity [89]. The cause of it lies in the transfer of longitudinal energy of particles across the magnetic field due to their drift in the crossed fields.

The question arises of the possibility of development of similar instabilities in a gravitating medium, which, as the plasma, refers to the number of systems with Coulomb interaction.

2.1 Cylinder of Constant Density with Radius-Dependent Temperature. Hydrodynamical Instability

The theoretical possibility of gradient instabilities in the gravitating media was proved in [28, 114] independently.

To begin with, consider the simplest model: a cylinder, *uniform* in density with the radial temperature gradient. The general equation (14), §3, Chapter II, describing small perturbations of the nonuniform cylinder with circular orbits of particles, in this case becomes somewhat simplified:

$$\frac{1}{r}\frac{d}{dr}\left(r\varepsilon_\perp\frac{d\Phi}{dr}\right) - \frac{m^2}{r^2}\varepsilon_\perp\Phi - k_z^2\varepsilon_\parallel\Phi - \frac{4m}{r}\frac{dI}{dr}\Phi = 0, \tag{1}$$

$$\varepsilon_\parallel = 1 + 2\int\frac{f_0\,dv_z}{\omega'^2}, \qquad \varepsilon_\perp = 1 + 2\int\frac{f_0\,dv_z}{\omega'^2 - 4};$$

$$I = \int\frac{f_0\,dv_z}{\omega'[\omega'^2 - 4]}, \qquad \Omega_0^2 = 2\pi G\rho_0. \tag{2}$$

The *local* dispersion equation corresponding to (1) is of the form (below we assume $\Omega_0 = 1$)

$$\varepsilon_\perp k_\perp^2 + \varepsilon_\parallel k_z^2 + \frac{4m}{r}\frac{dI}{dr} = 0, \tag{3}$$

where $k_\perp^2 = k_r^2 + m^2/r^2$. Calculating the integrals in (2) for the Maxwellian distribution function $f_0 = \exp(-v_z^2/v_T^2)/\sqrt{\pi}v_T$, we obtain instead of (3) for

$$\omega_* \equiv \omega - m \gg k_z v_T \tag{4}$$

(i.e., in the hydrodynamical limit)

$$k_z^2 + k_\perp^2 \frac{\omega_*^2 - 2}{\omega_*^2 - 4} + \frac{2k_z^2}{\omega_*^2} - \frac{k_\perp^2}{4} \frac{k_z^2 v_T^2}{(\omega_* + 2)^3} + \frac{k_\perp^2}{4} \cdot \frac{k_z^2 v_T^2}{(\omega_* - 2)^3}$$

$$= \frac{m}{2r} \frac{k_z^2 v_T^2}{v_T} \frac{\partial v_T}{\partial r} \left[\frac{1}{(\omega_* + 2)^3} + \frac{1}{(\omega_* - 2)^3} - \frac{2}{\omega_*^3} \right]. \tag{5}$$

For low-frequency oscillations satisfying the condition $\omega_* \ll 1$, i.e., close to the frequency of the rotational ("cyclotron") resonance, we have from (5)

$$\omega_*^3 + 4\frac{k_z^2}{k_\perp^2}\omega_* + \frac{2m}{r}\frac{k_z^2 v_T^2}{k_\perp^2}\frac{1}{v_T}\frac{\partial v_T}{\partial r} = 0. \tag{6}$$

Under condition $4\omega_* \ll (m/2r)v_T \, \partial v_T/\partial r$, dispersion equation (6) describes unstable oscillations with the growth rate

$$\gamma = \frac{\sqrt{3}}{2} \left| \frac{2m}{r} \frac{k_z^2 v_T^2}{k_\perp^2} \frac{1}{v_T} \frac{\partial v_T}{\partial r} \right|^{1/3}. \tag{7}$$

On the limit of the applicability of this approximation, we obtain the growth rate

$$\gamma \sim \left(\frac{m}{r} v_T \frac{\partial v_T}{\partial r} \right)^{1/3}. \tag{8}$$

It should be noted that these results are coincident with the results known from plasma physics: dispersion equation (6) can be obtained directly from the respective plasma equation (cf. [89]) by substituting $k_y \to m/r$, $\omega_B \to 2$. The physical meaning of the obtained instability is also similar to the case of plasma. The drift of particles leading to convection of heat in the radial direction is caused by Coriolis acceleration due to the appearance of perturbations of the azimuthal velocity.

It is important to emphasize that the gradient-temperature instability may evidently exist also under conditions when the Jeans instability is practically suppressed (for which it is necessary that there be $v_T > v_0$).

The instability considered above has a *hydrodynamical* character [cf. condition (4)].[3] There is also kinetic instability; however, it turns out that, in order to obtain it, the WKB approximation is insufficient. The simplest model in which one can most simply make sure that there is kinetic instability is treated in the following subsection.

---

[3] Accordingly, this instability could have been obtained [28] by using the equations of anisotropic hydrodynamics for a rotating gravitating medium.

2.2 Cylinder of Constant Density with a Temperature Jump. Kinetic Instability

One is able to get an answer in explicit form only in the case when the temperature jump (from T_2 at $0 < r < r_0$ to T_1 at $r_0 < r < R$) is experienced by only a small part of particles with the density $\alpha\rho_0$, $\alpha \ll 1$. As far as the remaining part is concerned, we shall assume that it has a fixed temperature T. Let the temperature jump be sufficiently great: $T_2 \gg T_1$, so that there are such k_z for which the condition (4) is valid only for the cold medium, while the inverse inequality takes place in a hot one, i.e.,

$$k_z v_{T1} \ll \omega_* \ll k_z v_{T2}. \tag{9}$$

Under such conditions, the hot medium should be considered kinetically (while the cold one, again hydrodynamically).

Using the inequality $\Omega \gg k_z v_{T2}$, for low-frequency perturbations ($\omega_* \ll \Omega$), we obtain $\varepsilon_\perp = \frac{1}{2}$. Let us seek solutions localized near the temperature jump and decreasing according to the law

$$\Phi \sim e^{-\varkappa(r-r_0)} \qquad (\varkappa > 0). \tag{10}$$

Substituting (10) into Eq. (1), we find, under condition $m^2 \gg 2k_z^2 r_0^2 \omega_0^2/\omega_*^2$,

$$\varkappa^2 \simeq \frac{m^2}{r_0^2}, \qquad \varkappa = \left|\frac{m}{r_0}\right|. \tag{11}$$

We now use the boundary conditions on the jump:

$$\Phi|_{r_0-0}^{r_0+0} = 0, \tag{12}$$

$$\varepsilon_\perp \frac{d\Phi}{dr} - \frac{2m\Phi}{r_0}\omega_0^2\Omega \int \frac{f_0}{\omega'[\omega'^2 - 4\Omega_0^2]} dv_z \Big|_{r_0-0}^{r_0+0} = 0. \tag{13}$$

In calculating the integral in (13) for the cold medium, it is enough to restrict oneself to the basic term, while, for the hot medium, let us write the integral completely. As a result, we obtain the dispersion equation

$$1 + \alpha\Omega \operatorname{sgn} k_\varphi \left(\int \frac{f_0^{(2)} dv_z}{\omega_* - k_2 v_z} - \frac{1}{\omega_*} \right) = 0 \qquad \left(k_\varphi = \frac{m}{r_0} \right). \tag{14}$$

If $\omega_* \ll k_z v_{T2}$, then (14) takes the form

$$1 - \frac{\alpha\Omega \operatorname{sgn} k_\varphi}{\omega_*} \left[1 + \frac{i\pi\omega_*}{|k_z|} f_0^{(2)}\left(\frac{\omega_*}{|k_z|}\right) \right] = 0. \tag{15}$$

We write $\omega_* = \omega^{(0)} + i\gamma$. At $\omega_* \ll k_z v_{T2}$, the inequality $\omega^{(0)} \gg \gamma$ is valid. Therefore, we find the solution in the form

$$\omega^{(0)} = \alpha\Omega \operatorname{sgn} k_\varphi, \qquad \gamma = \omega^{(0)2} \frac{\pi}{|k_z|} f_0^{(2)}\left(\frac{\omega^{(0)}}{|k_z|}\right), \tag{16}$$

i.e., there is kinetic instability. The oscillations are low-frequency ones ($\omega_* \ll \Omega_0$), if $\alpha \ll 1$. To have the condition (9), it is necessary that

$$k_z v_{T1} \ll \alpha\Omega \ll k_z v_{T2},$$

and for the validity of (11) it is necessary that

$$m^2 \gg 4\alpha^2 k_z^2 r_0^2.$$

2.3 Cylinder with Inhomogeneous Density and Temperature [28]

The density gradient $d\rho_0/dr$ exerts, as we shall see, a stabilizing effect on instability whenever it is sufficiently large and coincides in direction with the temperature gradient, but per se for $T = \text{const}$ and in the absence of Jeans instability does not lead to gradient instability.

Instability arises for $k_z \to 0$, $m \neq 0$ and $k_r r \gg 1$ (perturbation $\sim e^{i(k_z z + m\varphi + k_r r - \omega t)}$), i.e., perturbations with a large wavelength along the z-axis, depending on angle, and with a small wavelength on r, are unstable; therefore, quasiclassical consideration is possible for $k_r r \gg 1$. We shall be interested in low-frequency ($\omega_* \ll \Omega$) oscillations without restricting ourselves beforehand to perturbations of the hydrodynamical type with $\omega_* \gg k_z v_T$. Then we obtain the following dispersion equation:

$$k_\perp^2 + \left(1 + \frac{\omega_0^2}{2\Omega^2}\right)\frac{2\omega_0^2}{v_T^2}\left[\omega_* \int \frac{f_0\, dv_z}{\omega_* - k_z v_z} - 1\right] + \frac{\omega_0^2 \omega_T}{v_T^2}$$

$$\times \left[\frac{\omega_*}{k_z^2 v_T^2} + \left(\frac{1}{2} - \frac{\omega_*^2}{k_z^2 v_T^2}\right)\int \frac{f_0\, dv_z}{\omega_* - k_z v_z}\right]$$

$$- 2\frac{\Omega^2\omega_0^2}{2\Omega^2 + \omega_0^2}\frac{\omega_n}{v_T^2}\int \frac{f_0\, dv_z}{\omega_* - k_z v_z} = 0, \tag{17}$$

$$\omega_T = \frac{m}{r\Omega}\frac{dv_T^2}{dr}, \tag{18}$$

$$\omega_n = k_\varphi\left[\frac{d\ln\rho}{d\ln r} + \frac{(\omega_0^2 - 2\Omega^2)}{4\Omega^4}\right]\frac{v_T^2}{r\Omega}. \tag{19}$$

In the absence of Jeans instability $v_T \gg R\Omega$, one can neglect the second term in (17) (as in the uniform cylinder). To define the instability boundary, let us proceed in the following manner (similarly to the corresponding case in the plasma [89]). Assume in (17) ω^* to be real and then equate the real and imaginary parts to zero; then on the stability boundary we have

$$|k_z|_{cr} = \frac{\omega_0^2\,\omega_T}{v_T^2\,k_\perp^2}\left(\frac{1}{2} - \frac{2\Omega^2}{2\Omega^2 + \omega_0^2\,\eta}\frac{1}{\eta}\right)^{1/2}, \qquad \eta = \frac{\omega_T}{\omega_n},$$

$$\omega^* = -\frac{\omega_0^2\,\omega_T}{v_T^2 k_\perp^2}\left(\frac{1}{2} - \frac{2\Omega^2}{2\Omega^2 + \omega_0^2\,\eta}\frac{1}{\eta}\right) = -|k_z|_{cr}v_T\left(\frac{1}{2} - \frac{2\Omega^2}{2\Omega^2 + \omega_0^2\,\eta}\frac{1}{\eta}\right)^{1/2}.$$

$$\tag{20}$$

The condition that ω_* be real may be satisfied if

$$\eta > \frac{4\Omega^2}{2\Omega^2 + \omega_0^2} \quad \text{or} \quad \eta < 0. \tag{21}$$

Consider the case of "weak" inhomogeneity when the density variations are small, but large gradients ρ are admissible.

In this case

$$\omega_0^2 = 2\Omega^2, \quad \omega_n = k_\varphi \frac{d \ln \rho}{d \ln r} \frac{v_T^2}{r\Omega}, \quad \eta = \frac{d \ln v_T^2}{d \ln \rho},$$

$$|k_z|_{\mathrm{cr}} v_T \frac{k_\perp^2 v_T^2}{\omega_0^2 \omega_T} = \frac{1}{\sqrt{2}} \left(1 - \frac{1}{\eta}\right)^{1/2}. \tag{22}$$

Then the conditions of (21) will take the form

$$\eta < 0 \quad \text{or} \quad \eta > 1. \tag{23}$$

In a similar plasma case [89], the second instability boundary lies at $\eta = 2$. The cause of the difference is in the fact that in the gravitating medium, due to the equilibrium conditions, the angular velocity of rotation is uniquely linked with the density, and therefore it is variable. This yields the contribution to the criterion of (21) and (23) even for weak inhomogeneity, because it is necessary to take into account the second derivatives $d^2\Omega/dr^2$, which in their order of magnitude are coincident with $d\rho/dr$.[4]

The stability boundaries on the plane x, η, where

$$x = 2 \frac{k_z^2 v_T^2 k_\perp^4 v_T^4}{\omega_0^4 \omega_T^2}, \tag{24}$$

for the case of (22) are given in Fig. 88 ($x = 1 - 1/\eta$). From (20), it follows that on the stability boundary

$$\frac{\omega^*}{k_z v_T} = \left(\frac{1}{2} - \frac{2\Omega}{2\Omega^2 + \omega_0^2} \frac{1}{\eta}\right) \sim 1$$

---

[4] In [28], attention is paid to the fact that one gradient of density does not lead to instability. A contrary statement is contained in the papers of M. N. Maksumov [80–83]. In spite of the fact that the correctness of the approximations used in [80–83] is dubious, the question itself of the possibility of development of drift instabilities in gravitating systems, advanced by Maksumov, is, of course, of interest.

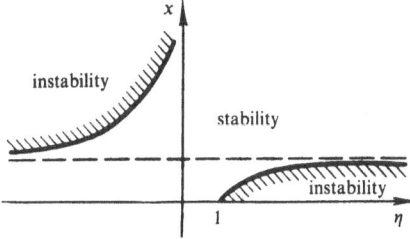

Figure 88. Stability and instability regions of the temperature- and density-inhomogeneous collisionless cylinder [28].

for $\eta \neq 4\Omega^2/(2\Omega^2 + \omega_0^2)$. Therefore, it was impossible, for its determination, to restrict oneself to the hydrodynamical consideration and expand in parameter $\omega^*/k_z v_T$; a general consideration with arbitrary $\omega^*/k_z v_T$ is needed.

§ 3 Hydrodynamical Instabilities of a Gravitating Medium with a Growth Rate Much Greater than that of Jeans [98–100]

All the instabilities considered by us so far, have their growth rates less than, or of the order of that of Jeans. A typical feature of instabilities of the non-Jeans type (beam, gradient types) is the difficulty of their application to real objects. Indeed, these instabilities considered in an infinitely long cylinder exist under condition that the limiting wavelength of the perturbation greatly exceeds the radius of the cylinder, $\lambda_z \gg R$.

Such a situation stimulates the search for instabilities of a gravitating medium qualitatively different from those mentioned. Of great interest are instabilities, whose development may proceed with a growth rate much greater than the Jeans growth rates, as well as instabilities not subjected (like those of Jeans) to the stabilizing influence of thermal dispersion or lacking such exotic conditions of existence, unlike the beam and gradient types.

Such instabilities involve the Kelvin–Helmholtz (KH) instability and the flute-like instability [98–100].

3.1 Hydrodynamical Instabilities in the Model of a Flat Parallel Flow

The simplest model for the investigation of the KH and flutelike instabilities is provided by the flat-parallel flow of gravitating fluid with varying (in the direction perpendicular to the flow velocity) values of velocity and density. Here it should be mentioned that the most general criteria of stability of the model described were obtained in [129, 186] in the approximation of an idealized noncompressible liquid. The need to account for compressibility complicates significantly the analysis and does not allow the general stability criteria to be obtained.

Stability of a flat tangential discontinuity has earlier been investigated in the approximation of incompressible fluid in the external gravitational field [67, 186], in compressible fluid and in the MHD approximation [126] in the absence of a gravitational field.

In [99], the effect of the gravitating properties of the medium on the stability of a flat tangential discontinuity (Fig. 89) is investigated.

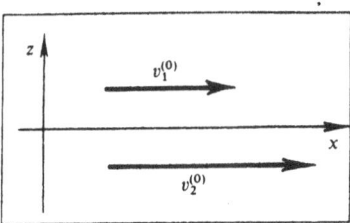

Figure 89. The plane tangential discontinuity.

1. Write the initial system of linearized equations in the form

$$\frac{\partial \mathbf{v}_1}{\partial t} + (\mathbf{v}_0 \nabla)\mathbf{v}_1 + v_{z1}\frac{d\mathbf{v}_0}{dz} = -\frac{\nabla P_1}{\rho_0} - \nabla \Phi_1, \tag{1}$$

$$\frac{\partial v_{z1}}{\partial t} + (\mathbf{v}_0 \nabla)v_{z1} = -\frac{1}{\rho_0}\frac{\partial P_1}{\partial z} + \frac{\rho_1}{\rho_0^2}\frac{dP_0}{dz} - \frac{\partial \Phi_1}{\partial z}, \tag{2}$$

$$\frac{\partial \rho_1}{\partial t} + (\mathbf{v}_0 \nabla)\rho_1 + \rho_0(\nabla \mathbf{v}_1) + \frac{\partial}{\partial z}(\rho_0 v_{z1}) = 0, \tag{3}$$

$$\frac{\partial s_1}{\partial t} + (\mathbf{v}_0 \nabla)s_1 + v_{z1}\frac{ds_0}{dz} = 0, \tag{4}$$

$$\Delta_2 \Phi_1 + \frac{\partial^2 \Phi_1}{\partial z^2} = 4\pi G \rho_1. \tag{5}$$

Here the vector notations are two-dimensional (for the x, y components), the surface of the discontinuity coincides with the $z = 0$ plane, G is the gravitational constant, $s_0 = P_0/\rho_0^\gamma$ and $s_1 = (P_1 - c^2\rho_1)/\rho_0^\gamma$ are the unperturbed and perturbed entropies, and $c^2 = \gamma P_0/\rho_0$ is the speed of sound. In considering perturbations of the type $\exp[i(\mathbf{k}\mathbf{r} - \omega t)]$, we reduce the system (1)–(5) to the following:

$$\zeta' = \frac{k^2}{\omega_*^2}\left(\frac{P_1}{\rho_0} + \Phi_1\right) - \frac{P_1 + \xi P_0'}{c^2\rho_0}, \tag{6}$$

$$\frac{P_1'}{\rho_0} + \Phi_1' = \xi\left[\omega_*^2 - \frac{P_0'\rho_0'}{\rho_0^2} + \frac{P_0'}{\rho_0^2 c^2}\right] + \frac{P_0' P_1}{\rho_0^2 c^2}, \tag{7}$$

$$\Phi_1'' = k^2\Phi_1 + 4\pi G\left[\frac{P_1 + \xi P_0'}{c^2} - \xi\rho_0'\right], \tag{8}$$

where $\omega_* = \omega - k v_0(z)$, $\xi = i v_{z1}/\omega_*$, and the prime denotes differentiation over z.

Let ρ_0 and v_0 be discontinuous in the $z = 0$ plane. The boundary conditions on the discontinuity are readily obtained from (6)–(8) by the familiar procedure of integration over the layer:

$$[\xi] = \xi(z = +0) - \xi(z = -0) = 0, \tag{9}$$

$$[\Phi'_1] = -4\pi G\xi[\rho_0], \tag{10}$$

$$[\Phi_1] = 0, \tag{11}$$

$$[P_1] = \xi g[\rho_0]. \tag{12}$$

Here $[g \equiv d\Phi_0/dz|_{z=0}]$, while (10) denotes that, in the $z = 0$ plane, a simple layer is formed of surface density $\sigma = -\xi[\rho_0]$.

Let us assume that, within the ranges $z > 0$ and $z < 0$, the unperturbed densities and velocities are constant (though different). Solving then the system (6)–(8) separately for $z > 0$ and for $z < 0$ and matching the solutions thus obtained according to (9)–(12), we obtain the dispersion equation for the frequencies of small oscillations. The coefficients of the system (6)–(8) may be assumed to be independent of z only for sufficiently short-wave perturbations

$$\lambda \ll \min(\lambda_1, \lambda_2), \tag{13}$$

where $\lambda_1 = g/4\pi G\rho_0$, $\lambda_2 = c^2/g$ or (for $g = 0$) for the wavelengths

$$\lambda^2 \ll \lambda_j^2 = \frac{c^2}{4\pi G\rho_0}. \tag{13'}$$

With these limitations, the dispersion equation has the form

$$\begin{vmatrix} k, & k, & \omega_1^2[k^2\omega_{01}^2 + (\chi_1^2 - k^2)(k^2c_1^2 - \omega_1^2)], \\ \omega_1^2 + kg, & kg - \omega_2^2, & \omega_1^2\omega_{01}^2(\chi_1\omega_1^2 + k^2g), \\ \omega_1^2 + \omega_{01}^2 + kg, & \omega_2^2 + \omega_{02}^2 - kg, & kg\omega_{01}^2(\chi_1\omega_1^2 + k^2g), \\ 0, & 0, & -\rho_{01}c_1^2\omega_1^2(\chi_1^2 - k^2)(\chi_1\omega_1^2 + k^2g), \end{vmatrix}$$

$$\begin{array}{c} \omega_2^2[k^2\omega_{02}^2 + (\chi_2^2 - k^2)(k^2c_2^2 - \omega_2^2)] \\ \omega_2^2\omega_{02}^2(k^2g - \chi_2\omega_2^2) \\ kg\omega_{02}^2(k^2g - \chi_2\omega_2^2) \\ \rho_{02}c_2^2\omega_2^2(\chi_2^2 - k^2)(\chi_2\omega_2^2 - k^2g) \end{array}$$

$$= 0 \tag{14}$$

Here the index "1" denotes the quantities referring to the range $z > 0$, while the index "2," the range $z < 0$; $\omega_{1,2} = \omega - kv_{01,2}$, $\omega_0^2 = 4\pi G\rho_0$,

$$\chi_{1,2}^2 = k^2 - \frac{\omega_{1,2}^2 + \omega_{01,2}^2}{c_{1,2}^2} + \frac{k^2g^2}{\omega_{1,2}^2c_{1,2}^2}, \qquad \text{Re } \chi_{1,2} > 0. \tag{15}$$

2. To begin with, consider the effects connected with the velocity discontinuity (Kelvin–Helmholtz instability, Fig. 90), assuming in (14) that $\rho_{01} = \rho_{02}$, $c_1^2 = c_2^2$, $v_{01} = -v_{02} \equiv v_0$, $g = 0$, and we describe the solution

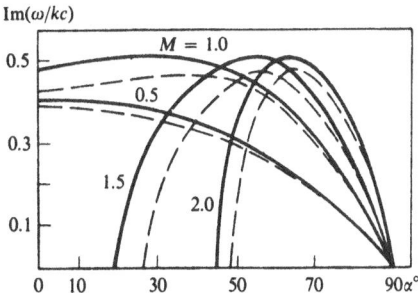

Figure 90. The dependence of the instability growth rate of the plane tangential discontinuity on "Mach number" $M = v_0/c$ as well as on the direction of the wave vector α. Solid lines show the growth rates for $v^2 = \omega_0^2/k^2c^2 = 0$ (incompressible fluid); dotted lines mark the growth rates for $v^2 = 0.2$.

with the aid of the following dimensionless parameters:

$$M = \frac{|\mathbf{v}_0|}{c}, \qquad \beta = M \cos \alpha, \qquad \cos \alpha = \frac{(\mathbf{k}\mathbf{v}_0)}{|\mathbf{k}||\mathbf{v}_0|}, \qquad v = \frac{\omega_0}{kc}$$

In the limit of short-wave perturbations $\omega_0 \ll kc$, in the zeroth approximation from (15) we have

$$\omega = ikc\beta\gamma, \qquad \gamma = \frac{[(1 + 4\beta^2)^{1/2} - (1 + \beta^2)]^{1/2}}{\beta}. \tag{16}$$

From (16), it is seen that the tangential discontinuity in this approximation is unstable for any $M \neq 0$; however, for $M < \sqrt{2}$ perturbations with arbitrary direction of the wave vector (apart from the purely transversal ones), are unstable, and for $M > \sqrt{2}$ perturbations prove to be unstable only when $\cos^2 \alpha < 2/M^2$. The growth rate is maximal

$$(\omega = ikc[(1 + 4M^2)^{1/2} - (1 + M^2)]^{1/2})$$

for $M < \sqrt{\frac{3}{2}}$ for purely longitudinal ($k \| v_0$) perturbations, while, for $M > \sqrt{\frac{3}{2}}$, the maximum of the growth rate ($\omega = ikc/2$) is reached at $\cos^2 \alpha = 3/4M^2$.

In the following approximation with respect to $v = \omega_0/kc$ we have

$$\omega = ikc\beta[1 - v^2 A(\gamma)], \qquad A(\gamma) = \frac{(1 - \gamma)(1 + \gamma^2)[2\gamma + (1 + \gamma)^2]}{4\gamma(1 + \gamma)(3 - \gamma^2)}. \tag{17}$$

It is easy to see that $A(\gamma) > 0$ and is a monotonically increasing β function. Thus, perturbations of the discontinuity surface with increasing wavelength suffer additional stabilization connected with taking into account the gravitating properties of the medium. However, one should bear in mind that according to (13'), the results of (17) are applicable only to wavelengths $\lambda \ll \lambda_J = c/\omega_0$. Here, as seen from (17), the growth rate of instability proves to be much greater than that of Jeans: Im $\omega \gg \omega_0$.

3. Consider now the effects due to density discontinuity. By assuming in

(14) that $\bar{v}_{01} = \bar{v}_{02} = 0$ and assuming the value of density discontinuity to be not very small in comparison with ρ_0, we obtain

$$\omega = \left(kg \frac{\rho_{02} - \rho_{01}}{\rho_{02} + \rho_{01}}\right)^{1/2}$$

$$\times \left[1 + \frac{\pi G}{kg}(\rho_{01} - \rho_{02}) + \frac{\rho_{01}\rho_{02}g(\rho_{01}c_1^2 + \rho_{02}c_2^2)}{kc_1^2c_2^2(\rho_{01} - \rho_{02})(\rho_{01} + \rho_{02})^2}\right]. \quad (18)$$

In the approximation $c \to \infty$, from (18), we obtain the result known in incompressible fluid theory (10).

4. Thus, the growth rates of hydrodynamical instabilities of the gravitating medium can essentially exceed the growth rate of Jeans instability.

Jeans instability is stabilized by thermal dispersion within the range of short ($k^2c^2 \gtrsim \omega_0^2$) wavelengths (a critical wavelength exists). Hydrodynamical instabilities, unlike the gravitational instability, are not stabilized by thermal dispersion in the short-wave range.[5] Moreover, according to (16) and (18), the growth rates of hydrodynamical instabilities increase with decreasing wavelength of perturbation.[6] This unique property of instabilities of KH and the flutelike instability distinguish them from the hydrodynamical instabilities of the gravitating medium investigated earlier.

It is easy to see that gravitation does not exert any influence at all on the short-wave part of the oscillation spectrum. If one assumes the gravitating medium to be in equilibrium, $\nabla P_0 + \rho_0\nabla\Phi_0 = 0$, then from the initial system of equations, by taking into account the gradients of unperturbed values, it is easy to see that $|\nabla P_1|$ is kL times larger than $|\rho_1\nabla\Phi_0|$ (L is the characteristic size of the inhomogeneity, $kL \gg 1$), while $|\rho_0\nabla\Phi_1|$ is less than $|\rho_1\nabla\Phi_0|$ also kL times. Thus, the influence of the "external" gravitational field may be considered as a small correction to the hydrodynamical effects. The influence of "self-gravitation" is even more negligible.

As is seen from expressions (16) and (18), compressibility has different influences on each of the instabilities considered. In case of the tangential discontinuity of the velocities, with increasing parameter M, the oscillations excited at small angles toward the direction of the velocity of the medium are stabilized, and the maximum growth rate of instability is displaced toward the region of larger angles between the direction of the velocity of the medium and the wave vector.

The growth rate of the flutelike instability increases with due regard for compressibility (as well as with due regard for "self-gravitation"), which is especially essential for long-wave oscillations.

---

[5] This fact becomes obvious if one remembers that the model of discontinuity considered in item 2 is unstable also in the approximation of incompressible fluid [67], where the value of thermal dispersion is, according to definition, infinite.

[6] This statement is valid, at least, for wavelengths greater than, or of the order of, the size of the transition layer $ka \ll 1$.

3.2 Hydrodynamical Instabilities of a Gravitating Cylinder

In the previous section, the hydrodynamical instabilities were dealt with in the case of flat geometry, when the gravitating system is nonstationary. Since the growth rates of KH and flutelike instabilities, as already mentioned, are much larger than the Jeans growth rate, such a consideration is correct because the deviation from the stationary state occurs for the time $1/\omega_0$ (ω_0 is the Jeans frequency, $\omega_0 = \sqrt{4\pi G\rho_0}$, ρ_0 is the density of the medium, G is the gravitational constant), which is much more than the time of instability development $1/\gamma$.

Nonetheless, the theory of gravitational instabilities presents many examples, when the gravitational instability investigated in flat geometry does not develop in a stationary gravitating system. This is due to the stabilizing role of the centrifugal forces (or forces of pressure). Below, we treat the possibility of development of hydrodynamical instabilities with the growth rate much more than that of Jeans, in the simplest gravitating system of cylindrical geometry.

1. Consider the cylindrical tangential discontinuity (Fig. 91) assuming that equilibrium is provided by equality of the centrifugal and gravitational forces ($P_0 = \text{const}; \Omega^2 = \omega_0^2/2$ is the square of the angular velocity of rotation of the cylinder, and R is the radius of discontinuity):

$$\Omega(r > R) = -\Omega(r < R).$$

To investigate perturbation stability of the discontinuity surface of the form $\exp[i(m\varphi + kz - \omega t)]$, it is easy to obtain the following dispersion equation:

$$\det (a_{ik}) = 0. \tag{1}$$

Here

$$a_{1,k} = 1, \qquad a_{2,k} = J_k, \qquad a_{3,k} = \mu_k^2 - 1 + a_{4,k},$$

$$a_{4,l} = v_l \frac{J_l - 2m/(x - m)}{(x - m)^2 - 4} \qquad (l = 1, 3),$$

$$a_{4,l} = v_0 \frac{J_l + 2m/(x + m)}{(x + m)^2 - 4} \qquad (l = 2, 4),$$

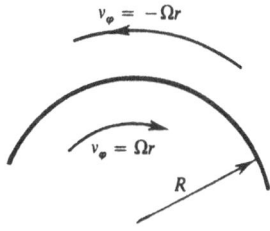

Figure 91. The cylindrical tangential discontinuity.

where

$$v_k^2 = 2\beta^2 + \mu_k^2 - 1,$$

$$J_{1,3} = \varkappa\mu_{1,3}\frac{I_m'(\varkappa\mu_{1,3})}{I_m(\varkappa\mu_{1,3})},$$

$$J_{2,4} = \varkappa\mu_{2,4}\frac{K_m'(\varkappa\mu_{2,4})}{K_m(\varkappa\mu_{2,4})},$$

$$\mu_{1,3}^2 = \frac{1}{2}\left[(1 + \varepsilon_1^2) \pm \sqrt{(1 - \varepsilon_1^2)^2 - \frac{32\beta^2}{(x - m)^2}}\right],$$

$$\mu_{2,4}^2 = \frac{1}{2}\left[(1 + \varepsilon_2^2) \pm \sqrt{(1 - \varepsilon_2^2)^2 - \frac{32\beta^2}{(x + m)^2}}\right],$$

$$\varepsilon_{1,2}^2 = 1 - \frac{4}{(x \mp m)^2} - \beta^2[(x \mp m)^2 - 2],$$

$$x = \frac{\omega}{\Omega}, \qquad \varkappa = kR, \qquad \beta = \frac{M}{\varkappa}, \qquad M = \frac{\Omega R}{c}.$$

Here, I_m is the Bessel function of the imaginary argument, K_m is the MacDonald function, and the prime denotes differentiation over the argument.

Consider a series of limiting cases for the most interesting modes $m \geqslant 2$. In the limit of incompressible fluid from (1) we have

$$\omega = \omega^{(0)} = \Omega(-1 + i\sqrt{m^2 - 1}). \tag{2}$$

In a weakly compressible case ($M \ll 1$) and in the long-wave ($kR \ll m$) approximation

$$\omega = \omega^{(0)} + \frac{\varkappa^2\Omega}{m^2 - 1} - \frac{i\Omega}{\sqrt{m^2 - 1}}\left(\frac{\varkappa^2}{m^2 - 1} + m^2M^2\right). \tag{3}$$

We see that compressibility, as in the flat case, partly stabilizes instability. In the inverse limiting case ($kR \gg m$; no restrictions are imposed on M), by using the asymptotics of the Bessel I_m and K_m functions, it is easy to obtain the stability condition for such perturbations:

$$\frac{m^2M^2}{k^2R^2} \gtrsim 2 - \frac{4}{m^2}, \tag{4}$$

coinciding for $m \gg 1$ with the stability condition of perturbation of a flat tangential discontinuity (in the same approximation $k_\parallel \ll k_\perp$)

$$\beta^2 = \frac{k_\parallel^2}{k_\perp^2}M^2 \geqslant 2. \tag{5}$$

In the limit of very short waves ($kR \gg m$), the instability growth rate tends asymptotically to the quantity $\Omega\sqrt{m^2 - 2}$.

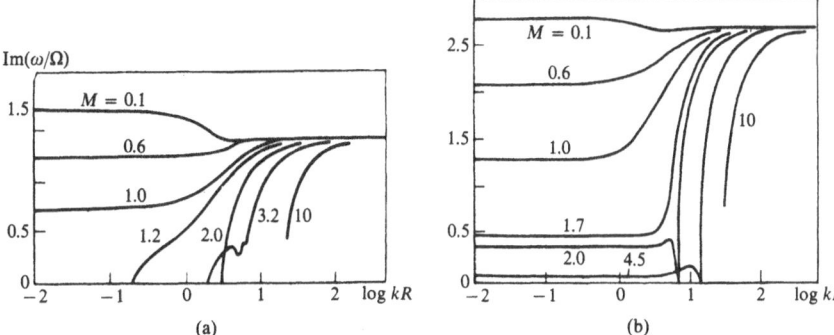

Figure 92. The dependence of the instability growth rate of the cylindrical tangential discontinuity on "Mach number" M (the number is written near each curve) and wavelength (kR) for the modes $m = 2$ (a) and $m = 3$ (b).

The investigation of dispersion equation (1) for the largest-scale modes $m = 2$ and $m = 3$ has been undertaken numerically. The results are: the dependence of the instability growth rate (in Ω units) on the Mach number M and the dimensionless wavelength kR are depicted in Fig. 92(a) and (b). One should emphasize a rather complicated dependence of the instability growth rate in the region $1 \lesssim kR \lesssim 10$ for supersonic discontinuities.

2. Investigate now the possibility of excitation of the flutelike instability in a gravitating cylinder. For that purpose, let us consider the model of an infinitely long cylinder with the discontinuity of the $\rho_0(r)$ density at a distance R from the cylinder axis, by assuming that equilibrium is provided by the resulting action of the centrifugal and gravitational forces and pressure force, so that $g = d\Phi_0/dr - \Omega^2 r \neq 0$. Considering the short-wave (in comparison with the Jeans scale) oscillations, for which the influence of the perturbed gravitational potential is negligible, we obtain the following equations for perturbations of the pressure P and the displacement $\zeta = iv_r/(\omega = m\Omega)$ of the discontinuity surface (the prime denotes differentiation over r):

$$P_1' = \frac{2m\Omega}{r\omega_*} P_1 - g\left[\frac{P_1 + \zeta P_0'}{c^2} - \zeta\rho_0'\right] - (\varkappa^2 - \omega_*^2)\rho_0\zeta, \tag{6}$$

$$\zeta' = \frac{\tilde{k}^2 P_1}{\omega_*^2\rho_0} - \frac{(\omega_* + 2m\Omega)}{\omega_* r}\zeta - \frac{P_1 + \zeta P_0'}{\rho_0 c^2}, \tag{7}$$

where $\omega_* = \omega - m\Omega$, $\varkappa^2 = 4\Omega^2(1 + r\Omega'/2\Omega)$, $\tilde{k}^2 = k^2 + m^2/r^2$, and $c^2 = \gamma P_0/\rho_0$ is the speed of sound. From (6) and (7) follows the boundary conditions for matching of solutions at $r = R$:

$$[\zeta] = \zeta(R + 0) - \zeta(R - 0) = 0, \tag{8}$$

$$[P_1] = \zeta g[\rho_0]. \tag{9}$$

In solving the systems (6)–(9) in the limit $m^2 \gg k^2 R^2$, we obtain the following dispersion equation

$$\sum_{n=1}^{2} \rho_{0n}(-1)^n \left\{ \frac{[(-1)^n \alpha_n + (\omega_* + 2m\Omega)/\omega_* R - g/c_n^2]}{(k^2/\omega_*^2 - 1/c_n^2)} - g \right\} = 0, \quad (10)$$

where

$$\alpha_n^2 = k^2 \left(1 + \frac{g^2}{c_n^2 \omega_*^2} \right) - \frac{\omega_*^2}{c_n^2} - \frac{g}{c_n^2} \frac{\omega_* + 4m\Omega}{R\omega_*} + \frac{2m\Omega(\omega_* + 2m\Omega)}{R^2 \omega_*^2}. \quad (11)$$

Since we are interested only in principal possibility of excitation of the flutelike instability, consider the case of a sufficiently hot medium. Then from (10), for $\rho_{01} = \rho_0 \ (r > R) \neq \rho_{02} = \rho_0(r < R)$, we obtain the growth rate of instability

$$\gamma \approx \left(kgA + \frac{(2 - A^2)m^2\Omega^2}{k^2 R^2} \right)^{1/2}, \quad (12)$$

where

$$A = \frac{\rho_{01} - \rho_{02}}{\rho_{01} + \rho_{02}}.$$

As follows from expression (12), the necessary condition of instability is

$$gA > 0. \quad (13)$$

This implies that for $g = \partial\Phi_0/\partial r - \Omega^2 r > 0$, the flutelike instability develops for $\rho_{01} > \rho_{02}$, while, in the case $g < 0$, for $\rho_{02} < \rho_{01}$. The second summand in (12) is much less than the first one and plays a role of small correction on account of the cylindrical symmetry. Since $|A| < 1$, the effect of curvature exerts a destabilizing action on the flutelike perturbations. Note that the same effect of curvature, as follows from formula (2), in the case of tangential discontinuity of the velocity, exerts a stabilizing influence.

Now, we consider the case opposite to that considered above [cf. (12)] $\lambda \ll a$. For perturbations of the type $\exp[i(kr + m\varphi - \omega t)]$, instead of (12), we obtain the following growth rate of the flute instability:

$$\gamma = \sqrt{g \frac{d \ln \rho_0}{dr} \frac{m^2}{k^2 r^2}}. \quad (14)$$

Of course, the instability condition is similar to (13). The instability growth rate in (14) is much greater than the Jeans growth rate at $m/kr \gg 1$.

For the simplest model of a gaseous uniformly rotating cylinder of radius R, having at $r = 1 \ (R \gg 1)$ density discontinuity ($\rho = \rho_1$, if $r < 1$; $\rho = \rho_2$, if $r > 1$), it is possible to obtain the following expression for the oscillation frequencies ($m > 0$):

$$\omega - m\Omega$$

$$= \frac{\Omega(\rho_2 - \rho_1) \pm \sqrt{\Omega^2(\rho_2 - \rho_1)^2 + m((1/\rho_0) dP_0/dr)_{r=1}(\rho_2 - \rho_1)(\rho_2 + \rho_1)}}{(\rho_1 + \rho_2)}.$$

Hence, in particular, it follows also that the necessary condition of instability is the inequality $(\nabla \rho_0)(\nabla P_0) < 0$.

§ 4 General Treatment of Kinetic Instabilities [35^{ad}]

The problem of kinetic instabilities in a collisionless gravitating system appears if parameters of the system (density, velocity dispersion of stars, rotation, etc.) are such that the main "hydrodynamical" instabilities are absent.[7] Below we shall mainly speak of instabilities of the "beam" type, which are connected with the composite ("heterogeneous") character of the system, say, with the presence of a "beam" in the velocity distribution of particles. Accordingly, in this case one assumes also that the beam instability of the hydrodynamical type produced by all particles of the beam is suppressed too. Then there is only the possibility of kinetic instability connected with the interaction of certain resonance groups of particles with the waves.

In Section 4.1, the heterogeneous system is considered, which consists of a spherical component at rest and of a rotating disk component. Such systems have earlier been treated by a number of authors [84, 109] as a model of the Galaxy; also some possible kinetic instabilities of the "beam" type in such models have been studied in [84]. However, the finite character of the movement of the particles was disregarded (cf. Section 4.1 [35^{ad}]); this has led to incorrect expressions for the growth rate of the kinetic beam instability. In this connection it should be noted that the simple models of real systems used in the investigation of the kinetic instabilities may be divided into two classes. The models of the flat layer of finite thickness or a cylinder with an infinite generatrix (Section 4.3) are not limited in at least one direction. These models have much in common with the homogeneous plasma. In particular, the different kinetic effects in such systems usually have the respective plasma analogies to be found easily. On the other hand, in the models of disks, spherical systems, or ellipsoids, the movement is finite in all directions. The kinetic effects in such systems are due to the resonances between the waves and the orbital movement of the particles.[8]

The solution of the problem under Section 4.1 as well as similar problems for the systems with a different geometry (Section 4.3) may be given in the general form, and it is separated into two independent stages: (1) derivation of the formulae describing the effect of the interaction of the resonance particles with the wave and (2) calculation of the wave energy. For a qualitative solution (i.e., to answer the question whether either instability takes place or the wave damps) it is sufficient to solve two simpler

---

[7] As "hydrodynamical" we understand instabilities in whose excitation all the particles take part and not a certain singled-out group of particles.

[8] Recently, in plasma physics, such effects have also begun to be studied in connection with investigations of the plasma instabilities in closed traps: tokamaks, Earth's magnetosphere, etc.

problems: to determine *the sign* of the wave energy and *the sign* of the variation rate of this energy due to the interaction with resonance stars.

For the heterogeneous model described above, one may, in the investigation of the beam instability, use the results attained by Lynden-Bell and Kalnajs, who have derived general formulae for the wave energy and its variation rate in the disk system [289].

Similar formulae for other systems are derived [35^{ad}] in Section 4.2 (sphere) and Section 4.3 (cylinder, flat layer). Section 4.3 investigates the beam instability. It has already been treated earlier on the simplest model of the cylinder with circular orbits of particles (in §1, [88]). Below we suggest the most general approach to the investigation of this point, and, in addition, we consider the beam instability in a gravitating layer.

In Section 4.2, on the basis of the general formulae to be derived therein, we briefly consider the effects connected with a possible presence in the center of the stellar system of a massive formation of the black hole type. Interaction with the excited waves leads in this case to falling of part of resonance stars onto "the black hole." Such a process may in principle serve as one of the possible mechanisms of luminosity of such objects. Excitation of the waves in the systems with stellar orbits rather extended in radius may be due to, say, Jeans instability (cf. §5–7, Chapter III).

It should be noted that the problem of the stars falling on the black hole, due to diffusion of stars caused by collisions, has earlier been considered in a number of papers; note here, for instance, the theory of Dokuchaev and Ozernoy [8^{ad}]. We shall (under section 3) briefly discuss the role of some *collective* effects; further (§8, Chapter IX) we shall again turn to this problem and investigate it in some detail.

4.1 Beam Effects in the Heterogeneous Model of a Galaxy

In this subsection we shall use, as given, some formulae and conclusions which were obtained by Lynden-Bell and Kalnajs in their work [289], devoted to analysis of resonance interactions between the stars of a disk galaxy and a spiral density wave. A detailed account of this important work can be found below, in §2, Chapter XI.

Let us consider the galactic model in the form of superposition of the flat and spherical subsystems. We shall assume that dispersion properties of disturbances are determined by the flat component. In this case, as was shown by Lynden-Bell and Kalnajs [289], the wave energy in epicyclic approximation is given by the formula

$$\delta E \simeq - \frac{\Omega_p}{16\pi^2} \iint dI_1 \, dI_2 \sum_{l=1}^{\infty} \frac{4l^2 m^2 \Omega_1 (\Omega_2 - \Omega_p)(-\partial F_d/\partial I_1)}{|l^2\Omega_1^2 - m^2(\Omega_2 - \Omega_p)^2|^2} \cdot |\psi_{lm}|^2, \quad (1)$$

in which $F_d(I_1, I_2)$ is the star distribution function of the flat component. The remaining notations and other details are given in §2, Chapter XI; here it is only important for us that in the case when the corotation radius

$(\Omega_2 = \Omega_p)$ lies on the galactic periphery, the sign of the wave energy is negative: $\delta E < 0$. The rate of wave energy change due to the resonance interactions with stars is given by formula (16), §2, Chapter XI (for the derivation, see the same place), which we write in the following convenient form:

$$\frac{dE}{dt} = \frac{dE_s}{dt} + \frac{dE_d}{dt}$$

$$= \iint \frac{dEdL}{\Omega_1} \sum_l \left(\omega^2 \frac{\partial F}{\partial E} + \omega m \frac{\partial F}{\partial L} \right) \delta(\omega - l\Omega_1 - m\Omega_2) \cdot |\psi_{lm}|^2, \qquad (1')$$

where $F = F_s + F_d$ is the total distribution function of the system in the disk plane and L is the angular moment.

We may separately consider the interaction of the wave with stars of the spherical component (dE_s/dt). In particular, for the case of the isotropic distribution of these stars $(\partial F_s/\partial L = 0)$ under the natural condition $\partial F_s/\partial E < 0$ we get $dE_s/dt < 0$. This corresponds to the "beam" instability of the wave with negative energy (we assume so far that $|dE_d/dt| < |dE_s/dt|$).

Note that the solution of the problem in the local approximation leads to the resonance of the kind $\omega = kv$ where k is the radial wave number $(kr \gg m)$ and v is the radial velocity of stars. In reality, however, the interaction of stars of the spherical component with the wave has a nonlocal character, and the correct condition of the resonance $\omega = l\Omega_1 + m\Omega_2$ [cf. (1')] involves the frequencies $\Omega_1(E, L)$ and $\Omega_2(E, L)$ rather than velocities. The expression for the growth rate of instability $\gamma = -\dot{E}_s/2\delta E$ may be obtained from formulae (1) and (1'). The local growth rate $\gamma(r)$ can be derived in a natural way only in the description of the interaction of a tightly twisted wave with the stellar subsystems, whose orbits are close to circular (cf. [290a]). Namely for such subsystems it is easy to determine, as in [290a], the local density of energy sources (apart from the local density of the wave energy).

The fact of instability (or damping) of the wave is defined not only by the wave resonance with the stars of the spherical component but also with stars of the flat constituent. The resonance interactions of the wave and stars of the disk subsystem in the epicyclic approximation are comprehensively[9] investigated by Lynden-Bell and Kalnajs [289] (see above). The main resonances in usual galaxies are of two types: the inner Lindblad resonance $(\omega = l\Omega_1 + m\Omega_2$ for $l = -1$; in the epicyclic approximation $\Omega_2 = \Omega(r)$ is the angular rate of rotation, $\Omega_1 = \varkappa = 2\Omega(1 + r\Omega'/2\Omega)^{1/2}$ is the epicyclic frequency), and the corotational resonance $\omega = m\Omega(r)$. The waves of negative energy are amplified at the corotational resonance and damping at the inner Lindblad resonance. The wave–star interaction effects of spherical and flat subsystems should, generally speaking, be considered jointly; moreover, the resulting effect may be of any sign. In this respect, the situation here is essentially different from that to be discussed below (under

---

[9] Note in particular that the kinetic "drift" instabilities which are treated by M. N. Maksumov (e.g., [81]) are nothing other than a result of the wave–star interaction in the region of the corotation resonance.

section 3), where the resonance effects in the main subsystem may be made negligible (so that the effects of the "beam-likeness" are manifested in the pure form) if the velocity of the beam is high as compared to the thermal velocity of the stars of the "background."

Nevertheless, even in case, say, the eigenoscillations in the heterogeneous system, with due regard for the joint influence of the resonance particles of both subsystems, are damping, the effect of the wave amplified by the stars of the spherical subsystem may be of some interest, if the wave itself is originally excited by some external action. For example, it may be initiated, in the region of the corotational radius, by a bar or barlike excitation at the galactic center. With its further propagation toward the center, such a wave is amplified by giving out its energy to the stars of the spherical component of the galaxy. For the system of the type of our Galaxy, the conditions of amplification of the wave appear to be too artificial: it is more probable that the influence of the inner Lindblad resonance of the disk component would be decisive and the wave would damp in its propagation. It is possible that here it is necessary, as suggested in $[72^{ad}, 79^{ad}]$, to take into account the dissipative phenomena in the gaseous layer of the Galaxy. The effect of the wave–star interaction of the spherical constituent is of real interest in case of colder flat systems, with very narrow resonance regions, or in those cases close to the uniform rotation of the galaxy when the inner resonance is absent (here, however, the role of the corotational resonance in the disk subsystem should be of importance).

The conclusion on excitation of the waves by the stars of the spherical constituent is obtained above, strictly speaking, only for the isotropic distribution $(\partial F_s/\partial L = 0)$. In the case $F_s = F_s(E, L)$, one needs a more detailed investigation, which is impossible to make in the general form and should be performed for concrete distribution functions or their series (one can simply write only some sufficient conditions of instability). No principal difficulties are presented in this, but one has to do a considerable amount of technical work. The same is valid for different generalizations of the consideration given above (for the investigation of the "open" waves in the disk system, for an accurate quantitative analysis of the wave interaction effect with resonance stars in the Galaxy with due regard for the competition between the spherical and disk subsystems, etc.).

Let us concern ourselves with the problem of the nonlinear generalization of the theory. The quasilinear theory may be constructed in the standard way; this theory gives the answers to the standard questions. This is first of all the question of the relaxation of the resonance stars' distribution function. The resonance conditions, i.e., the equalities $\omega = l\Omega_1(E, L) + m\Omega_2(E, L)$, determine (at $l = 0, \pm 1, \cdots$) at the plane E, L, some family of curves. It was to be expected that in the statement of the problem[10] usual for the

---

[10] The problem of the self-consistent relaxation of distribution functions of both subsystems, taking into account the nonlinear drift of barlike disturbances' frequency in the galactic center, which may serve as the source of spiral waves, is more interesting to our minds.

quasilinear theory, "a plateau" will form in the vicinity of these curves so that the expression in the brackets in formula (1') goes to zero.

Probably the most interesting question from the point of view of possible applications is the second question—on the adiabatic relaxation of non-resonance stars' distribution function of the flat subsystem under an influence of the instability [21[ad]]. In the paper [21[ad]], which was specially devoted to the problem of relaxation under acting the spiral structure, in reality, an influence onto the star distribution ("adiabatic heating") of the gravitational noises with an isotropic spectrum, but not of the spiral density waves, was considered. The latter are most probably narrow one-dimensional wave packets ($m = 2$, $k_r \in [k_{r0}, k_{r0} + \Delta k_r]$, $\Delta k_r \ll k_{r0}$). A deformation of the distribution function (which has a more complicated character) under the influence of such spiral disturbances may be determined from the formulae given in the work [32[ad]] (see Section 1.3, Chapter VII).

4.2 Influence of a "Black Hole" at the Center of a Spherical System on the Resonance Interactions Between Stars and Waves

Derive first of all the general formulae for the variation rate of the angular momentum and the stellar energy of a spherically symmetrical stellar system, due to their resonance interaction with the waves of a given frequency ω [34[ad]], 35[ad]]. These formulae are quite similar to the respective disk formulae (1) and (1') and may be derived by the "Lagrange" method used in [289] (see §2, Chapter XI). We shall obtain them, however, by a somewhat different method (more formal). By writing the linearized kinetic equation in the action-angle variables [cf. (7) in §4, Appendix] and integrating it along the path of the unperturbed movement of the particle, we readily find the perturbation of the distribution function (formula (10), §4, Appendix):

$$f_1 = -\frac{1}{(2\pi)^3} \sum_{l_1 l_2 l_3} \Phi_{l_1 l_2 l_3} \frac{e^{i(l_1 w_1 + l_2 w_2 + l_3 w_3)}}{\omega - (l_1 \Omega_1 + l_2 \Omega_2 + l_3 \Omega_3)}$$

$$\times \left(l_1 \frac{\partial f_0}{\partial I_1} + l_2 \frac{\partial f_0}{\partial I_2} + l_3 \frac{\partial f_0}{\partial I_3} \right) + \text{c.c.}, \tag{2}$$

where $f_0(I_1, I_2, I_3)$ is the distribution function of stars, I_1, I_2, I_3 are the actions due to the coordinates r, θ, φ, respectively, $\Omega_j = \partial E/\partial I_j$ are the frequencies, and

$$\Phi_{l_1 l_2 l_3}(I_1, I_2, I_3) = \int_0^{2\pi} \int_0^{2\pi} \int_0^{2\pi} dw_1 \, dw_2 \, dw_3 \, \Phi_1(I_i, w_i) e^{-i(l_1 w_1 + l_2 w_2 + l_3 w_3)}.$$

Φ_1 is the perturbed potential; w_1, w_2, w_3 are the angular variables.

Using the equation of motion of the particle, we find

$$\frac{dL}{dt} = -\left(\frac{\partial}{\partial w_2} + \sigma \frac{\partial}{\partial w_3}\right)\Phi_1, \tag{3}$$

$$\frac{dE}{dt} \equiv \frac{d}{dt}\left(\frac{v^2}{2} + \Phi\right) = -\frac{\partial \Phi_1}{\partial t}, \tag{4}$$

where $\sigma = \mathrm{sgn}(I_3)$. For the angular momentum variation average over time, as well as for energy, we have

$$\frac{d\langle L \rangle}{dt} = \int d\Gamma \, f_1 \frac{dL}{dt}, \qquad \frac{d\langle E \rangle}{dt} = \int d\Gamma \, f_1 \frac{dE}{dt}, \tag{5}$$

where $d\Gamma = dI_1 \, dI_2 \, dI_3 \, dw_1 \, dw_2 \, dw_3$ is the element of the phase volume of the system. By Eqs. (2)–(5) one may obtain the sought-for expressions:

$$\frac{d\langle L \rangle}{dt} = -\frac{2\gamma}{(2\pi)^3} e^{2\gamma t} \int dI_1 \, dI_2 \, dI_3 \sum_{l_1 l_2 l_3} \left(l_i \frac{\partial f_0}{\partial I_i}\right)$$

$$\times (l_2 + \sigma l_3) \frac{|\Phi_{l_1 l_2 l_3}|^2}{|l_1\Omega_1 + l_2\Omega_2 + l_3\Omega_3 - \omega|^2}, \tag{6}$$

$$\frac{d\langle E \rangle}{dt} = -\frac{2\gamma}{(2\pi)^3} e^{2\gamma t} \mathrm{Re}(\omega) \int dI_1 \, dI_2 \, dI_3 \sum_{l_1 l_2 l_3} \left(l_i \frac{\partial f_0}{\partial I_i}\right)$$

$$\times \frac{|\Phi_{l_1 l_2 l_3}|^2}{|l_1\Omega_1 + l_2\Omega_2 + l_3\Omega_3 - \omega|^2}, \tag{7}$$

where $\gamma = \mathrm{Im}(\omega)$. In case the instability growth rate is small, $\gamma \to 0$, from (6) and (7) we get

$$\frac{d\langle L \rangle}{dt} = -\frac{1}{(2\pi)^2} \int dI_1 \, dI_2 \, dI_3 \sum_{l_1 l_2 l_3} l_i \frac{\partial f_0}{\partial I_i} (l_2 + \sigma l_3)$$

$$\times |\Phi_{l_1 l_2 l_3}|^2 \delta(\omega - l_1\Omega_1 - l_2\Omega_2 - l_3\Omega_3), \tag{8}$$

$$\frac{d\langle E \rangle}{dt} = -\frac{\omega}{(2\pi)^2} \int dI_1 \, dI_2 \, dI_3 \sum_{l_1 l_2 l_3} l_i \frac{\partial f_0}{\partial I_i}$$

$$\times |\Phi_{l_1 l_2 l_3}|^2 \delta(\omega - l_1\Omega_1 - l_2\Omega_2 - l_3\Omega_3). \tag{9}$$

From (8) and (9) it is easy to see that the exchange of energy and moment between the particles and the wave occurs on the resonances $\omega = l_1\Omega_1 + l_2\Omega_2 + l_3\Omega_3$. We turn now to the variables E, L_z, L from the action variables

I_1, I_2, I_3 in Eq. (8), then for the distribution function $F = F(E, L)$ we obtain

$$\frac{d\langle L \rangle}{dt} = - \frac{1}{(2\pi)^2} \iint \frac{dE \, dL \, dL_z}{\Omega_1} \sum_{l_1 l_2 l_3} \left[\omega \frac{\partial F}{\partial E} (l_2 + \sigma l_3) + (l_2 + \sigma l_3)^2 \frac{\partial F}{\partial L} \right]$$

$$\times |\Phi_{l_1 l_2 l_3}|^2 \delta(\omega - l_1 \Omega_1 - l_2 \Omega_2 - l_3 \Omega_3). \tag{10}$$

With respect to the frequency spectrum of waves, which may be excited in the system, one may make some assumptions which appear to be natural. We deal with the spherically symmetrical stellar systems (galaxies of the type $E0^{11}$ or spherical clusters of stars) in which, according to observations the stellar orbits are very much elongated in the radial direction. Such a system may easily be unstable "by Jeans" with respect to "merging" of the neighboring nearly radial orbits of stars: the velocity dispersion in the tangential directions, which make the system stable, in this case may turn out to be insufficient. It is clear that the development of this instability will "heat up" the system in the transversal (toward the radius) directions, so that finally we probably will have a spherical system brought to the stability boundary with the waves excited in it with the frequencies[12] $\omega \simeq 0$. Assuming in formula (10) that $\omega = 0$, we have

$$\frac{d\langle L \rangle}{dt} = - \frac{1}{2\pi^2} \iint \frac{dE \, dL \, dL_z}{\Omega_1} \sum_{l_1 l_2 l_3} (l_2 + \sigma l_3)^2 \frac{\partial F}{\partial L}$$

$$\times |\Phi_{l_1 l_2 l_3}|^2 \delta(l_1 \Omega_1 + l_2 \Omega_2 + l_3 \Omega_3) > 0 \quad \text{if} \frac{\partial F}{\partial L} > 0. \tag{11}$$

The location of the resonances for $\omega = 0^{13}$ is determined by the frequencies $\Omega_{1, 2, 3}(E, L)$, i.e., by the equilibrium model of the system; the values $\Phi_{l_1 l_2 l_3}$ may be found by the formulae derived in §4, Appendix.

Let us now take into account the effects caused by the presence of a "black hole" at the system center. Let us suppose that a "hole" has really formed at the center of a galaxy. It must be formed by stars with small angular momenta, which leave the system of stars in the vicinity of the central body. As a result, the immediate vicinity of the "hole" is enriched with stars with near-circular orbits; i.e., in the vicinity, in the region of sufficiently small stellar angular momenta, we have $\partial F/\partial L > 0$. The latter condition is the necessary condition for the kinetic loss-cone instability (of course, for waves with negative energy, the necessary condition of the instability has an opposite

---

[11] The qualitative conclusions are valid also for other galaxies of the elliptical type, as well as for spherical components or nuclei of spiral galaxies.

[12] In the general case, the eigenfrequencies of a spherically symmetrical system should be determined by the numerical solution of dispersion equation (23), §4, Appendix.

[13] Note that in the Coulomb potential $\Omega_1(E) = \Omega_2(E)$ so that in this case all the stars formally are resonant (for corresponding l_1, l_2, l_3).

sign), leading to an abnormally rapid (compared with collisions) filling of the loss cone in momentum space. This means that a stellar flow onto the "black hole" can considerably exceed a similar flow caused by star–star collisions. Since the accretion stellar flow onto the central body is responsible for the observed radiation flux from the central region, the existence of a loss-cone instability imposes an upper limit on the central body mass. Readers interested in more detail of the physics of the kinetic loss-cone instability, its consequences, and in some figures connected with applications are referred to the paper [39[ad]].

4.3 Beam Instability in the Models of a Cylinder and a Flat Layer

It is easy to show (for example, by the method described in §4, Appendix) that the variation rate of the energy of resonance particles in a gravitating cylinder with an infinite generatrix is given by the formula

$$\dot{E} = -\frac{\omega}{2\pi} \iiint dI_1\, dI_2\, dv_z \sum_l \left(l\frac{\partial F}{\partial I_1} + m\frac{\partial F}{\partial I_2} + k_z\frac{\partial F}{\partial v_z} \right)$$

$$\times |\Phi_{k,l,m}|^2 \delta(\omega - k_z v_z - l_1\Omega_1 - l_2\Omega_2), \tag{12}$$

where the wave with a frequency ω and the wave number k_z along the generatrix (z) is considered; $F = F(I_1, I_2, v_z)$ is the distribution function; the remaining notations have the same meaning as those in Section 4.2. On the other hand, the wave energy

$$\delta E = -\frac{\omega}{2\pi^2} \iiint dI_1\, dI_2\, dv_z \sum_l \left(l\frac{\partial F}{\partial I_1} + m\frac{\partial F}{\partial I_2} + k_z\frac{\partial F}{\partial v_z} \right)$$

$$\times |\Phi_{k,l,m}|^2 \frac{1}{|\omega - k_z v_z - l\Omega_1 - m\Omega_2|^2}. \tag{13}$$

Formulae (12) and (13) are valid for arbitrary cylindrical models; as $k_z \to 0$, (12) and (13) coincide with the respective formulae in Section 4.1. On the basis of (12) and (13), the kinetic instabilities in these models may be investigated in the general form (the instability growth rate or the wave damping decrement is $\gamma = -\dot{E}/2\delta E$).

The case of axially symmetrical perturbations ($m = 0$) of a cylinder with stellar orbits close to circular (epicyclic approximation) is very simple. Then, considering the perturbations with sufficiently small k_z, the wave energy may be calculated approximately by omitting in (13) the term proportional to k_z, i.e., by assuming the perturbations to be "flutelike":

$$\delta E \simeq -\frac{\omega^2}{2\pi^2} \iint dI_1 \, dI_2 \sum_l \frac{\partial f_0}{\partial I_1} |\Phi_{k,l,m}|^2 \frac{4l^2\Omega_1}{|\omega^2 - l^2\Omega_1^2|^2},$$

(14)

$$f_0 = \int F \, dv_z.$$

With the natural requirement that $\partial f_0/\partial I_1 < 0$ from (14) it follows that $\delta E > 0$. Consequently, the wave will be unstable only when the resonance stars, on the average, lose their energy in their interaction with it, i.e., if $\dot{E} < 0$. The distribution function is $F = F_0 + F_b$, where F_0 corresponds to the basic plasma ("background") and F_b, to the beam. It is evident that if the velocity of the beam V is high as compared to the thermal velocity of the stars of the background V_{T0}, then the contribution to $\dot{E}$ of the resonance interaction of the waves with the phase velocities $\omega/k_z \sim V$[14] with the stars of the background may be neglected (because there are practically no stars of this kind). After that, there remains only the effect of the interaction with the stars of the beam, and it is evident that such a situation appears to be possible only due to the formally infinite extension of the cylinder along the z-axis; in case of finite systems, the situation is quite different, cf. Section 4.2.

Consider, for the sake of simplicity, the case of a sufficiently narrow beam in the velocity space, $k_z|\Delta v_z| < \max(\Omega_1)$.[15] In this case, any one resonance practically "works," with a fixed l (dependent on the value k_z). For the Cherenkov resonance $l = 0$ ($\omega = k_z v_z$) we get

$$\dot{E} = -\frac{1}{2\pi}\left(\frac{\omega}{k_z}\right)k_z^2 \iiint dI_1 \, dI_2 \, dv_z \frac{\partial F_b}{\partial v_z} |\Phi_{k,0,0}|^2 \delta(\omega - k_z v_z), \quad (15)$$

so that $\dot{E} < 0$ (which means instability) under the condition that for the wave in question $(\partial F_b/\partial v_z)|_{v_z=\omega/k_z} > 0$. For other resonances ($l \neq 0$), a definite contribution is made by the term [in (12)] proportional to $\partial F/\partial I_1$, and the calculation gets somewhat complicated; in the general case, one has to specify the distribution function of the beam F_b, to know the equilibrium state determining $\Omega_1, \Omega_2(E, L)$ and the values of the frequency ω of the flutelike mode in question.

---

[14] Note that among the eigenfrequencies ω of flutelike oscillations of cylindrical models there is always $\omega \sim \Omega_1$ (cf. §5, Chapter II). For such waves, the phase velocity ω/k_z may be arbitrarily high (for respectively small k_z).

[15] The Jeans instability in both subsystems may here be practically completely suppressed (cf. §2, Chapter II).

For the model of a flat gravitating layer the formulae similar to (12) and (13) have the form

$$\dot{E} = -\omega \iint dI \, dv_x \sum_{n=0}^{\infty} \left(n\frac{\partial F}{\partial I} + k_z \frac{\partial F}{\partial v_x} \right)$$

$$\times |\Phi_{k,n}|^2 \delta(\omega - k_x v_x - n\Omega), \tag{16}$$

$$\delta E = \frac{\omega}{2\pi} \iint dI \, dv_x \sum_{n=0}^{\infty} |\Phi_{k,n}|^2$$

$$\times \frac{(\partial F/\partial E)n\Omega + k_x(\partial F/\partial v_x)}{|\omega - kv_x - n\Omega|^2}, \tag{17}$$

where the distribution function $F = F(I, v_x)$, I is the action associated with the coordinate perpendicular with the layer, $\Omega = \partial E/\partial I$, the wave $\sim e^{ik_x x}$. The analysis of the beam instability for $\partial F/\partial I < 0$ in this case is performed (very simply for the Cherenkov resonance $n = 0$) practically in the same way as for the cylinder, and the same results are obtained. Therefore, we shall not consider this point separately but consider, on the basis of formulae (16) and (17) a somewhat distinguished case of a homogeneous flat layer; for the case of a homogeneous cylinder with circular orbits of all particles (§1) one may consider by the analogy.

The distribution function of a homogeneous layer with the thickness $h = 1$ and density $\rho_0 = 1$ may be represented in the form (§1, Chapter I)

$$F_0 = \varphi_0(v_x)(1 - z^2 - v_z^2)^{-1/2}\theta(1 - z^2 - v_z^2),$$

where θ is the unit step function, $4\pi G\rho_0 = 1$, G is the gravitational constant; for the homogeneous cylinder with circular orbits $F_0 = \varphi_0(v_x)\delta(E - L\Omega)$, Ω is the angular velocity of rotation of the particles. The derivative $\partial F_0/\partial E$ in both cases does not have a definite sign everywhere; in case of the layer, moreover, the integrals in formulae (16) and (17) are formally divergent and, for example, formula (17) for the oscillative energy at $k_x = 0$ should be written in the form

$$\delta E = -\frac{\omega}{2\pi} \int dE \, f_0(E) \sum_{n=0}^{\infty} \frac{n\Omega}{|\omega - n\Omega|^2} \frac{\partial |\Phi_n|^2}{\partial E}. \tag{18}$$

Since, below, in the consideration of the wave–beam interaction we restrict ourselves by the Cherenkov resonance ($n = 0$), formula (16) again yields

$$\dot{E} = -\left(\frac{\omega}{k_x}\right)k_x^2 \iint dI \, dv_x |\Phi_0|^2 \frac{\partial F_b}{\partial v_x} \delta(\omega - kv_x). \tag{19}$$

The sign of the wave energy from formula (18) is not determined without calculations; however, from (19) it is easy to see that instability is present for any sign of δE: if $\delta E > 0$, then all the waves with such k_x, for which $(\partial F_b/\partial v_x)|_{v_x = \omega/k_x} > 0$, are unstable; if $\delta E < 0$, then those k_x are unstable for which $(\partial F_b/\partial v_x)|_{v_x=\omega/k_x} < 0$. By the way, in the simple case in question, we know (cf. Problem 2, Chapter I) all the eigenfrequencies and eigenfunctions for all "perpendicular" oscillations $(k_x = 0)$: $\Phi_1^{(n)}(z) = P_{n+2}(z) - P_n(z)$ and $P_n(z)$ are the Legendre polynomials, $n = 0, 1, 2, \cdots$. Therefore, analytical calculations here may be performed to an end, and, for example, the growth rates of the beam instability may be obtained. The values of $\Phi_l(E)$ involved in (18) and (19) are defined as the coefficients at e^{ilw} (w is the angular variable) in the expansion of the potential $\Phi_1(z)$. Since $z = \sqrt{2E}\cos w$, $E = I$, and according to the addition theorem for the Legendre polynomials [42],

$$
P_n(\sqrt{2E}\cos w) = P_n(\sqrt{1-2E})P_n(0) + \sum_{k=1}^{\infty} \frac{\Gamma(n-k+1)}{\Gamma(n+k+1)}
$$

$$
\times\, P_n^k(\sqrt{1-2E})P_n^k(0)(e^{ikw} + e^{-ikw}), \tag{20}
$$

(the P_n^m are the associated functions), then

$$
\Phi_k = \frac{\Gamma(n+3-k)}{\Gamma(n+3+k)}\, P_{n+2}^k(0)P_{n+2}^k(\sqrt{1-2E})
$$

$$
-\frac{\Gamma(n-k+1)}{\Gamma(n+k+1)}\, P_n^k(0)P_n^k(\sqrt{1-2E}), \qquad k \neq 0,
$$

$$
\Phi_0 = P_{n+2}(\sqrt{1-2E})P_{n+2}(0) - P_n(\sqrt{1-2E})P_n(0), \tag{21}
$$

where $|k| < n + 2$. Take, for example, the case $n = 0$ $(k = 0, \pm 2)$ corresponding to homogeneous extensions–compressions of the layer; $\omega = \sqrt{3}$, and determine the sign of the wave energy. Since $\Phi_0 = -\frac{3}{2}(1 - E)$ and $\Phi_2 = \frac{3}{4}E$, then

$$
\delta E \sim \tfrac{9}{4}\omega \int dE\ EF(E)\left(\frac{1}{(\omega-2)^2} - \frac{1}{(\omega+2)^2}\right) = \frac{1}{2\pi}\cdot\frac{2}{3}\cdot 54 > 0.
$$

The beam instability in the model of the homogeneous cylinder with circular orbits of all the particles was earlier considered in §1. So we calculate, for comparison, by the method described above, the increment of the beam instability for this case. As the disturbed potential $\Phi_1 = J_0(k_\perp R)$ and $R = \sqrt{a - b\cos w_1}$, $a = I_2 + 2I_1$, $b = 2\sqrt{I_1 I_2 + I_1^2}$, then, using the addition theorem for cylindrical functions [42],

$$
J_0(k_\perp R) = J_0(k_\perp\rho)J_0(k_\perp r) + \sum_{l=1}^{\infty} J_l(k_\perp\rho)J_l(k_\perp r)(e^{ilw_1} + e^{-ilw_1}),
$$

where $r = (\sqrt{a+b} + \sqrt{a-b})/2$ and $\rho = (\sqrt{a+b} - \sqrt{a-b})/2$, we obtain $\Phi_l = \Phi_{-l} = J_l(k_\perp\rho)\cdot J_l(k_\perp r)$. In the limiting case of the cylinder with the circular orbits

$$
I_1 \to 0, \quad a \to I_2, \quad b \to 0, \quad r \to \sqrt{a} \to \sqrt{I_2}, \quad \text{and} \quad \rho \to 0.
$$

In the sum for δE (as well as for $\dot{E}$), then there is only one term (we consider Cherenkov resonance $l = 0$):

$$\delta E = \rho_0 \pi^3 k_\perp^2 \frac{4\omega^2}{(\omega^2 - 4)^2} \int_0^1 dI_2 \, J_1^2(k_\perp\sqrt{I_2}), \tag{22}$$

$$\dot{E} = - 4\pi^4 \rho_0 \omega F_b'|_{v=\omega/k_z} \cdot \int_0^1 dI_2 \, J_0^2(k_\perp\sqrt{I_2}). \tag{23}$$

Consider some rotational oscillatory branch with $k_\perp \gtrsim 1$. For $k_z \to 0$, $k_\perp$ is determined from $J_0(k_\perp) = 0$ (see §2, Chapter II). In this case the integrals in (22) and (23) are identical $[=J_1^2(k_\perp)]$ [42], so that the increment of instability is equal to

$$\gamma = - \frac{\dot{E}}{2\delta E} = - \frac{4\alpha\sqrt{\pi}}{k_z k_\perp^2 v_T^2} \left(1 - \frac{k_z V}{\sqrt{2}}\right) \exp\left[- \frac{(\sqrt{2} - k_z V)^2}{k_z^2 v_T^2}\right]$$

($\omega^2 = 2$, v_T is the thermal velocity of the beam with Maxwellian particle distribution, and α is the ratio of densities of the beam and the main system), which is coincident with the results of calculations in §1.

CHAPTER VII

Problems of Nonlinear Theory

Either the well was very deep,
or she fell very slowly, for she had plenty of time
as she went down to look about her
and to wonder what was going
to happen next...
L. CARROLL, *Alice in Wonderland.*

In this chapter we shall touch upon some important problems of nonlinear theory of density wave evolution in a gravitating medium. Nonlinear theory is in its first stage only. It faces, for example, the development of gravitating medium turbulence theory, and now the tasks set forth in this chapter are its unfinished foundation.

§ 1 Nonlinear Stability Theory of a Rotating, Gravitating Disk

1.1 Nonlinear Waves and Solitons in a Hydrodynamical Model of an Infinitely Thin Disk with Plane Pressure [90a, 20[ad], 89a, 31[ad]]

1.1.1. Statement of the Problem and Initial Equations. The question of the possibility of stationary solutions in the form of solitons travelling in the gravitating disk plane is extremely interesting in itself and possibly as one bearing a relation to the galactic spiral structure problem (cf. Chapter XI).

In this section we shall investigate nonlinear processes in the infinitesimally thin, rotating disk near its stability boundary $[|\omega^2| \ll \Omega_0^2$, where ω is a frequency of the perturbation wave and $\Omega_0(r)$ is an angular disk velocity]. For the sake of simplicity we restrict ourselves to consideration of a short-wavelength range of spectrum.

37

Let us proceed from the system of equations commonly used in the analysis of perturbation located at the plane of a gravitating disk ($v_z = 0$) (see Chapter V):

$$\frac{\partial v_r}{\partial t} + v_r \frac{\partial v_r}{\partial r} + \frac{v_\varphi}{r}\frac{\partial v_r}{\partial \varphi} - \frac{v_\varphi^2}{r} = -\frac{\partial \psi}{\partial r} - \frac{1}{\sigma}\frac{\partial p}{\partial r}, \tag{1}$$

$$\frac{\partial v_\varphi}{\partial t} + v_r \frac{\partial v_\varphi}{\partial r} + \frac{v_\varphi}{r}\frac{\partial v_\varphi}{\partial \varphi} + \frac{v_r v_\varphi}{r} = -\frac{1}{r}\frac{\partial \psi}{\partial \varphi} - \frac{1}{r\sigma}\frac{\partial p}{\partial \varphi}, \tag{2}$$

$$\frac{\partial \sigma}{\partial t} + \frac{1}{r}\frac{\partial}{\partial r}(rv_r\sigma) + \frac{1}{r}\frac{\partial}{\partial \varphi}(v_\varphi\sigma) = 0, \tag{3}$$

$$\frac{1}{r}\frac{\partial}{\partial r}\left(r\frac{\partial \Phi}{\partial r}\right) + \frac{1}{r^2}\frac{\partial^2 \Phi}{\partial \varphi^2} + \frac{\partial^2 \Phi}{\partial z^2} = 4\pi G\sigma\delta(z). \tag{4}$$

$\Phi = \Phi(r, \varphi, z, t)$ is the gravitational potential, $\psi = \psi(r, \varphi, 0, t)$ is the gravitational potential at the disk plane, and p is the plane pressure (a usual pressure we denote by a letter P). We shall consider the oscillations (and the process of instability itself) to be adiabatic. Then for a thin disk, as was shown by Hunter [233] (see Section 1.3), one may write

$$p = \frac{c_s^2}{\varkappa}\sigma_0\left(\frac{\sigma}{\sigma_0}\right)^\varkappa, \qquad c_s^2 \equiv \varkappa\frac{P_0}{\sigma_0}, \qquad \varkappa = 3 - \frac{2}{\gamma}, \tag{5}$$

where γ is an adiabatic index ($P/\rho^\gamma = K \equiv$ const). Stationary quantities have an index 0.

Let us represent quantities $X = (v_r, v_\varphi, \psi, \sigma)$ in the form

$$X = X^0 + X,$$

where X^0 describes a slowly varying part of X and $\tilde{X}$ corresponds to a quickly varying part. Assuming that $|\tilde{X}| \ll |X^0|$ ($v_r^0 = 0$) and also that $r\,\partial \ln \tilde{X}/\partial r \gg m^2$ (the last inequality corresponds to tightly wound spirals, where m is a number of an azimuthal mode), we obtain from (2) and (3)

$$\tilde{v}_r = -\frac{1}{\varkappa_0}\hat{L}\tilde{v}_\varphi, \tag{6}$$

$$\sigma = \frac{\sigma_0}{\varkappa_0}\frac{\partial \tilde{v}_\varphi}{\partial r}, \tag{7}$$

where the operator $\hat{L} = \partial/\partial t + \Omega_0\,\partial/\partial\varphi$ [$\varkappa_0 = (1/r)(\partial/\partial r)(rv_\varphi^0)$] is introduced. Substituting (5)–(7) into (1), we get

$$\hat{L}^2\tilde{v}_\varphi = \left(c_s^2\frac{\partial^2}{\partial r^2} - 2\Omega_0\varkappa_0\right)\tilde{v}_\varphi + \varkappa_0\frac{\partial \tilde{\psi}}{\partial r}$$

$$+ \frac{\varkappa - 2}{2\varkappa_0}c_s^2\frac{\partial}{\partial r}\left[\left(\frac{\partial \tilde{v}_\varphi}{\partial r}\right)^2 + \frac{\varkappa - 3}{3\varkappa_0}\left(\frac{\partial \tilde{v}_\varphi}{\partial r}\right)^3\right] - \varkappa_0\frac{v_\varphi^2}{r}. \tag{8}$$

In deriving this equation we have used the condition of a slow change in the values $\tilde{X}$ in comparison with the disk rotation frequency Ω_0: $\partial \ln \tilde{X}/\partial t \ll \Omega_0$. On the basis of the last inequality, the nonlinear term $v_r \, \partial v_r/\partial r$ in (1) is omitted. The appearance of nonlinear summands in square brackets is associated with the term describing the pressure force.

1.1.2. Nonlinear Dispersion Equation Near the Stability Boundary.
Restricting ourselves in (8) only to linear terms and taking into account the relation between $\tilde{v}_\varphi$ and $\tilde{\psi}$,

$$\tilde{\psi} = -i \frac{2\pi G \sigma_0}{\varkappa_0} \tilde{v}_\varphi \operatorname{sgn} k, \qquad \operatorname{sgn} k = \frac{k}{|k|}, \tag{9}$$

for perturbations of the form

$$\tilde{v}_\varphi(r, t), \tilde{\psi}(r, t) \backsim e^{-i\omega t + ikr}, \tag{10}$$

we arrive at the local Toomre dispersion equation

$$(\omega - m\Omega_0)^2 = \varkappa^2 - 2\pi G \sigma_0 |k| + c_s^2 k^2. \tag{11}$$

Let us consider,[1] first of all, the range of disk equilibrium parameters corresponding to dispersion curves close to the curve which touches the abscissa axis (see Fig. 93). Obviously, the two following conditions, $\omega^2(k_*) = 0$ and $\partial \omega^2(k_*)/\partial k = 0$, are satisfied at the touching point $k = k_*$. Hence one may obtain $k_* = \varkappa/c_s$ and $\varkappa = \pi G \sigma_0/c_s$.

We shall investigate nonlinear evolution of the small-amplitude wave with the wave number k_0 near k_*, $|(k_0 - k_*)/k_*| \ll 1$. Deformation of such a wave is due as is known to the production of multiple harmonics, i.e., of waves with wave number nk_0 ($n = 0, \pm 1, \cdots$). Consequently, let us represent disturbed quantities in the form of Fourier expansions:

$$\tilde{v} = \sum_n v_n \exp(ink_0 r), \qquad \tilde{\psi} = \sum_n \psi_n \exp(ink_0 r). \tag{12}$$

Substituting (12) into (8), taking into account (9), and introducing the notation

$$\omega_l^2 = 2\Omega_0 \varkappa_0 + l^2 k_0^2 c_s^2 - 2\pi G \sigma^0 l |k_0|, \tag{13}$$

---

[1] The general case is considered at the end of this subsection and in Problem 1.

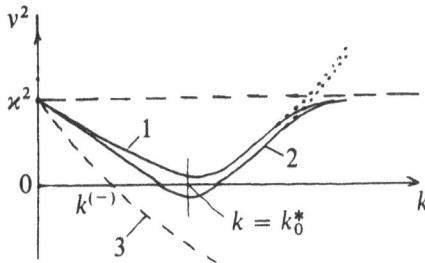

Figure 93. Dispersion curves for rotating collisionless disks; 1 and 2—for nearly marginally-stable disks, 3—for a "cold" disk. Pointed lines correspond to different behavior of $v^2(k)$ for gas disks at large k.

we obtain the equation

$$\sum_l \hat{L}^2 v_l(t, \varphi) \exp(ilk_0 r) = -\sum_l \omega_l^2 v_l(t, \varphi) \exp(ilk_0 r)$$

$$-\frac{\varkappa - 2}{2\varkappa_0} \sum_l lk_0 c_s^2 \left\{ i \sum_n nk_0^2(l + n)v_l(t, \varphi)v_n(t, \varphi) \exp[i(l + n)k_0 r] \right.$$

$$-\frac{\varkappa - 3}{3\varkappa_0} \sum_{n, m} k_0^3 nm(l + n + m)v_l(t, \varphi)v_n(t, \varphi)v_m(t, \varphi)$$

$$\left. \times \exp[i(l + n + m)k_0 r] \right\} + \frac{\varkappa_0}{r} \sum_{l, n} v_l(t, \varphi)v_n(t, \varphi) \exp[i(l + n)k_0 r].$$

(14)

Equation (14) may be written in the form $\sum_l A_l \exp(ilk_0 r) = 0$. The equation $A_0 = 0$ describes the linear approximation considered by Toomre [333]. The equations $A_1 = 0$ and $A_2 = 0$ are the following:

$$\hat{L}^2 v_1(t, \varphi) = -\omega_{k_0}^2 v_1(t, \varphi) + \frac{\varkappa - 2}{2\varkappa_0} k_0^2 c_s^2 [4iv_2(t, \varphi)v_{-1}(t, \varphi)$$

$$-\frac{\varkappa - 3}{\varkappa_0} k_0 |v_1(t, \varphi)|^2 v_1(t, \varphi)] + 2\frac{\varkappa_0}{r} v_2(t, \varphi)v_{-1}(t, \varphi), \quad (15)$$

$$\hat{L}^2 v_2(t, \varphi) = -\omega_{2k_0}^2 v_2(t, \varphi) - \frac{\varkappa - 2}{2\varkappa_0} k_0^3 c_s^2 2iv_1^2(t, \varphi) + \frac{\varkappa_0}{r} v_1^2(t, \varphi), \quad (16)$$

where

$$\omega_{2k_0}^2 = \varkappa_0^2, \quad (17)$$

$$v_2(t, \varphi) = \frac{1}{\hat{L}_{k_0}^2 + \omega_{2k_0}^2} \left(i\frac{2 - \varkappa}{\varkappa_0} k_0^3 c_s^2 + \frac{\varkappa_0}{r} \right) v_1^2(t, \varphi). \quad (18)$$

Taking into account that $\gamma^2 \ll \omega_{2k_0}^2$, $\varkappa_0^2 = k_0^2 c_s^2$, $k_0 r \gg 1$, let us transform (18) into the form

$$v_2(t, \varphi) = \frac{2i(2 - \varkappa)\Omega_0}{\omega_{2k_0}^2} k_0 v_1^2(t, \varphi). \quad (19)$$

Substituting this expression into (15), we finally obtain the following non-linear equation of velocity oscillations of a gaseous gravitating rotating disk with a small but finite amplitude:

$$\hat{L}^2 v_1(t, \varphi) = \gamma_{k_0}^2 v_1(t, \varphi) + \frac{(2 - \varkappa)k_0^2 \Omega_0}{\varkappa_0} \left[\frac{8(2 - \varkappa)\Omega_0 \varkappa_0}{\omega_{2k_0}^2} - (3 - \varkappa) \right]$$

$$\times |v_1(t, \varphi)|^2 v_1(t, \varphi). \quad (20)$$

The nonlinear dispersion equation corresponding to (20) has the form

$$(\omega - m\Omega_0)^2 = -\gamma_{k_0}^2 - \tfrac{3}{2}k_0^2(2 - \varkappa)(\tfrac{5}{3} - \varkappa)|v_1|^2. \quad (21)$$

From (20) or (21) it is seen that the growth of the perturbed velocity amplitude due to instability stops at the level

$$|v_1|^2 = \frac{\gamma_{k_0}^2 \varkappa_0}{3(2 - \varkappa)k_0^2 \Omega_0 (\varkappa - \frac{5}{3})},$$ (22)

while the amplitude of the perturbed density increases up to the value

$$\frac{|\sigma_1|^2}{\sigma_0^2} = \frac{k_r^2}{k_0^2} \frac{\gamma_{k_0}^2}{3(2 - \varkappa)\Omega_0 \varkappa_0 (\varkappa - \frac{5}{3})}.$$ (23)

1.1.3. Solitons. Let us show that Eq. (20) has a stationary solution of the soliton type [90a].

Let a narrow wave packet be excited in the vicinity of the point k_0, $\Delta k/k_0 \ll 1$. Taking into account the dispersion of wave numbers, it is not difficult to represent the function $\gamma^2(k)$ in the form of a series in the vicinity of the point k_0, where this function has a maximum

$$\gamma_k^2 = \gamma_{k_0}^2 + \frac{1}{2} \frac{\partial^2 \gamma_k}{\partial k^2}\bigg|_{k = k_0} (k - k_0)^2 + \cdots.$$ (24)

As follows from (11), $\frac{1}{2}\partial^2 \gamma_k/\partial k^2 = -c_s^2$, $k - k_0 = k_r - k_{0r} \equiv k_{1r}$ since $k_r \gg k_\varphi = m/r$. Hence, instead of (24) we have

$$\gamma_k^2 = \gamma_{k_0}^2 - k_{1r}^2 c_s^2.$$ (25)

Substituting in (20) $\gamma_{k_0}^2$ for γ_k^2 and making use of the expansion (25), proceed in (20) to the coordinate representation. For this purpose, we multiply term by term (20) by $\exp(ik_1 r)$ and integrate over k_1, assuming that

$$\int v_1(k_1, t) \exp(ik_1 r) \, dk_1 = v_1(r, t).$$

As a result, we obtain the equation

$$\hat{L}^2 v_1(r, t) = (\gamma_{k_0}^2 + c_s^2 \Delta_r) v_1(r, t) + 3 \frac{(2 - \varkappa)k_0^2 \Omega_0}{\varkappa_0}$$

$$\times (\tfrac{5}{3} - \varkappa)|v_1(r, t)|^2 v_1(r, t), \quad (26)$$

where $\Delta_r \equiv \partial^2/\partial r^2$.

We turn now to the local rotating frame of reference and select the co-ordinate-temporal dependence $v_1(r, t)$ in the form $v_1(r, t) = V(\xi) = V(r - ut)$. In (26) we proceed to the variable ξ. Finally, we obtain the equation

$$\frac{\alpha^2 d^2 V}{d\xi^2} = \gamma_0^2 V - \beta^2 V^3,$$ (27)

where

$$\alpha^2 = u^2 - c_s^2, \qquad \beta^2 = 3(2 - \varkappa)(\varkappa - \tfrac{5}{3})\frac{\Omega_0}{\varkappa_0} k_0^2.$$ (28)

The solution of Eq. (27) is

$$V(\xi) = \sqrt{2}\,\frac{\gamma_0}{\beta}\,\text{sech}\,\frac{\gamma_0 \xi}{\sqrt{u^2 - c_s^2}}. \tag{29}$$

Thus, as is seen from relations (27)–(29), in a rotating gravitating disk, the existence of two types of solitons is possible.

(1) The supersound solition, $\alpha^2 > 0$, which can be produced and propagated in a weakly unstable disk, $\gamma_{k_0}^2 > 0$ (Fig. 93). Here the equation of state of the disk must satisfy the condition $\beta^2 > 0$. The condition $\beta^2 > 0$ corresponds to the adiabatic index γ, which lies in the range $\frac{3}{2} < \gamma < 2$.

(2) The subsonic soliton, $\alpha^2 < 0$, that can propagate in a stable (according to Jeans; cf. Fig. 93) disk, $\gamma_{k_0}^2 < 0$. Then, the equation of state of the disk must satisfy the condition $\beta^2 < 0$, which corresponds to the adiabatic index $\gamma < \frac{3}{2}$.

1.1.4. Nonlinear Dispersion Equation in the General Case. In concluding let us note that in Problem 1, in the same model we shall derive [31[ad]] the non-linear equation generalizing (21) for the case of an arbitrary location of the wave number k (not necessarily corresponding to the minimum of the curve $\omega^2 = \omega^2(k)$ and not necessarily to the system being at the stability boundary). The derivation is performed by another, possibly simpler and more obvious method (in the Lagrange approach by using an accompanying locally Cartesian coordinate system x, y). Let us write here this equation for the Fourier image $\xi_k(t)$ of one-dimensional displacements of particles

$$\xi(x_0, t) = \sum_k \xi_k(t)e^{ikx_0}$$

(x_0 is the Lagrange coordinate of the particle)

$$\left(\frac{d^2}{dt^2} + \omega_k^2\right)\xi_k = i\sum_{k_1,k_2} F_{k,k_1,k_2}\xi_{k_1}\xi_{k_2} + \sum_{k_1,k_2,k_3} Q_{k,k_1,k_2,k_3}\xi_{k_1}\xi_{k_2}\xi_{k_3}, \tag{30}$$

where $\omega_k^2 = \varkappa_0^2 + k^2 c_s^2 - 2\pi G\sigma_0|k|$, $\varkappa_0$ is the epicyclic frequency, c_s is the sound velocity, σ is the surface density, G is the gravity constant, and

$$F_{k,k_1,k_2} = \delta_{k,k_1+k_2}\left[2\pi G\sigma_0\left(-\frac{k|k|}{2} + k_1|k_1|\right)\right.$$

$$\left. + c_s^2\left(\frac{k^3}{2} - k_1^3 + \frac{\varkappa - 2}{2\varkappa}kk_1k_2\right)\right],$$

$$Q_{k,k_1,k_2,k_3} = \delta_{k,k_1+k_2+k_3}\left\{2\pi G\sigma_0(-\tfrac{1}{6}k^2|k| + \tfrac{1}{2}|k_1 + k_2|^3 - \tfrac{1}{2}|k_1|^3)\right.$$

$$+ c_s^2\left[\tfrac{1}{6}k^4 + \frac{\varkappa - 2}{2\varkappa}kk_1(k_2 + k_3)^3 - \frac{\varkappa - 2}{3\varkappa^2}kk_1k_2k_3\right.$$

$$\left.\left. - \tfrac{1}{2}(k_1 + k_2)^4 + \tfrac{1}{2}k_1^4 - \frac{\varkappa - 2}{2\varkappa}(k_1 + k_2)^2k_1k_2\right]\right\}.$$

Hence, in particular, we find the following generalization of Eq. (21): the nonlinear dispersion equation which can be used in the consideration of one-dimensional wave packets located near the arbitrary wave number k:

$$\omega^2 = \omega_k^2 - \frac{k^2}{\varkappa^2(\omega_{2k}^2 - 4\omega_k^2)} \left\{ 8[(2\varkappa - 1)c_s^2 k^2 - \varkappa\pi G\sigma_0|k|]^2 \right.$$
$$\left. - [(2\varkappa - 1)(5\varkappa - 2)c_s^2 k^2 - 4\varkappa^2\pi G\sigma_0|k|(\omega_{2k}^2 - 4\omega_k^2)] \right\}|\xi_k|^2.$$

Let us point out the simplest partial case of a nonrotating disk at the stability boundary:

$$\omega^2 = -\frac{c_s^2 k^4}{\varkappa^2}(\varkappa - 1)(\varkappa - 2)|\xi_k|^2.$$

Here one also can construct a solition of the envelope, with an arbitrary wave number of "filling" k_0. For real $\omega(k_0)$, these solitons must evidently displace along the disk (in the radial direction) with a group speed $c_g = (d\omega/dk)_{k_0}$. For k_0 corresponding to the curve minimum $\omega^2 = \omega^2(k)$, we arrive (by continuity) at solitons at rest ($c_g = 0$), which are a partial case of the above-treated soliton solutions (for $u = 0$). For $d\omega/dk = 0$, there is degeneration: there are solutions with arbitrary velocities of propagation u.

1.2 Nonlinear Waves in a Gaseous Disk [31[ad]]

In the previous section we described some results of the basic work [90a], which for the first time considers the effects of weak nonlinearity in a rotating gravitating medium. The model of an infinitely thin disk ($kh \ll 1$; h is the thickness) adopted in this work is justified in the consideration of a gaseous layer, which is within a more massive (stellar) constituent[2]: the dispersion curve $\omega^2 = \omega^2(k)$ has a minimum at $kh \sim \rho_g/\rho_s$ (ρ_g and ρ_s are the volume densities of gas and stars). However, for purely gaseous systems (for example, the protoplanetary cloud of the Solar system) being at the stability boundary, the infinitesimally thin disk model is not valid: from the equilibrium condition in the vertical direction it follows that $4\pi G\rho_0 h^2 = 2\pi G\sigma_0 h \sim c_s^2$. Since at the stability boundary the wave number $k = k_0 \sim \pi G\sigma_0/c_s^2$, then it turns out that $kh \sim 1$.

The construction of the consistent theory of nonlinear oscillations of a gaseous disk with a finite thickness presents a significantly more complex

---

[2] For detail see §5.1, Chapter VII. One may also consider that the infinitesimally thin disk with plane pressure ($p_z = 0$, $p_\perp \neq 0$) leads to a model description of a real strongly flattened stellar system of the type of spiral galaxy. Here, however, the question arises of the magnitude of the "effective" adiabatic index γ (cf. Section 1.3). Finally it is necessary to say that, in accordance with the notation of Churilov and Shukhman, for the gas disk immersed into more massive halo, the relation between the plane ($\varkappa$) and volume (γ) adiabatical exponents changes: $\varkappa = 3 - 4/(\gamma + 1)$. This formula may be obtained from the dimension analysis if we take into account that the gravitational constant G must occur in the given case only in the combination ($G\rho_s$).

task, which must be solved by numerical methods. Analytical solution happens to be possible only for some partial values of the adiabatic index γ, for example, for $\gamma = 1$ and $\gamma = 2$ (for more detail, cf. Problems 2–4).

Consider now small oscillations of a flat gaseous layer with the adiabatic index γ. Corresponding equilibrium models as well as linear oscillations for three partial values of the adiabatic index $\gamma = 1, 2, \infty$ are considered analytically by Goldreich and Lynden-Bell [210]. Below we construct, now basically by numerical methods, a nonlinear theory of oscillations of these models for any values of γ.

Initially we have the system of hydrodynamical equations and the Poisson equation

$$\frac{\partial \rho}{\partial t} + \text{div}(\rho V) = 0,$$

$$\frac{\partial V}{\partial t} + (V\nabla)V - 2[V\Omega] = -\nabla\lambda,$$

$$\lambda = \varphi + R, \qquad \Delta\varphi = 4\pi G\rho,$$

$$(A = \text{const}) \qquad \text{quantity } R = \frac{\gamma A}{\gamma - 1}\rho^{\gamma-1} \equiv \frac{c_s^2}{\gamma - 1}\left(\frac{\rho}{\rho_c}\right)^{\gamma-1}, \qquad \gamma > 1,^3$$

(1)

ρ_c and c_s are the density and the sound velocity in the middle plane of the layer.

Let the $0z$ axis be directed perpendicularly to the plane of the layer and all the values be dependent on x and z (in the stationary state, however, only on z). We introduce the dimensionless quantities, ξ, ζ, τ, σ, c, Φ, $\varkappa$, Λ, v in the following way: $z = a\zeta$, $x = a\xi$, $\tau = \omega_0 t$, $\sigma = \rho/\rho_c$, $c^2 = c_s^2/a^2\omega_0^2$, $\varphi = \varphi/a^2\omega_0^2$, $\varkappa = 2\Omega/\omega_0$, $\Lambda = \lambda/a^2\omega_0^2$, $v = (u, v, w) = V/\omega_0 a$, where $\omega_c^2 = 4\pi G\rho_c$, a is the semi-thickness of the layer. In the stationary state, from the z-component of the Euler equation and the Poisson equation, we obtain the equation [210]:

$$\frac{c^2}{\gamma - 1}\frac{d^2\sigma_0^{\gamma-1}}{d\zeta^2} + \sigma_0 = 0,$$

whose integration, with due regard for the fact that, at $\zeta = 0$, $\sigma_0 = 1$, $d\sigma_0/d\zeta = 0$, yields [$\Gamma(x)$ is the gamma function]

$$\frac{d}{d\zeta}\sigma_0^{\gamma-1} = -\frac{\gamma - 1}{C}\sqrt{\frac{2}{\gamma}}\sqrt{1 - \sigma_0^\gamma}, \qquad C = \sqrt{\frac{2\gamma}{\pi}}\frac{\Gamma(\frac{3}{2} - 1/\gamma)}{\Gamma(1 - 1/\gamma)}. \qquad (2)$$

Since, as follows from (2), asymptotically, as $\zeta \to \pm 1$, $\sigma_0 \sim (1 - \zeta^2)^{1/(\gamma-1)}$, σ_0 may be represented in the form $\sigma_0 = f(\zeta)(1 - \zeta^2)^{1/(\gamma-1)}$, where $f(\zeta)$ is the function regular on the segment $[-1, 1]$ together with its derivatives.

---

[3] For $\gamma \leq 1$, the thickness of the disk is infinite; note also that the adiabatic link between the equilibrium pressure and density is generally not necessary and is assumed (same as, e.g., in [210]) for the sake of simplicity.

Let us consider the perturbations symmetrical with respect to the plane $z = 0$ (of Jeans type) near the stability boundary $\omega \simeq 0^4$ (ω is the perturbation frequency). Assuming that $\delta\Phi \sim \varepsilon$, $\partial/\partial\tau \sim \varepsilon^2$ ($\delta\Phi$ is the potential perturbation, ε is the small parameter),[5] we shall obtain from (1) the following estimates of the orders of magnitudes: $\delta\sigma \sim \varepsilon$, $u \sim \varepsilon^3$, $v \sim \varepsilon$, $w \sim \varepsilon^3$, the displacement of the disk boundary $\eta_1 \sim \varepsilon$. System (1) then gets simplified:

$$\int_0^1 d\zeta\left(\frac{\partial\sigma}{\partial\tau} + \frac{\partial(\sigma u)}{\partial\xi}\right) + \eta\left(\frac{\partial\sigma}{\partial\tau} + \frac{\partial(\sigma u)}{\partial\xi}\right)\bigg|_{\zeta=1} + \frac{\eta^2}{2}\left(\frac{\partial^2\sigma}{\partial\tau\partial\zeta} + \frac{\partial^2(\sigma u)}{\partial\zeta\partial\xi}\right)\bigg|_{\zeta=1} = 0,$$
(3)

$$u = -\frac{1}{\varkappa^2}\left(\frac{\partial^2\Lambda}{\partial\tau\partial\xi} + u\frac{\partial^2\Lambda}{\partial\xi^2}\right), \qquad \frac{\partial\Lambda}{\partial\zeta} = 0,$$
(4)

$$\frac{\partial^2\Phi}{\partial\xi^2} + \frac{\partial^2\Phi}{\partial\zeta^2} = \sigma,$$
(5)

$$\frac{\delta\sigma}{\sigma_0} = -\frac{\sigma_0^{1-\gamma}}{c^2}(\Phi - \Lambda) + \frac{2-\gamma}{2c^4}\sigma_0^{2(1-\gamma)}(\Phi - \Lambda)^2$$

$$-\frac{(3-2\gamma)(2-\gamma)}{6c^6}\sigma_0^{3(1-\gamma)}(\Phi - \Lambda)^3.$$
(6)

We derive the boundary conditions for the values of jumps of the potential Φ and its derivative $\partial\Phi/\partial z$ at the unperturbed boundary $z = a$. We deal here with a common—for such a kind of problem—representation of the total perturbed potential as due to, first, "local" density changes $\rho_1(x, z) = \rho(x, z) - \rho_0(z)$ within unperturbed boundaries ($|z| < a$) and, second, by induced, at the old boundary $z = a$ (and, symmetrically, at $z = -a$), a simple and double layers. In essence, we determine below the powers of the latter. We write for that purpose the total (dimensionless) density σ with due regard for perturbation symmetry, in the form

$$\sigma = \sigma(\zeta)[\theta(\zeta + 1 + \eta) - \theta(\zeta - 1 - \eta)],$$
(7)

where θ is unit step, η is the boundary displacement. Expanding the θ function near $z = 1$ in a series in degrees η up to third order inclusively and then substituting σ into the Poisson equation, we obtain

$$\frac{\partial^2\Phi}{\partial\zeta^2} + \frac{\partial^2\Phi}{\partial\xi^2} = \sigma(\zeta)[1 - \theta(\zeta - 1) + \eta\delta(\zeta - 1) - \tfrac{1}{2}\eta^2\delta'(\zeta - 1)$$

$$+ \tfrac{1}{6}\eta^3\delta''(\zeta - 1)], \quad (5')$$

where δ, δ', δ'' are the δ function and its derivatives.

---

[4] Goldreich and Lynden-Bell in [210] have shown that ω^2 is real for the system of (1).

[5] Derivative $\partial/\partial\tau \curvearrowright \omega$, where ω is the "full" (taking into account nonlinear correction) frequency; the ordering scheme assumed here suggests that ω^2 is a value of greater order of smallness than ε^2 (for example, one may assume $\omega \sim \varepsilon^2$).

The first boundary condition (for the jump $\partial\Phi/\partial\zeta$) will be found by integrating (5') over ζ within the limits from $(1 - \varepsilon)$ to $(1 + \varepsilon)$, $\varepsilon \to +0$:

$$\left[\frac{\partial\Phi}{\partial\zeta}\right] = \eta\left(\sigma + \frac{1}{2}\eta\frac{\partial\sigma}{\partial\zeta} + \frac{1}{6}\eta^2\frac{\partial^2\sigma}{\partial\zeta^2}\right)\bigg|_{\zeta=1}. \tag{8}$$

Multiplying (5') by $(\zeta - 1)$ and integrating within the same limits, we easily find also the second boundary condition

$$[\Phi] = -\eta^2\left(\frac{\sigma}{2} + \frac{\eta}{3}\frac{\partial\sigma}{\partial\zeta}\right)\bigg|_{\zeta=1}. \tag{9}$$

The continuity condition of the pressure on the boundary of the disk $[p] = 0$ can be written in the form

$$\left(\sigma + \eta\frac{\partial\sigma}{\partial\zeta} + \frac{1}{2}\eta^2\frac{\partial^2\sigma}{\partial\zeta^2} + \frac{1}{6}\eta^3\frac{\partial^3\sigma}{\partial\zeta^3}\right)\bigg|_{\zeta=1} = 0. \tag{10}$$

Since as $|\zeta| \to 1, \partial\sigma_0/\partial\zeta \sim (1 - \zeta^2)^{(2-\gamma)/(\gamma-1)}, \partial^2\sigma_0/\partial\zeta^2 \sim (1 - \zeta^2)^{(3-2\gamma)/(\gamma-1)}$, and $\partial^3\sigma_0/\partial\zeta^3 \sim (1 - \zeta^2)^{(4-3\gamma)/(\gamma-1)}$, then within the interval of interest to us $1 < \gamma \lesssim 2$ the right-hand sides of (8)–(10) tend to infinity as $|\zeta| \to 1$ (correspondingly for $\gamma > \frac{3}{2}$ and $\gamma > \frac{4}{3}$). To eliminate these fictitious divergents (they result from the displacement of the boundaries of the disk), one may, for example, use the following approach. Assume that there is a small pressure P_h of the "halo" surrounding the disk. Then the right-hand side of (3) becomes

$$- \sigma_h(\partial\eta/\partial\tau + u\partial\eta/\partial\xi)|_{\zeta=1} \qquad [\sigma_h = (\gamma P_h/(\gamma - 1)c^2)^{1/(\gamma-1)}],$$

the right-hand side of (10) will be substituted for σ_h, and σ_0 will be presented in the form $\sigma_0 = f(\zeta)(1 + \alpha - \zeta^2)^{1/(\gamma-1)}, \alpha = \gamma[f(1)]^{1-\gamma}P_h/(\gamma - 1)c^2$. Then one has to perform all calculations and in the final formulae put that $P_h = 0$ (by preliminarily reducing them to the such form when this limiting process makes sense).

The perturbation periodical in ξ may be presented in the form (c.c. means complex conjugate)

$$\Lambda = (\Lambda^{(1)}e^{ik\xi-i\omega\tau} + \text{c.c}) + [(\Lambda^{(2)}e^{2ik\xi-2i\omega\tau} + \text{c.c}) + \Lambda^{(0)}] + \cdots,$$

where $\Lambda^{(0)}, \Lambda^{(1)}, \Lambda^{(2)}$ are the constants (since $\partial\Lambda/\partial\zeta = 0$). We choose $\Lambda^{(1)}$ as the amplitude of perturbation and seek the solution using the perturbation theory within an accuracy of $(\Lambda^{(1)})^3$.

1. *Linear approximation.* Expressing from (6) $\sigma^{(1)}$ through $(\Phi^{(1)} - \Lambda^{(1)})$, we find the solution for the Poisson equation (5) with the boundary conditions of (8) and (9):

$$\Phi^{(1)} = c_1(\zeta)\Lambda^{(1)}. \tag{11}$$

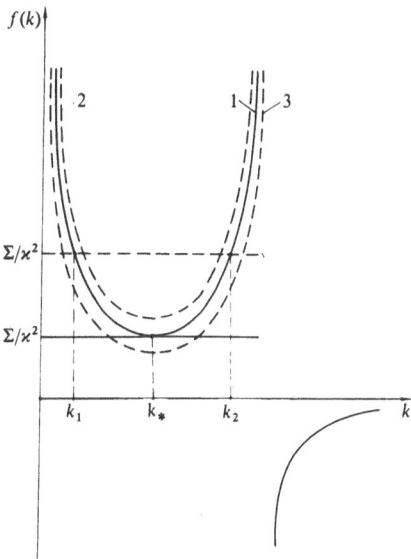

Figure 94. Nonlinear shift of the stability boundary determined by the linear problem (1): 2—stabilization; 3—destabilization.

Then, substituting (11) into (3), we obtain the linear dispersion equation (for $\omega = 0$) in the form

$$F_k - \frac{k^2}{\varkappa^2}\Sigma_0 = 0, \quad F_k = \int_0^1 d\zeta \frac{\sigma_0^{2-\gamma}}{c^2}(c_1 - 1), \quad \Sigma_0 = \int_0^1 \sigma_0 \, d\zeta = \sqrt{\frac{2}{\gamma}}\, C. \tag{12}$$

The behavior of the function $f(k) = F_k/k^2$ is qualitatively identical for all values of γ [210] and is shown in Fig. 94. From this figure it is seen that Eq. (12) $[f(k) = \Sigma_0/\varkappa^2]$ has solutions only for fairly small $\varkappa$ (angular velocities of rotation). If (12) has two solutions, then the region $k_1 < k < k_2$ corresponds to instability ($\omega^2 < 0$), and the regions $k < k_1$ and $k > k_2$, to stability, $\omega^2 > 0$. At some critical value of $\varkappa = \varkappa_*$, there is only one solution of (12) $k = k_*$, which corresponds to the minimum $f(k)$.

2. *Nonlinear corrections.* In taking into account the nonlinear terms (up to cubic), instead of (12), we have the nonlinear dispersion equation of the form

$$F_k - \frac{k^2}{\varkappa^2}\Sigma_0 = A|\Lambda^{(1)}|^2, \tag{13}$$

and the stability boundaries will be somewhat shifted. Nonlinearity exerts a destabilizing action if the equation $f_*(k) \equiv F_k/k^2 - A|\Lambda^{(1)}|^2/k^2 = \Sigma_0/\varkappa_*^2$ has two real solutions and a stabilizing action, if there are no real roots. Due to the smallness of $\Lambda^{(1)}$, one can write near $k = k_*$

$$\frac{F_k}{k^2} - \frac{\Sigma_0}{\varkappa_*^2} \simeq \frac{1}{2}\frac{\partial^2 f}{\partial k^2}\bigg|_{k=k*}(\delta k)^2 = \frac{A}{k_*^2}|\Lambda^{(1)}|^2.$$

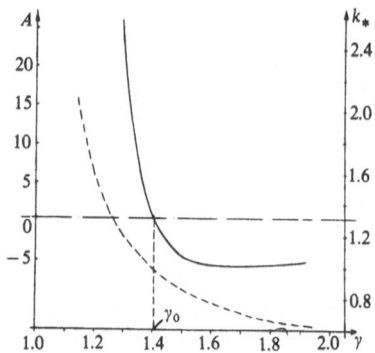

Figure 95. Dependence of the critical wave number k_* (dotted line) and the coefficient A in the nonlinear correction—see Eq. (13) (solid line) on the adiabatic exponent γ.

Since $(\partial^2 f / \partial k^2)_{k=k_*} > 0$, then the sign $(\delta k)^2$ is defined by the sign of A: for $A > 0$, nonlinearity plays a destabilizing role, while for $A < 0$, a stabilizing one.

The system of equations (2)–(10) was solved numerically. First of all, the stationary density distribution σ_0 was found. Then, in each order with respect to $\Lambda^{(1)}$ (up to third order inclusively) the solution to the Poisson equation (5) was found with the right-hand side of (6), and with boundary conditions of (8) and (9). Then, by using (6), the corresponding density perturbation was calculated and, together with the velocity perturbation u found from (4), was substituted into (3). As a result, we have the following equations:

First order:

$$F_k - \frac{k^2}{\varkappa^2} \Sigma_0 = 0. \tag{14}$$

Solving this equation, we find k_* and $\varkappa_*$ (for results, cf. Fig. 95):

Second order:

$$(F_{2k_*} - 4F_{k_*})\Lambda^{(2)} = B\Lambda^{(1)2}. \tag{15}$$

(The value $\Lambda^{(0)}$ is not necessary in further calculations.)

Third Order:

$$\left(F_k - \frac{k^2}{\varkappa^2} \Sigma_0\right) = D\Lambda^{(2)} - C\Lambda^{(1)2}. \tag{16}$$

The right-hand sides of (15) and (16) are calculated at $k = k_*$ (note that in all problems of this type $D = 2B$). Solving (15) with respect to $\Lambda^{(2)}$ and substituting into (16), we obtain the nonlinear dispersion equation in the form (13), where $A = BD/(F_{2k_*} - 4F_{k_*}) - C$. Computations (cf. Fig. 95) showed that in the range $1 < \gamma < 2$ for $1 < \gamma < \gamma_0$ there is destabilization ($A > 0$), while for $\gamma_0 < \gamma < 2$, stabilization ($A < 0$ and $\gamma_0 \simeq 1.404 \simeq \frac{7}{5}$).[6]

---

[6] Note that numerically this value proved to be close to the value $\gamma_0 = \frac{3}{2}$ obtained in the previous subsection, where the case of an infinitely thin disk is considered.

The cases $\gamma = 1$ and $\gamma = 2$ are peculiar ones for the numerical method described above. However, in these cases it is possible to investigate analytically, which was just performed with the aim of control (cf. Problems 2–4), For $\gamma = 2$, the nonlinear correction is expressed through elementary functions while for $\gamma = 1$ the answer is in quadratures (by the way, being rather cumbersome so that their computation demanded more computer time than the complete calculation described above). As was to be expected (cf. Fig. 95), nonlinearity has a destabilizing character for $\gamma = 1$; for $\gamma = 2$ stabilization takes place, so that the nonlinear correction changes sign (a second time after $\gamma = \gamma_0$) for $\gamma > 2$.

For a nonrotating layer (at its stability boundary) the calculations are performed for $\gamma = 1$ and $\gamma = 2$ (in more detail, cf. also Problems 2–4). In the first case, the nonlinear correction ($\sim \Lambda^{(1)2}$) happened to be zero. In the second case, stabilization takes place. It should be expected that also within the whole region $1 < \gamma < \gamma_1$ ($\gamma_1 > 2$) the character of the nonlinear correction remains stabilizing.

The main result of the investigation performed above is the determination of the critical value of the adiabatic index $\gamma_c \simeq 1.404$ for the rotating gaseous layer. Only for sufficiently low $\gamma < \gamma_c$, perturbations may increase up to large values. Note the closeness of γ_c to the value of the adiabatic index of the biatomic gas ($\gamma = 7/5$ in normal conditions). Some possibilities of applications of the results attained above are discussed in Chapter XI.

1.3 Nonlinear Waves and Solitons in a Stellar Disk [32^{ad}]

1.3.1. Derivation of the Equation for Nonlinear Waves. Consider an infinitesimally thin stellar disk, which rotates with an angular velocity Ω. Assume for the sake of simplicity that rotation is uniform: $\Omega = \text{const} \neq \Omega(r)$, r is the radius of the point in the disk plane (x, y). The distribution function of stars in a rotating reference system is assumed to be Maxwellian

$$f^{(0)} = \frac{\sigma_0}{2\pi T} \exp\left(-\frac{v^2}{2T}\right), \qquad v^2 = v_x^2 + v_y^2, \tag{1}$$

where T is the temperature. The linear dispersion equation describing small perturbations is conveniently written in the form of (23) of Section 4.1, Chapter V,

$$\frac{k_T}{k} = \frac{1}{x}\left(1 - \frac{v\pi}{\sin v\pi}\frac{1}{2\pi}\int_{-\pi}^{\pi} e^{-x(1+\cos s)}\cos vs\, ds\right). \tag{2}$$

We recall the definition of the values involved in (2): $k_T = x^2/2\pi G\sigma_0$ is the Toomre critical wave number, $x = 2\Omega$ is the epicyclic frequency, $v = (\omega - m\Omega)/x$ is the dimensionless frequency, m is the azimuthal number, k is the wave number that is assumed to be large, $kr \gg 1$, $x = k^2T/x^2 = k^2\rho^2$, and $\rho^2 = T/x^2$ is the square of the epicyclic linear size. The dispersion equation in the form of (2) is normally used in the theory of spiral structure of galaxies similar to our Galaxy [270, 271]. The qualitative behavior of the curves $v(k)$, defined by Eq. (2) is given in Fig. 93 for two different pairs of parameters k_T, T characterizing the equilibrium state. The disk is stable if the parameter $z = 2\pi G\sigma_0\sqrt{2/T}/\varkappa$ does not exceed the critical value $z^* \simeq 2.652$ (cf. Section 4.1, Chapter V). As $z > z^*$, there appears the instability region (cf. Fig. 93). If $z = z^*$, the dispersion curve touches the abscissa axis at a certain point $k = k_0^*$. The value k_0^* can be determined from (2): $k_0^* = 1.377\varkappa/\sqrt{2T}$.

The goal of further calculations is to obtain the nonlinear equation for the potential harmonic Φ_k. To begin with, we derive the nonlinear dispersion equation that generalizes Eq. (2), and this is done for an arbitrary value of the wave number k. The situation is considered in more detail when the disk is in the state close to the stability boundary, i.e., at $z \simeq z^*$ either in a stable $(z < z^*)$ or in an unstable $(z > z^*)$ region. Moreover, we shall mainly be interested in the wave numbers k close to the wave number k_0 corresponding to the minimum of the dispersion curve. In this case the small parameter of the problem may be considered the value $|\omega_{k_0}|^2/\varkappa^2 = |v_{k_0}|^2 \ll 1$. At such assumptions, we have the nonlinear equation of the form (20), Section 1.1.

In the local approximation used by us it is convenient to introduce [210, 334] the locally Cartesian system of coordinates with the origin at the center of the region of the disk under consideration rotating with an angular velocity Ω. The orientation of the axes (x, y) is arbitrary; for certainty one may, for example, assume that the x-axis is directed along the radius r, with the y-axis across the radius.

Let the potential perturbation have the form

$$\Phi(x, t) = \sum_k \Phi_k(t)e^{ikx} \qquad (\Phi_k = \Phi_{-k}^*).$$

Substituting Φ into the kinetic equation

$$\frac{\partial f}{\partial t} + v_x \frac{\partial f}{\partial x} + \varkappa\left(v_y \frac{\partial f}{\partial v_x} - v_x \frac{\partial f}{\partial v_y}\right) = \frac{\partial \Phi}{\partial x}\frac{\partial f}{\partial v_x} \tag{3}$$

(where it is assumed that in the x-direction, the potential Φ changes much more rapidly than along y), we can calculate the distribution function by the iteration method. Assume that

$$f = f^{(0)} + \sum_k f_k e^{ikx} \qquad (f_k = f_{-k}^*), \tag{4}$$

then

$$f_k = f_k^{(1)} + f_k^{(2)} + f_k^{(3)}, \tag{5}$$

where the values $f_k^{(1)}$, $f_k^{(2)}$, and $f_k^{(3)}$ are linear, quadratic, and cubic in Φ_k, respectively. Equation (3) is readily solved by the method of integration over angle, by transforming to the variables v and φ: $v_x = v \cos \varphi$, $v_y = v \sin \varphi$.

Integrating then (5) over velocities, we obtain the surface density perturbation

$$\sigma_k = \sigma_k^{(1)} + \sigma_k^{(2)} + \sigma_k^{(3)}, \tag{6}$$

where $\sigma_k^{(1)}$, $\sigma_k^{(2)}$, $\sigma_k^{(3)}$ are connected with Φ_k via relations of the form

$$\sigma_k^{(1)} = \sigma_0 A_k \Phi_k, \tag{7}$$

$$\sigma_k^{(2)} = \sigma_0 \sum_{k_1, k_2} B_{k, k_1, k_2} \Phi_{k_1} \Phi_{k_2} \delta_{k, k_1 + k_2} \tag{8}$$

$$\sigma_k^{(3)} = \sigma_0 \sum_{k_1, k_2, k_3} C_{k, k_1, k_2, k_3} \Phi_{k_1} \Phi_{k_2} \Phi_{k_3} \delta_{k, k_1 + k_2 + k_3}. \tag{9}$$

$A_k, B_{k, k_1, k_2}, C_{k, k_1, k_2, k_3}$ are the notations of the corresponding coefficients. This expression for σ_k is further to be substituted into the Toomre relation following from the Poisson equation (in the short-wave approximation) and connecting the values σ_k and Φ_k:

$$\Phi_k = -(2\pi G / |k|) \sigma_k. \tag{10}$$

We have

$$\Phi_k = -\frac{2\pi G \sigma_0}{|k|} \left(A_k \Phi_k + \sum_{k_1 k_2} B_{k, k_1, k_2} \Phi_{k_1} \Phi_{k_2} \delta_{k, k_1 + k_2} \right.$$

$$\left. + \sum_{k_1, k_2, k_3} C_{k, k_1, k_2, k_3} \Phi_{k_1} \Phi_{k_2} \Phi_{k_3} \delta_{k, k_1 + k_2 + k_3} \right). \tag{11}$$

For the harmonic $k = k_0$ (arbitrary so far) we have from (11) within an accuracy of cubic terms

$$\Phi_{k_0} = -\frac{2\pi G \sigma_0}{|k_0|} [A_{k_0} \Phi_{k_0} + (B_{k_0, 2k_0, -k_0} + B_{k_0, -k_0, 2k_0}) \Phi_{2k_0} \Phi_{k_0}^*$$

$$+ (C_{k_0, k_0, k_0, -k_0} + C_{k_0, k_0, -k_0, k_0} + C_{k_0, -k_0, k_0, k_0}) \Phi_{k_0}^2 \Phi_{k_0}^*]. \tag{12}$$

Eliminating the value Φ_{2k_0} with the help of Eq. (11), written for $k = 2k_0$,

$$\Phi_{2k_0} = -\frac{(2\pi G \sigma_0 / |2k_0|) B_{2k_0, k_0, k_0}}{1 + (2\pi G \sigma_0 / |2k_0|) A_{2k_0}} \Phi_{k_0}^2, \tag{13}$$

we obtain [32^{ad}]

$$\Phi_{k_0} = -\frac{2\pi G \sigma_0}{|k_0|} (A_{k_0} \Phi_{k_0} + R_{k_0} \Phi_{k_0} |\Phi_{k_0}|^2). \tag{14}$$

Here

$$R_{k_0} = C_{k_0} - \frac{2\pi G \sigma_0 B_{k_0} D_{k\theta}}{1 + (2\pi G \sigma_0 / |2k_0|) A_{2k_0}},$$

$$B_{k_0} \equiv B_{2k_0, k_0, k_0}, \qquad D_{k_0} \equiv B_{k_0, 2k_0, -k_0} + B_{k_0, -k_0, 2k_0},$$

$$C_{k_0} \equiv C_{k_0, k_0, k_0, -k_0} + C_{k_0, k_0, -k_0, k_0} + C_{k_0, -k_0, k_0, k_0}. \tag{15}$$

Equation (14) (if it is divided by Φ_{k_0}) is the sought-for nonlinear dispersion equation.

Some details of the derivation (rather cumbersome) of this equation are given in §8, Appendix; also given are the expressions for the coefficients

$A, \ldots, D$. Below, we shall need only their values at $z = z^*$ and $k_0 = k_0^* = 1.377$ (we assume the units in which $2T = 1$, $\varkappa = 1$). To calculate them, one has in the general formulae to switch to the limit $v \to 0$, opening the uncertainties to be presented. The expressions thus obtained are also given in §8, Appendix. The integrals in them are computed, which result in the following values for the coefficients of the nonlinear dispersion equation of interest:

$$A_{2k_0} = -1.574, \qquad B_{k_0} = 0.230, \qquad D_{k_0} = 2B_{k_0},$$
$$C_{k_0} = 0.206, \qquad R_{k_0} = -0.002, \qquad 2\pi G\sigma_o = 1.326. \tag{16}$$

From dispersion equation (14), one can switch to the corresponding nonlinear differential equation for Φ_{k_0} with $\partial/\partial t \neq 0$. This is easily done if one bears in mind that, in the linear approximation, the equation has the form

$$\frac{\partial^2}{\partial t^2} \Phi_k + v_k^2 \Phi_k = 0, \tag{17}$$

where v_k^2 is the square of the dimensionless frequency

$$v_k^2 = \frac{1 - I_0(k^2/2)e^{-k^2/2} - |k|/4\pi G\sigma_0}{(\pi^2/6)I_0(k^2/2)e^{-k^2/2} - \int_0^\pi e^{-k^2(1 + \cos x)/2}x^2 \, dx/2\pi} \tag{18}$$

Finally, we obtain [32^{ad}] the following equation:

$$\frac{\partial^2 \Phi_k}{\partial t^2} + v_k^2 \Phi_k = \mu_k |\Phi_k|^2 \Phi_k, \tag{19}$$

where

$$\mu_k = \frac{R_{k_0}}{2} \left[\frac{\pi^2}{6} I_0\left(\frac{k_0^2}{2}\right) e^{-k_0^2/2} - \frac{1}{2\pi} \int_0^\pi e^{-k_0^2(1 + \cos x)/2} x^2 \, dx \right]. \tag{20}$$

The numerical value of the quantity μ_k turns out to be small: $\mu_k \simeq 0.002$. Hence it follows that the essential role may be played even by a comparatively small in mass, gaseous component. Since the velocity dispersion of gaseous clouds in the Galaxy is small in comparison with the velocity dispersion of stars, it may be suggested that the gaseous disk is "cold." In this case, the coefficients A_{2k}, B_k, D_k, and C_k, for the gas similar to that used earlier for stars, are

$$A_{2k} = \frac{4k^2}{4v^2 - 1}, \qquad 2B_k = D_k = \frac{12v^2k^4}{(v^2 - 1)^2(4v^2 - 1)},$$
$$C_k = -\frac{6v^2k^6}{(v^2 - 1)^3(4v^2 - 1)}. \tag{21}$$

On the stability boundary $B = D = C = 0$, therefore, the role of the gas is reduced mainly to the change of the critical wave number k_0^*, which now must be determined from the equation

$$1 = -\frac{2\pi G\sigma_0}{|k|} \left\{ -2\left[1 - I_0\left(\frac{k^2}{2}\right) e^{-k^2/2} \right] - k^2 \frac{\sigma_g}{\sigma_0} \right\}, \tag{22}$$

where σ_0 and σ_g are the surface densities of stars and gas, respectively. The nonlinear correction R_k is determined by the expression (15), in which the substitution $A_{2k} \rightarrow A_{2k} - 4k^2(\sigma_g/\sigma_0)$ must be done. For $\varepsilon = \sigma_g/\sigma_0 = 0.1$, we have $R_{k_0} = -0.45$. For the arbitrary $\varepsilon \lesssim 0.05$, we have $k_0^* = 1.377 + 2\varepsilon$, $(4\pi G\sigma_0)^{-1} \simeq 0.377 + 0.7\varepsilon$, $R_{k_0} = -(0.002 + 0.85\varepsilon)$.

Equation (19) coincides in shape with the corresponding nonlinear equation derived in Section 1.1, where waves have been considered for an infinitesimally thin disk. For collisionless systems interesting to us, the "hydrodynamical" description is, of course, unfit. The reason is that, as is well known, the approximate hydrodynamical description of a stellar system (of the Chu–Goldberger–Low type,—see, for example, [46^{ad}]) is correct only for $k\rho \ll 1$ (ρ is the radius of epicycle), whereas in our region of interest $k\rho \simeq 1$.

The correct inclusion in the theory of gas is nontrivial, if arbitrary wavelengths are considered. The fact is that at wavelengths less than or comparable to the thickness h, the approximation of a gaseous infinitely thin disk is no longer valid. But just that very case is realized in a purely gaseous disk (see Section 1.2). Corresponding theory was constructed in Section 1.2 [31^{ad}].

At the same time, for the description of gaseous subsystems in a galaxy, the approximation of an infinitely thin disk is good since the wavelengths $\lambda = 2\pi/k \sim 2\pi/k_0^*$ of interest are much greater than the thickness of this subsystem. For instance, for the Galaxy $\lambda \simeq 2.5$ kps and $h \simeq 200$ pc. However, for the application of the theory to the Galaxy one must also take account of the finite thickness of the stellar component.

1.3.2. Soliton-like Solutions. The most interesting class of solutions which follow from Eq. (19) are solitons of the envelope. To obtain these solutions, we assume, similarly to Section 1.1, that this equation is valid in some vicinity of the wave numbers k near $k = k_0$. In order not to deal with the complex values, we can, without loss of generality, assume that $\Phi_k = \Phi_{-k} = \psi_k/2$, where ψ_k are the expansion coefficients of the potential in the Fourier series in $\cos(kx)$. We have, near $k = k_0$: $v_k^2 = v_{k_0}^2 + c_s^2 \varkappa_x^2$, $c_s^2 = 0.42$, $k - k_0 = \varkappa$, $\varkappa = (\varkappa_x, \varkappa_y)$, $|\varkappa| \ll k_0$. Integrating (19) over $\varkappa$ with the weight $\cos(\varkappa r)$, we obtain the equation for the envelope $\bar{\psi}(x, y, t)$ $(\Phi(r, t) = \bar{\psi}(r, t) \cos k_0 x]$:

$$\frac{\partial^2 \bar{\psi}}{\partial t^2} + v_{k_0}^2 \bar{\psi} - c_s^2 \frac{\partial^2 \bar{\psi}}{\partial x^2} = \frac{\tilde{\mu}}{4} \bar{\psi}^3, \qquad \tilde{\mu} = \mu_{k_0}(2\pi)^2. \tag{23}$$

Let us seek the solution for Eq. (23) in the form:

$$\bar{\psi} = \psi(x \cos \alpha + y \sin \alpha - ut) = \bar{\psi}(z - ut).$$

The solutions of the soliton type are possible if $v_{k_0}^2 > 0$, $u^2 - c_s^2 \cos^2 \alpha < 0$. We denote:

$$\frac{v^2}{c_s^2 \cos^2 \alpha - u^2} = \frac{1}{\Delta^2}, \qquad \frac{\tilde{\mu}}{4}\Big/(c_s^2 \cos^2 \alpha - u^2) = \frac{2}{A^2\Delta^2}. \tag{24}$$

Then the solution has the form:

$$\bar{\psi}_0(z, t) = \frac{A}{\cosh[(z - ut)/\Delta]} \qquad A^2 = \frac{8v_{k_0}^2}{\tilde{\mu}}. \qquad (25)$$

1.3.3. Stability of the Solitons. Investigate the stability of the soliton obtained. For that purpose, turn to the frame of reference, in which the soliton is at rest. In this system, Eq. (23) has the form

$$\frac{\partial^2 \bar{\psi}}{\partial t^2} - (c_s^2 \cos^2 \alpha - u^2)\frac{\partial^2 \bar{\psi}}{\partial z^2} = \frac{\tilde{\mu}}{4}\bar{\psi}^3 - v_{k_0}^2 \bar{\psi}. \qquad (26)$$

We linearize this equation assuming that

$$\bar{\psi} = \bar{\psi}_0(z) + \bar{\psi}_1(z)e^{-i\lambda t + iqw},$$

where w is the coordinate in the direction perpendicular to z. For ψ_1, we obtain the equation:

$$\Delta^2 \frac{\partial^2 \psi_1}{\partial z^2} + \left[\frac{6}{\cosh^2(z/\Delta)} - \left(1 - \frac{\lambda^2}{v_{k_0}^2}\right)\right]\psi_1 = 0. \qquad (27)$$

We make in (27) the substitutions $\zeta = \tanh(z/\Delta)$ and $m^2 = 1 - \lambda^2/v_{k_0}^2$; then we have

$$\frac{\partial}{\partial \zeta}(1 - \zeta^2)\frac{\partial \psi_1}{\partial \zeta} + \left(6 - \frac{m^2}{1 - \zeta^2}\right)\psi_1 = 0. \qquad (28)$$

Equation (28) must be solved with the boundary conditions: $|\psi_1(\pm 1)| < \infty$. The solution has the form

$$\psi_1(\zeta) = P_2^m(\zeta), \qquad (29)$$

where P_α^β is the associated Legendre function. The boundary conditions are satisfied by solutions of two types: (1) $m^2 < 0$; (2) $m_{1,2}^2 = 1,4$. The solutions of the first type correspond to a continuous spectrum $\lambda^2 = v_{k_0}^2(1 - m^2)$ and describe nonincreasing perturbations. The solutions of the second type correspond to discrete frequencies

$$\lambda_1^2 = v_{k_0}^2(1 - m_1^2) = 0, \qquad (30a)$$

$$\lambda_2^2 = v_{k_0}^2(1 - m_2^2) = -3v_{k_0}^2. \qquad (30b)$$

It is clear that the mode (30a) corresponds to the displacement of the soliton as a whole along the z-axis. The mode (30b) describes perturbations increasing with the growth rate $\mathrm{Im}(\lambda_2) = \sqrt{3}v_{k_0}$ small in comparison with the angular frequency of rotation of the disk. The spatial structure of this mode

$$\psi_1 = e^{iqw}P_2^2(\zeta) \sim \frac{e^{iqw}}{\cosh^2(z/\Delta)}. \qquad (31)$$

For solitons obtained in this section, the consideration just performed completely solves the question of their stability. In gaseous systems (Sections 1.1 and 1.2) for values of the adiabatic index $\gamma_0 < \gamma < \gamma_1$ there are soliton solutions of another type ("supersonic" according to the terminology of

Section 1.1). The problem of the stability of these solitons is solved by the same formulae as above, but in this case one must assume that $v_{k_0}^2 < 0$. Therefore, "supersonic" solitons are unstable with the growth rate $|v_{k_0}|\sqrt{1 - m^2}(m^2 < 0)$.

1.3.4. Influence of the Finite Disturbance Amplitude on the Jeans Instability in Homogeneous Systems. Consider briefly nonlinear generalizations of the criterion of Jeans instability in the simplest homogeneous collisionless systems: (1) in an infinite homogeneous space (Jeans' classic problem), (2) in an infinitely thin homogeneous nonrotating layer (the two-dimensional analog of the Jeans problem), and (3) of an infinitesimally thin "thread" (the one-dimensional analog). If, as the unperturbed distribution function, we assume for the sake of simplicity[7] the "step"

$$f(v_x) = \left(\frac{\rho_0}{2\pi v_0}\right) [\theta(v_x + v_0) - \theta(v_x - v_0)]$$

[$\theta(x)$ is the Heaviside function], so by the method similar to that used above we may obtain

$$(1) \qquad k_0' = k_0\left(1 + \frac{1}{3}\frac{|\Phi|^2}{v_0^4}\right), \qquad\qquad (32)$$

$$(2) \qquad k_0' = k_0\left(1 + \frac{|\Phi|^2}{v_0^4}\right), \qquad\qquad (33)$$

$$(3) \qquad k_0' = k_0\left(1 + \frac{|\Phi|^2}{c^4}\right), \qquad\qquad (34)$$

where k_0' is the critical wave number (separating the stability region for $k > k_0'$ and instability region for $k < k_0'$), k_0 corresponds to an infinitely small amplitude of perturbation, $c^2 = 8 \ln 2(G\rho')^2$, and $\rho' = \pi\rho R^2$ is the linear density of the "thread." Thus including the nonlinearity leads here to the destabilizing effect: to an expansion of the instability zone in the space of wave numbers k. Note that formulae (32) and (33) were earlier obtained by another method in [26[ad]] {more exactly, in [26[ad]], general formulae are derived, from which, in particular, one may obtain also (32) and (33)}.

1.4 Explosive Instability [20[ad]]

Since nonlinear equations for the stellar and gaseous disks for corresponding values of parameters have an identical form, both cases can be considered simultaneously. We take, for the sake of concreteness, the "gaseous" equation (26) of Section 1.1 and show that, under the condition

$$\varkappa < \tfrac{5}{3}, \qquad\qquad (1)$$

---

[7] For continuous distribution functions, of the Maxwellian type, there are some complications due to the appearance of trapped particles.

the value $v_1(t, \varphi)$ for a finite time may become arbitrarily large. From (5), subsection 1.1, it follows that the condition (1) corresponds to the inequality $\gamma < \frac{3}{2}$.

By using expression (1), Section 1.1, we rewrite equation (20), in a simpler form:

$$\frac{\partial^2 v_1(t)}{\partial t^2} = [\gamma_{ko}^2 + \tfrac{3}{2}(2 - \varkappa)(\tfrac{5}{3} - \varkappa)k_0^2 |v_1|^2] v_1(t). \tag{2}$$

We multiply, term by term, Eq. (2) by $\partial v_1/\partial t$ and integrate twice over t. Omitting further the index 1 in the v letter, we obtain

$$t - t_0 = \int_{v_0}^{v} \frac{dv}{\sqrt{w_0(r) + \gamma_{ko}^2 v^2 + Av^4}}, \tag{3}$$

where $A = \tfrac{3}{2}(2 - \varkappa)(\tfrac{5}{3} - \varkappa)k_0^2$, v_0 is the initial perturbation of the velocity at the time t_0, and $w_0(r)$ is the arbitrary function of r.

In the general case, the integral in (3) is expressed through the elliptical integral. To clear up the character of the solution, let us consider some partial case, in which the integral in (3) is easily taken. Let, for example, $w_0(r)$, $\gamma_{ko}^2 v_0^2 \ll Av_0^4$. Then

$$\frac{v}{v_0} = \frac{1}{1 - \sqrt{A}\,v_0(t - t_0)}. \tag{4}$$

From (4), it is seen that for a finite time $t - t_0 \to 1/\sqrt{A}\,v_0$ the velocity perturbation tends to infinity. Such an impetuous growth of perturbation characterizes the so-called "explosive" instability [20[ad]].

1.5   Remarks on the Decay Processes

The investigation of non-one-dimensional (in particular, isotropic[8]) spectra of perturbation encounters, in the case of interest of flat gravitating systems, an additional difficulty. It is associated with a possibility, in principle, of a decay instability. The decay instability for the waves with parallel wave vectors is considered in [70[ad]]. Of more interest, however, is the following possibility of a more general decay. It is easy to make sure that, for example, the 3-waves, whose wave vectors lie in the vicinity of the touching point of the dispersion curve and the k-axis (cf. Fig. 93) and form configurations close to the equilateral triangle may be connected via the decay conditions $\omega_1 = \omega_2 \pm \omega_3$, $k_1 = \pm k_3 + k_2$.[9] This leads to the necessity to modify the very statements of the problems in such cases: for example, in an attempt to seek nonlinear corrections to the frequency of some radial perturbation in the way used above, divergent expressions ensue. Note, however, that at

---

[8] The partial case of which in turn are the radial perturbations.

[9] In more detail, the 3-wave decay processes are investigated in §4, Chapter VII.

least in two interesting cases, the consideration performed by us has a meaning. First, these are localized small-scale perturbations (for example, ring-shaped); second, there are global, including radial, modes in *nonuniform* systems, where as is shown by the simplest examples (for example, of uniformly rotating disks), the satisfaction of the resonance conditions is difficult.

1.6 Nonlinear Waves in a Viscous Medium [52^{ad}]

Below we describe a scheme of deriving the nonlinear equation describing stationary waves of finite amplitude in a rotating gravitating disk in the presence of viscosity.

In Section 1.1 it was shown that when the rotating gravitating disk is near the stability boundary, two types of solitons may form and propagate in it: subsonic and supersonic. The type of soliton is dependent on whether the disk is stable or not. The medium, in which solitons propagate, was assumed to be dissipationless.

In a plasma medium, as shown by R. Z. Sagdeev [46^{ad}], taking dissipation into account causes solitons to be transformed into shock waves [Fig. 96(a), 96(b)]. Solitons considered in [46^{ad}] are described by the Korteweg–de Vries equation (if the amplitude of these solitons is small: the Mach number $M \approx 1$), the solution of which has the form $\infty 1/\cosh^2 \alpha\xi$, where α is a constant and $\xi = r - ut$. Solitons in a gravitating disk are described by the function of the form $\sim \sin k_0 r/\cosh \beta\xi$, where β is a constant and $k_0 \gg \beta$. Such a structure has received the name of the "envelope" soliton [Fig. 97(a)] in the plasma physics literature.

The Sagdeev collisionless shock wave [Fig. 96(b)] remains the basic property of the hydrodynamical shock wave: the jumps of the main characteristics of the medium on the wave front are not equal to zero. If one assumes that the envelope soliton [Fig. 97(a)] in a viscous medium is transformed into a structure similar to the Sagdeev one [Fig. 97(b)] but with small-scale

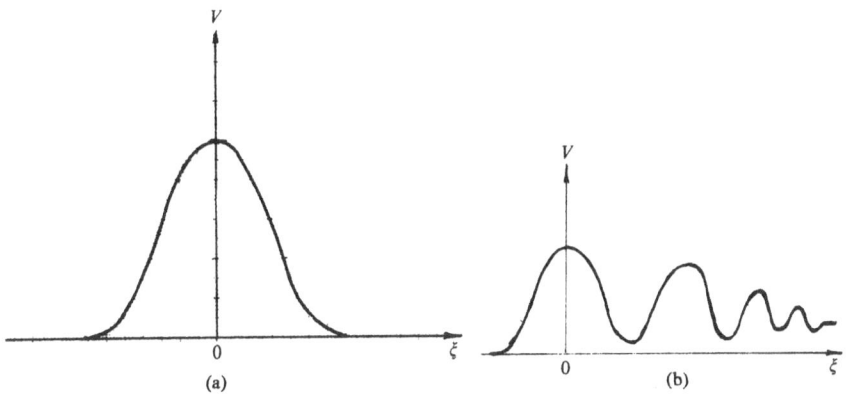

Figure 96. Soliton (a) and oscillatory profile of a shock's front (b).

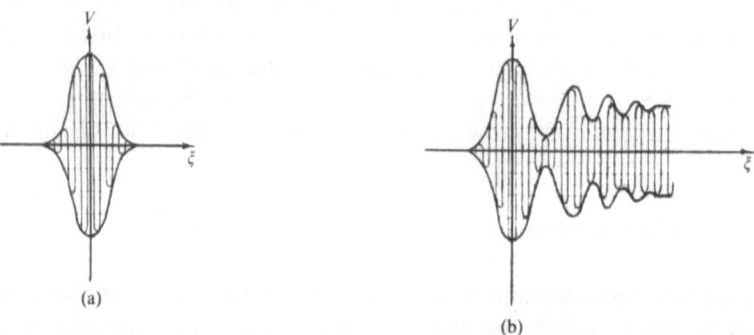

Figure 97. Envelope soliton (a) and corresponding profile of a shock's front (b).

oscillations, then the "classical" definition of the shock wave does not match such a structure: we shall not obtain a jump of the first moments (of density and velocity) on the front width.

Derive the nonlinear equation for density waves in a rotating, infinitesimally thin disk (the model of Section 1.1) with due regard for viscosity. In the initial system of equations used in the analysis of perturbations located in the plane of the gravitating disk, the continuity and Poisson equations have a standard form, while the Navier–Stokes equations will be written as [67]:

$$\frac{\partial v_r}{\partial t} + v_r \frac{\partial v_r}{\partial r} + \frac{v_\varphi}{r} \frac{\partial v_r}{\partial \varphi} - \frac{v_\varphi^2}{r} = -\frac{\partial \Phi}{\partial r} - \frac{1}{\sigma} \frac{\partial P}{\partial r}$$

$$+ v \left(\frac{1}{r^2} \frac{\partial^2 v_r}{\partial \varphi^2} - \frac{1}{r} \frac{\partial^2 v_r}{\partial \varphi \partial r} - \frac{1}{r^2} \frac{\partial v_\varphi}{\partial \varphi} \right)$$

$$+ \left(\frac{\zeta}{\rho} + \frac{4}{3} v \right) \frac{\partial}{\partial r} \left[\frac{1}{r} \frac{\partial}{\partial r} (r v_r) + \frac{1}{r} \frac{\partial v_\varphi}{\partial \varphi} \right]; \tag{1}$$

$$\frac{\partial v_\varphi}{\partial t} + v_r \frac{\partial v_\varphi}{\partial r} + \frac{v_\varphi}{r} \frac{\partial v_\varphi}{\partial \varphi} + \frac{v_r v_\varphi}{r} = -\frac{1}{r} \frac{\partial \Phi}{\partial \varphi} - \frac{1}{r\sigma} \frac{\partial P}{\partial \varphi}$$

$$+ v \left(\frac{\partial^2 v_\varphi}{\partial r^2} + \frac{1}{r} \frac{\partial v_\varphi}{\partial r} - \frac{v_\varphi}{r^2} - \frac{1}{r} \frac{\partial^2 v_r}{\partial r \partial \varphi} + \frac{1}{r^2} \frac{\partial v_r}{\partial \varphi} \right)$$

$$+ \left(\frac{\zeta}{\rho} + \frac{4}{3} v \right) \frac{1}{r} \frac{\partial}{\partial \varphi} \left[\frac{1}{r} \frac{\partial}{\partial r} (r v_r) + \frac{1}{r} \frac{\partial v_\varphi}{\partial \varphi} \right]. \tag{2}$$

We solve the initial set of equations by the method of perturbation theory in the same way as done in Section 1.1, where the nonlinear theory of stability of a gravitating, rotating disk is developed without taking account of dissipation. From Eq. (2), in the assumption $r\partial/\partial r \gg m$, we find the perturbed value of the radial velocity

$$\tilde{v}_r = -\frac{1}{\varkappa_0} \hat{L} \tilde{v}_\varphi + \frac{v}{\varkappa_0} \frac{\partial^2 \tilde{v}_\varphi}{\partial r^2},$$

$$\hat{L} \equiv \frac{\partial}{\partial t} + \Omega_0 \frac{\partial}{\partial \varphi}, \qquad \varkappa_0 = \frac{v_{\varphi 0}}{r} + \frac{\partial v_{\varphi 0}}{\partial r}, \tag{3}$$

($\varkappa_0$ is the epicyclic frequency and $\Omega_0 = v_0/r$), and, after substitution of (3) into the continuity equation, we obtain the expression for the perturbed value of the surface density:

$$\tilde{\sigma} = \frac{\sigma_0}{\varkappa_0} \frac{\partial \tilde{v}_\varphi}{\partial r} - \sigma_0 \frac{v}{\varkappa_0} \hat{L}^{-1} \frac{\partial^3 \tilde{v}_\varphi}{\partial r^3}. \tag{4}$$

By using the familiar expression of the link between the perturbed quantities of surface density, potential, and pressure [cf. formula (5) in 2.2, Chapter V]

$$\psi = -\frac{2\pi G\tilde{\sigma}}{k_r}, \qquad p_0 + \tilde{p} = \frac{c_s^2}{\varkappa} \sigma_0 \left(\frac{\sigma_0 + \tilde{\sigma}}{\sigma_0}\right)^\varkappa,$$

$$c_s^2 \equiv \varkappa \frac{p_0}{\sigma_0}, \qquad \varkappa = 3 - \frac{2}{\gamma}, \tag{5}$$

and, performing all the subsequent calculations, as made in Section 1.1, instead of Eq. (22), we obtain the expression [52^{ad}]

$$\hat{L}^2 v_1 = (\gamma_{k_0}^2 + c_s^2 \Delta_r)v_1 - \tfrac{3}{2}(2 - \varkappa)(\varkappa - \tfrac{5}{3})k_0^2 |v_1|^2 v_1$$
$$- (\mu - 4\Omega_0^2 v \hat{L}_0^{-2})k_0^2 \hat{L} v_1, \tag{6}$$

where $\mu = \tfrac{4}{3}v + \zeta/\sigma_0$; k_0 is the wave number corresponding to the maximum value of the growth rate of instability $\gamma_{k_0}(\partial \gamma_k/\partial k = 0)$ or in the absence of Jeans instability, the value of the argument of the function $\omega^2(k)$ at the point of its minimum; $v_1 \equiv v_\varphi$ (γ_{k_0} and $\varkappa_0$ should be distinguished from the adiabatic index γ and x!).

In the derivation of Eq. (6), we used the assumption about the smallness of the second summands on the right-hand sides of expressions (3) and (4) in comparison with the first summands. In fact, this means that the inequality

$$vk_0^2 \ll \gamma_{k_0} \tag{7}$$

holds true. Make sure in the example of a flat gaseous subsystem of the Galaxy that inequality (7) is satisfied with a large reserve. The viscosity coefficient

$$v \sim \frac{v_T}{\xi_0 n_0}, \tag{8}$$

where ξ_0 is the effective cross section of collisions of gas molecules, n_0 is the number of particles in 1 cm^3, and v_T is the thermal velocity. Assume for estimate $k_0 h \sim 3$ (h is the layer thickness); then $vk_0^2 \sim 10 \, \tfrac{2}{T}/\xi_0 n_0 h^2$. Substituting the characteristic parameters for the hot component of the gas

$$(T \sim 10^4 \, °K)v_T \sim 10^6 \, cm/c, \, \xi_0 \sim 10^{-16} \, cm^2, \, n_0 \sim 0.1 \, cm^{-3},$$

and $h \sim 200$ pc, we obtain $vk_0^2 \sim 2 \times 10^{-18} \div 2 \times 10^{-19} \, c^{-1}$. This value is at least by 3 orders of magnitude less than the galactic rotation frequency $\Omega_0 \sim 10^{-15} \, c^{-1}$. If one assumes that $\gamma_{k_0}/\Omega_0 \sim 10^{-1}$, then the inequality (8) is fulfilled at least with a 2-order-of-magnitude accuracy. The last circumstance allows one to take into account the "viscous" summands only in the third order of perturbation theory (in the parameter $\tilde{v}_\varphi/v_{\varphi_0} \ll 1$), which was used in the derivation of Eq. (6).

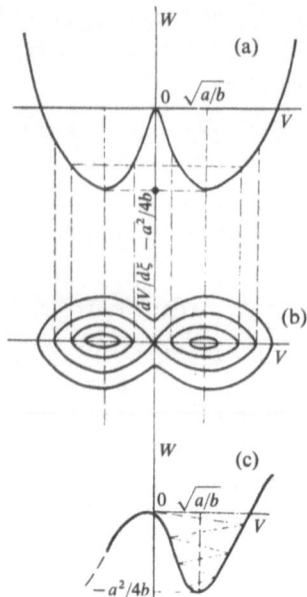

Figure 98. Potential energy $W(\xi)$ (a), structure of the phase plane for Eq. (12) without the viscous term (b), and trajectory of particle's gradual sliding down into the potential well in the presence of viscosity (c).

In the case when the coefficient of the first viscosity is by far more than the coefficient of the second viscosity, we obtain from (6) the following equation [52[ad]]:

$$\hat{L}\left(\hat{L}^2 - \gamma_{k_0}^2 + 3(2 - \varkappa)(\varkappa - \tfrac{5}{3})\frac{\Omega_0 k_0^2}{\varkappa_0}|v_1|^2\right)v_1 = 2\Omega_0\varkappa_0 vk_0^2 v_1. \quad (9)$$

In the opposite limiting case we have from (6) the equation[10] [52[ad]]

$$\hat{L}^2 v_1 = (\gamma_{k_0}^2 + c_s^2\Delta_r)v_1 - 3(2 - \varkappa)(\varkappa - \tfrac{5}{3})\frac{\Omega_0 k_0^2}{\varkappa_0}|v_1|^2 v_1 - \frac{\xi k_0^2}{\sigma_0}\hat{L}v_1. \quad (10)$$

Turning to the local rotating coordinate system $(r, \varphi, t) \to (r, \varphi', t)$, where $\varphi' = \varphi - \Omega_0 t$, and then introducing the variable $\xi = r - ut$, we have, instead of Eqs. (9) and (10),

$$\frac{d^3 V}{d\xi^3} - \left(\frac{\gamma_{k_0}^2}{\alpha^2} - \frac{3\beta^2}{\alpha^2}V^2\right)\frac{dV}{\partial\xi} + \frac{\varkappa_0^2 vk_0^2 V}{u\alpha^2} = 0, \quad (11)$$

$$\frac{d^2 V}{d\xi^2} - \frac{\xi}{\sigma_0\alpha^2}uk_0^2\frac{dV}{d\xi} = -\frac{\partial W}{\partial V}, \quad (12)$$

where

$$W = -\tfrac{1}{2}aV^2 + \tfrac{1}{4}bV^4, \qquad V \equiv v_1(\xi),$$

$$a = \frac{\gamma_0^2}{\alpha^2}, \qquad b = \frac{\beta^2}{\alpha^2}, \qquad \alpha^2 = u^2 - c_s^2, \qquad \beta^2 = 3(2 - \varkappa)(\varkappa - \tfrac{5}{3})\frac{\Omega_0 k_0^2}{\varkappa_0}.$$

[10] More exactly, provided that $\xi/\sigma_0 \gg v\Omega_0\varkappa_0/\gamma_{k_0}^2$.

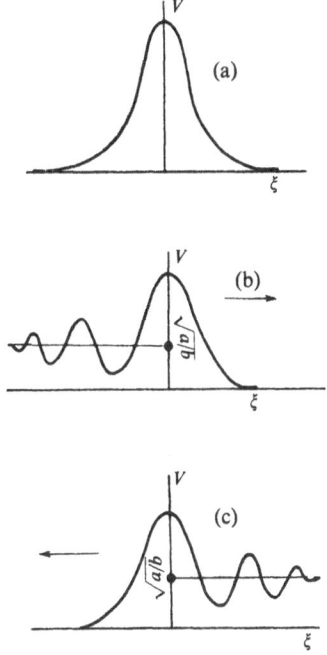

Figure 99. Soliton-like solution without the viscosity (a) and with the viscosity for the cases: $u < 0$ (b) and $u > 0$ (c).

In the derivation of Eqs. (11) and (12), we used the inequality

$$\Omega_0^2 \gg \gamma_{k_0}^2. \tag{13}$$

Equation (12) without the "viscous" term has the solution describing the soliton [90a, 89a]:

$$V(\xi) = \sqrt{2}\,\frac{\gamma_0}{\beta}\,\text{sech}\,\frac{\gamma_0\,\xi}{\sqrt{\alpha}} \tag{14}$$

in the case when the coefficients a and b are positive, which corresponds to two types of solitons: (1) subsound ($u^2 < c_s^2$, $\gamma_0^2 < 0$, $\beta^2 < 0$) and (2) supersound ($u^2 > c_s^2$, $\gamma_0^2 > 0$, $\beta^2 > 0$).

Equation (12) without the "viscous" term is Duffing's equation, which is well known in the theory of nonlinear oscillations. The structure of the phase plane for Duffing's equation is represented in Fig. 98(b). The separatrix which separates a region of periodical motions from other ones corresponds to the solution of the soliton-like wave [Fig. 99(a)]. Let us assume, for the sake of definition, $u > 0$. Then, considering the coordinate ξ as the time τ, one can interpret Eq. (12) as the equation of the nonlinear pendulum with damping.

The analogy considered may be used for the construction of the solution in all the region ξ (Kadomtsev [15^{ad}]). If, in the absence of damping, the "particle" is at some level of the potential well (in the case of the soliton this level passes through the point 0), then, in the presence of damping, the "particle" falls down to the bottom of the potential well. If the coefficient of the viscosity is sufficiently small, the "particle" has time for several oscillations during the fall [Fig. 98(c)]. This corresponds to deformation of the symmetrical profile of the soliton [Fig. 99(a)] into the nonsymmetrical oscillatory profile of the shock front [Fig. 99(b), (c)]. As the viscous coefficient v increases, the number of oscillations decreases and finally for a certain value v_{max} the "particle" falls down to the bottom of the potential well without reflections, which corresponds to the usual monotonically increasing profile of the shock wave.

The fact of the variation of the medium's state after the passage of the wave with the oscillatory profile (behind the wave front the medium moves with the velocity $V_0 = \sqrt{a/b}$) means that the wave is a shock.

In the case $u < 0$ one obtains the profile of the shock wave depicted in Fig. 99(c). Figures 99(b) and 99(c) correspond to a fall of the "particle" into the right potential well $W(V)$ [Fig. 98(a)]. Under a fall into the left potential well the pictures in Figs. 99(b) and 99(c) must be turned over the angle π.

All the above-mentioned corresponds to the envelope since the gravitational soliton is the "envelope" soliton. In this it differs, for example, from the ion-sound soliton in the plasma which bears, in the presence of dissipation, the oscillatory profile of the shock front (Sagdeev, [46^{ad}]). The oscillatory profile of the shock front of the envelope is filled with high-frequency oscillations with the wave number k_0, $k_0 \Delta \gg 1$, where Δ is the width of the soliton. Generally speaking, it is not necessary that we obtain a large-scale density jump under the averaging over the small-scale density oscillations (the latter supposition demands more detailed test) while in considering in terms of energy we understand that we are dealing with the shock wave. In this meaning, the gravitational shock waves produced by the gravitational soliton in the viscous medium principally differ from the shock waves in a gas and a plasma investigated earlier.

Here we do not touch upon the question of collisionless shock waves. Paper [63^{ad}] was the first to describe the model of a stellar system, in which collisionless shock waves can originate. In more detail, the problem of the existence of collisionless shock waves in stellar systems is considered in [64^{ad}].

§2 Nonlinear Interaction of a Monochromatic Wave with Particles in Gravitating Systems

2.1 Nonlinear Dynamics of the Beam Instability in a Cylindrical Model [85]

In real astrophysical objects, the distribution functions of particles (of stars, gas), with respect to their velocity, have sometimes a beam character. These involve all galaxies with heterogeneous structure, where the flat subsystems are rotating with respect to elliptical and spherical subsystems; regions of active centers characterized by ejections of large masses of gas, etc.

In §1, Chapter VI, the possibility was shown of excitation of the beam instability in gravitating systems, which leads to the growth of the amplitude of density waves of interacting subsystems. This effect was studied in the example of a gravitating cylinder. In [108a, 93], the role of beam effects was investigated in more complex systems of two interacting disks, a sphere and an ellipsoid of Freeman [201–204]. Of most importance here is the question whether the nonlinear stabilization of such a growth of the amplitude occurs or the instability process advances, leading to collapse of different clusters of density. Note that, for the problem of the galactic spiral structure, of particular interest is the interaction with the particles (stars) of the *monochromatic* density wave.

In this paragraph, we pay attention to the fact that in gravitating systems, an important role may be played by the nonlinear stabilization mechanism of the monochromatic density wave similar to that investigated in the collisionless plasma by Mazitov [78a] and O'Neill [295a]. The causes of the analogy between the mechanisms of collective processes in a gravitating and plasma medium are investigated in [87, 88, 100]. In particular, it is noted that the kinetic equation for small oscillations in a simple model of the gravitating system (rotating cylinder) by redetermining the characteristic parameters, proves to coincide with the kinetic equation for the collisionless magnetized plasma. With this is connected the fact that, in a gravitating cylinder, there may develop a beam instability described by the same relations as the beam instability in the plasma with a magnetic field (§1, Chapter VI).

Replacing the double frequency of rotation of the cylinder by the cyclotron frequency, $2\Omega \rightarrow \omega_B$, and the square of the Jeans frequency, by the negative square of the Langmuir frequency, $\omega_0^2 \rightarrow -\omega_p^2$, let us consider the upper hybrid branch of oscillations, $\omega^2 = \omega_p^2 + \omega_B^2$. In the case of a gravitating cylinder, this branch was called rotational. Here, it is characterized by the frequency $\omega^2 = \omega_0^2$, which, due to the equilibrium condition $\omega_0^2 = 2\Omega^2$, may be represented also in the form $\omega^2 = 2\Omega^2$.

In the presence of a beam moving along the generatrix of the cylinder, the rotational branch is excited with the linear growth rate

$$\gamma_L \sim \alpha \left(\frac{v}{v_T}\right)^2 \left(\frac{k_z}{k}\right)^2 \omega_0, \tag{1}$$

where α is the ratio of the beam density to the density of the medium, v, v_T are the directed and the thermal velocities of the beam, and k_z, k are the longitudinal and full wave numbers. Expression (1) is valid if the Cherenkov resonance $\omega \simeq k_z v$ dominates over the cyclotron one $\omega \pm 2\Omega \simeq k_z v$, i.e., under condition $2\Omega/k_z \gg v_T$. An estimate for the growth rate similar to (1) takes place also in case of the beam instability in a plasma. It remains in force also in the absence of the magnetic field, i.e., in the limit $\omega_B \rightarrow 0$, when the upper hybrid branch transforms to the branch of electron plasma oscillations. In the case of a gravitating cylinder, such a limiting transition is prohibited by the above conditions of equilibrium.

From the plasma theory, it is known that (cf. [39]) one may use the expression of the form of (1) for the linear growth rate in the problem of excitation via a beam of plasma oscillations only for small amplitudes of the wave, namely, for

$$\tau\gamma_L \gg 1, \tag{2}$$

where $\tau^2 = m/ek_z^2\psi$ (m and e are the mass and charge of the electron). This condition means that the inverse action of the wave field on the resonant particles is negligible. Otherwise, i.e., for $\tau\gamma_L \ll 1$, the wave field leads to trapping of the resonant particles, due to which the expression for the growth rate of the form of (1) is replaced for

$$\gamma(t) \sim \gamma_L F\left(\frac{t}{\tau}\right), \tag{3}$$

where $F(t/\tau)$ is the function, whose explicit form is given in [295a]. Then

$$\int_0^\infty \gamma(t)\, dt \sim \gamma_L\tau. \tag{4}$$

It was shown earlier that excitation of the plasma waves via a beam ceases as the amplitude of the field reaches the values corresponding to τ, such that

$$\tau\gamma_L \sim 1. \tag{5}$$

These results refer to the plasma without a magnetic field, $\omega_B \to 0$, and to perturbations propagating along the beam $k \simeq k_z$.

However, it may be shown that both for the plasma with a magnetic field and for a gravitating medium for $\omega_B \sim \omega_p$ and $k_z \ll k$ (just this case is interesting for the problem in question of a gravitating cylinder) the ordinal relations (1)–(5) remain in force. This allows one to continue the analogy between the plasma and gravitating media onto the region of nonlinear phenomena.

This section deals with the investigation of the nonlinear stage of instability in a gravitating cylinder and a disk. In Section 2.1.1, we give the necessary results of linear theory. Section 2.1.2 studies the movement of the particles in a gravitation field corresponding to the eigen (monochromatic) mode of oscillations of the cylinder. In the frame of reference of a rotating cylinder, the action of inertial forces on the gravitating particle is similar to the action of the longitudinal magnetic field on the probe charge. Then the particles of the cylinder turn out to be "magnetized," due to which (approximately) they retain their distance from the cylinder axis. For that reason, as shown in Section 2.1.2, the equation of the longitudinal movement of the particles is reduced to the equation of the type of the mathematical pendulum, which is solved similarly [78a, 295a], in elliptical functions.

Section 2.1.3 treats the nonlinear evolution of the distribution function of particles, while Section 2.1.4 finds the densities of kinetic energy of the particles and the energy of the monochromatic wave (averaged over the cylindrical layer). Using the method of energy balance (after averaging over the radius), we define the time dependence of the nonlinear growth rate. Section 2.1.5 shows the region of applicability of the theory constructed. Section 2.1.6 makes estimates of the steady-state amplitude of oscillations for different values of the parameters of the configuration.

2.1.1. Statement of the Problem. Consider the stationary collisionless system of gravitating particles in the form of a rotating uniform-in-density cylinder of infinite length, of radius R. In the frame of reference rotating along with the cylinder (at the angular frequency $\Omega = \sqrt{2\pi G\rho}$ where G is the gravitational constant and ρ is the density) the particles are moving only along the axis, which we assume to be the axis of the cylindrical frame of reference, (r and φ) will denote respectively the radial and angular coordinates).

Thus, the stationary radial and azimuthal velocities for all particles are zero,

$$v_r = v_\varphi = 0.$$

The distribution in longitudinal velocities $f^F(v_z)$, which is not restricted by the equilibrium conditions, will be assumed to have a beam form

$$f^F(v_z) = f^M(v_z) + f(v_z),$$

with the Maxwellian distribution of the basic component $f^M(v_z)$

$$f^M(v_z) = \frac{1}{\sqrt{\pi}V_T} e^{-v_z^2/V_T^2} \tag{6}$$

and the distribution function of the beam $f(v_z)$

$$f(v_z) = \frac{\alpha}{\sqrt{\pi}v_T} \exp\left[-\left(\frac{v_z - v}{v_T}\right)^2\right], \tag{7}$$

where the conditions

$$\alpha \ll 1, \tag{8}$$

$$v \gg v_T, \tag{9}$$

$$||v| - v_T| \gg V_T \tag{10}$$

are adopted.

Let us assume, in addition, that the thermal dispersion V_T of the basic component is sufficiently large

$$V_T \gg V = \Omega R. \tag{11}$$

Under these conditions, according to Section 2, Chapter II, in the system, there can propagate the axial-symmetrical oscillations of the structure

$$\Phi(r, z, t) = \Phi_0 J_0(k_\perp r) e^{-i\omega_0 t + ik_z z} e^{\gamma F(t)}, \qquad r \leq R, \tag{12}$$

(inside the cylinder) and

$$\Phi(r, z, t) = \tilde{\Phi}_0 K_0(k_z r)e^{-i\omega_0 t + ik_z z}e^{\gamma_F(t)}, \qquad r \geq R \qquad (13)$$

(outside). Here $\Phi(r, z, t)$ denotes the perturbation of the gravitational potential, t is the time variable, and Φ_0, $\tilde{\Phi}_0$ are the constants satisfying the conditions of matching on the boundary of the cylinder, in particular the continuity condition of the potential

$$\Phi_0 J_0(k_\perp R) = \tilde{\Phi}_0 K_0(k_z R), \qquad (14)$$

where J_0 and K_0 are conventional notations (cf. [42, 157]) of cylindrical functions; ω_0 is the Jeans frequency linked with the angular frequency Ω via the equilibrium condition

$$\omega_0^2 = 2\Omega^2, \qquad (15)$$

and the longitudinal wave parameter k_z must satisfy the conditions

$$k_z R \ll 1 \qquad (16)$$

and

$$k_z V_T \ll \omega_0, \qquad (17)$$

while the transversal one $k_\perp$, to the conditions

$$J_0(k_\perp R) \simeq 0, \qquad k_\perp R \gtrsim 1. \qquad (18)$$

The quantity γ_F in (12) and (13) is equal to the sum

$$\gamma_F = \gamma_M + \gamma$$

of the decrement γ_M of wave damping interacting with the basic component of the medium and the increment γ of instability caused by excitation of the waves via a beam. In turn, γ consists of the sum of two summands:

$$\gamma = \gamma_b + \gamma_c. \qquad (19)$$

Here γ_b is the consequence of the Cherenkov resonance

$$\gamma_b = \frac{\pi}{2}\frac{\omega_0^3}{k^2}\left[f'\left(\frac{\omega_0}{k_z v_T}\right)\operatorname{sgn} k_z\right], \qquad (20)$$

where $k^2 = k_z^2 + k_\perp^2$, while the prime denotes the derivative with respect to the argument; γ_c is the consequence of the cyclotron resonance,

$$\gamma_c = -\frac{\pi}{4}\frac{\omega_0^3}{|k_z|\Omega}\left[f\left(\frac{\omega_0 - 2\Omega}{k_z v_T}\right) - f\left(\frac{\omega_0 + 2\Omega}{k_z v_T}\right)\right], \qquad (21)$$

Apart from the beam instability (19), in the conditions described there is only the Jeans instability with an exponentially small (if $V_T^2 \gg V^2$) growth rate γ_J,

$$\gamma_J \simeq \frac{V_T}{V}e^{-V_T^2/V^2}. \qquad (22)$$

As will be shown below, the parameters of the configuration may be chosen to be such that the beam growth rate γ will be the largest and, moreover, due mainly to the Cherenkov resonance

$$\gamma_F \simeq \gamma \simeq \gamma_b, \qquad \gamma_c \ll \gamma_b, \tag{23}$$

$$\gamma_M \ll \gamma, \tag{24}$$

$$\gamma_g \ll \gamma. \tag{25}$$

We shall restrict ourselves just to this case.

2.1.2. Particle Motion in the Wave Field.
At some initial time moment $t_0 = 0$, in the above system, let the gravitational potential of the form [cf. (12) and (13)]

$$\Phi(r, z, t) = -\Phi_0(t)J_0(k_\perp r) \cos(-\omega_0 t + k_z z), \qquad r < R, \tag{26}$$

$$\Phi(r, z, t) = -\tilde{\Phi}_0(t)K_0(k_z r) \cos(-\omega_0 t + k_z z), \qquad r > R, \tag{27}$$

be "switched-on."

In a similar way [86], for the determination of the movement of the particles, we shall restrict ourselves to the zero order of perturbation theory in the small parameter $\delta\Phi_0/\Phi_0$, i.e., assume that

$$\Phi_0(t) = \text{const} = \Phi_0, \tag{28}$$

$$\tilde{\Phi}_0(t) = \text{const} = \tilde{\Phi}_0. \tag{29}$$

Let us assume that the potential in (26) and (27) satisfies the conditions in (14), (16)–(18).

Within the range of the wave vectors, where $\gamma \sim \gamma_{\max}$, the condition of the Cherenkov resonance may be represented in the form

$$\omega - k_z v \sim k_z v_T. \tag{30}$$

Let us study the movement of the particles arising due to the "switch-on" of the potential, in the frame of reference moving along with the wave on z (again rotating with a frequency Ω). In such a system, the potential of (26) and (27) has the form

$$\Phi(r, z) = -\Phi_0 J_0(k_\perp r) \cos(k_z z), \qquad r < R, \tag{31}$$

and the corresponding form for $r > R$.

The movement of the particles obeys the law

$$\frac{d\mathbf{v}}{dt} = 2[\mathbf{v}\Omega] - \nabla\Phi, \tag{32}$$

where $\Omega = \Omega\hat{z}$ and $\hat{z}$ is the ort along the axis of the cylinder. Such a movement would have been performed by a particle with a single charge and a mass in the constant and uniform magnetic field $\mathbf{H} = 2\Omega$ and in the electric potential Φ (the speed of light, via the selection of the system of units, is reduced to unity).

All the particles of the cylinder can be divided into three groups according to the value of their stationary velocity v_z:

(1) "slowly" moving (in the wave system) particles, which for the "cyclotron" period $T = 2\pi/2\Omega = \pi/\Omega$ are displaced on z by a distance much less than the longitudinal wavelength $\lambda = 2\pi/k_z$,

$$|v_z| \ll \frac{2\Omega}{|k_z|}; \tag{33}$$

(2) particles with a displacement of the order of the wavelength,

$$|v_z| \sim \frac{2\Omega}{|k_z|}; \tag{34}$$

(3) "fast" particles,

$$|v_z| \gg \frac{2\Omega}{|k_z|}. \tag{35}$$

The transversal movement of "slow" particles is similar to the movement of a charged particle under the action of a longitudinal constant "magnetic" field and slowly varying radial "electric" field i.e., is the drift in azimuth; the radial size of the orbit then is of the order of the "cyclotron" radius

$$r_{\text{rot}} \sim \frac{k_\perp \Phi_0}{4\Omega^2}. \tag{36}$$

For a comparatively weak potential Φ_0, the particle displaces little in radius; the corresponding condition $r_{\text{rot}} \ll R$ can be, with due regard for (18), written in the form

$$\frac{\Phi_0}{V^2} \ll 1, \tag{37}$$

and denotes the smallness of perturbation of the particles, which of course, also is suggested, with necessity, by linear theory. For the developed nonlinear theory, let us also assume the inequality (37) to be satisfied; below it will be seen that the condition of applicability of the nonlinear theory gives a restriction of the potential from below, but this constraint may not contradict (37).

Taking into account that the initial transversal velocity is zero, we find that the trajectory of a slow particle has the shape of an epicycloid, while the longitudinal movement occurs thus if the particle remained all the time at the same distance r from the axis (r is the coordinate of the particle before the switch-on of the additional gravitational field).

The transversal movement of fast particles is the movement in a rapidly oscillating (with a frequency much higher than the frequency of rotation, $|k_z v_z| \gg \Omega$) gravitational field. The oscillation in r with the amplitude

$$\Delta r \sim \frac{k_\perp \Phi_0}{(k_z v_z)^2} \ll \frac{k_\perp \Phi_0}{\Omega^2} \sim r_{\text{rot}} \tag{38}$$

will not lead, however, to a significant change in the radial position of the particle. Thus, the change in the radial location of fast particles is still less than the slow ones, and, in the study of their longitudinal movement, their radial coordinate can the more be considered as constant.

One should not neglect the change in radial coordinate only for the rotational-resonant particles (34). But the portion Δf of such particles can be estimated as

$$\Delta f \sim f_0\left(\frac{2\Omega}{k_z}\right)\Delta v_z \sim f_0\left(\frac{2\Omega}{k_z}\right)\frac{\Omega}{k_z},$$

and is exponentially small, if

$$\frac{2}{k_z r} \gg \frac{v_T}{V}. \tag{39}$$

This last condition is easily satisfied; therefore, we eliminate these particles from consideration. Thus, in the study of the longitudinal movement we assume the radial coordinate of each particle to be fixed in its initial value.

2.1.3. Nonlinear Evolution of the Distribution Function.
Consider now the evolution of the distribution function in its longitudinal velocity.

The longitudinal field of the potential (31) is

$$E_z(r, z) = -\frac{\partial}{\partial z}\Phi(r, z) = -\varepsilon_z(r)\cos(k_z z), \tag{40}$$

where the amplitude ε_z is given by the expression

$$\varepsilon_z = k_z \Phi_0 J_0(k_\perp r). \tag{41}$$

It is seen that the amplitude ε_z is dependent only on the radial variable (remaining, as established above, at the longitudinal movement of the particle). The amplitude changes from its maximum value on the cylinder axis to zero on the edge [cf. (18)].

The field (40) leads to the equation of longitudinal movement (cf. [295a]).

$$\dot{v}_z = \ddot{z} = -\varepsilon_z(r)\sin k_z z. \tag{42}$$

Conservation of energy of the longitudinal movement is written in the form

$$\frac{v_z^2}{2} - \frac{\varepsilon_z}{k_z}\cos(k_z z) = \text{const} \equiv W. \tag{43}$$

In a similar way [39], by the method of integration over trajectories, we arrive at the following distribution functions.

The distribution function of the particles trapped by the wave in its longitudinal movement has the form (in the wave system)

$$f_r(z, v_z, t) = f(0) + \frac{\partial}{\partial v_z} f(0)\sigma\sqrt{2\left[W(r, z, v_z) + \frac{\varepsilon_z(r)}{k_z}\right]}$$

$$\times \text{cn}\left\{F\left[\zeta(z), \frac{1}{\varkappa(z, v_z)}\right] - \frac{t}{\tau_r}, \frac{1}{\varkappa}\right\}, \qquad \varkappa > 1. \tag{44}$$

For the transit particles, the distribution function is of the form

$$f_r(z, v_z, t) = f(0) + \frac{\partial}{\partial z} f(0)\sigma \sqrt{2\left[W(r, z, v_z) + \frac{\varepsilon_z(r)}{k_z}\right]}$$

$$\times \, dn\left\{F\left[\frac{kz}{2}, \varkappa(z, v_z)\right] - \frac{t}{\varkappa\tau_r}, \varkappa\right\}, \qquad \varkappa < 1. \tag{45}$$

The notations here are as follows:

$$\sigma = \text{sgn } v_z, \qquad W(r, z, v_z) = \frac{v_z^2}{2} - \frac{\varepsilon_z(r)}{k_z} \cos(k_z z), \tag{46}$$

$F(\varphi, k) = \int_0^\varphi d\alpha/\sqrt{1 - k^2 \sin^2 \alpha}$ is the elliptical integral of the first kind, $cn(u, k)$ is the elliptical cosine, and $dn(u, k)$ is the delta of the amplitude—the function defined by the relation $dn[F(\varphi, k), k] = \sqrt{1 - k^2 \sin^2 \varphi}$;

$$\varkappa^2(z, v_z) = \frac{2\varepsilon_z(r)}{k_z W(z, v_z) + \varepsilon_z(r)}, \tag{47}$$

$$\zeta(z, v_z) = \sin\left(\arcsin\frac{k_z z}{2}\right), \tag{48}$$

$$\tau_r^2 = \frac{1}{k_z \varepsilon_z(r)} = \frac{1}{k_z^2 \Phi_0 J_0(k_\perp r)}, \tag{49}$$

i.e.,

$$\tau_r^2 = \frac{\tau_0^2}{J_0(k_\perp r)}, \tag{50}$$

where

$$\tau_0^2 = \frac{1}{k_z^2 \Phi_0}. \tag{51}$$

Similarly to [39], we arrive at the conclusion that, in the region of the phase space (z, v_z), corresponding to the trapped particles, a plateau is produced, and one can write the distribution function of nontrapped particles averaged over time. The difference from the evolution in problems [38, 78a] is in the fact that the evolution of our configuration proceeds at a variable rate at different distances from the axis. The oscillation period of the particles trapped by the wave, according to (50), increases from τ_0 on the axis up to (formally) an infinite quantity on the edge of the cylinder.

2.1.4. Nonlinear Evolution of the Monochromatic Wave. To find the growth rate of the field variation in the method applied here we make use of the equation of energy balance

$$\frac{dQ}{dt} = -2\gamma W, \tag{52}$$

where Q is the average (over the volume of the cylinder) density of kinetic energy of the particles and W is the mean density of the wave energy {i.e., the sum of the field energy and the energy of the nonresonance particles (c.f., e.g., [10])}.

Equation of motion (42) coincides with the corresponding equation, found by O'Neill [295a]; therefore, for the variation rate of density Q_r of the kinetic energy of the particles in a circular cylinder $(r, r + dr)$, averaged over the volume of the circular cylinder, we obtain (cf. (30) in [295a])

$$
\frac{dQ_r}{dt} = -\frac{\pi \omega_0^3}{2 k_z^2} f' \frac{\varepsilon_z^2(r)}{4\pi G} \sum_{n=0}^{\infty} \frac{64}{\pi} \int_0^1 d\varkappa
$$

$$
\times \left\{ \frac{2n\pi^2 \sin(\pi n t/\varkappa K \tau_r)}{\varkappa^5 K^2(1 + q^{2n})(1 + q^{-2n})} + \frac{(2n + 1)\pi^2 \varkappa \sin[(2n + 1)\pi t/2K\tau_r]}{K^2(1 + q^{2n+1})(1 + q^{-2n-1})} \right\},
$$

(53)

where

$$
q = \exp\left(\frac{\pi K'}{K}\right), \qquad K = F\left(\frac{\pi}{2}, \varkappa\right), \qquad K' = F\left(\frac{\pi}{2}, (1 - \varkappa^2)^{1/2}\right).
$$

This expression has to be further averaged over the radii of annular cylinders

$$
\frac{dQ}{dt} = \int_0^R \frac{dQ_r}{dt} \frac{2\pi r \, dr}{\pi R^2}.
$$

(54)

Note that the radius dependence in (53) enters only through $\varepsilon_z(r)$ and τ_r.

To calculate the value W entering into (52), one should take the integral [the external field is very small, according to (14), (18) even on the edge of the cylinder and drops rapidly with increasing r, so that we neglect its contribution to the energy]:

$$
\frac{1}{\pi R^2} \int_0^R \frac{\varepsilon_z^2 + \varepsilon_r^2}{8\pi G} 2\pi r \, dr.
$$

(55)

Here, similarly to (41)

$$
\varepsilon_r(r) = k_\perp \Phi_0 J_1(k_\perp R),
$$

(56)

so that the radial field intensity $E_r(z, r)$ is

$$
-\frac{\partial \Phi(r, z)}{\partial r} = E_r(z, r) = -\varepsilon_r(r) \cos(k_z z).
$$

(57)

Taking into account (56) and (41), the calculation of the integral in (55) is reduced to taking the integrals

$$
I_1 = \frac{2}{R^2} \int_0^R J_0^2(k_\perp r) r \, dr, \qquad I_2 = \frac{2}{R^2} \int_0^R J_1^2(k_\perp r) r \, dr.
$$

(58)

Using the familiar formula [157]

$$
\int x J_\mu^2(\alpha x) \, dx = \frac{x^2}{2} [J_\mu^2(\alpha x) - J_{\mu-1}(\alpha x) J_{\mu+1}(\alpha x)]
$$

and taking into account (18), we find that

$$I_1 = I_2 = J_1^2(k_\perp R). \tag{59}$$

Finally, the mean energy density of the wave is

$$W = \frac{1}{8\pi G} k^2 \Phi_0^2 J_1^2(k_\perp R), \tag{60}$$

where $k^2 = k_\perp^2 + k_z^2$. We shall not explicitly write the cumbersome general expression for the growth rate $\gamma(t)$. We show only that, in the limiting case $t \ll \tau_0$, there ensues an accurately linear growth rate. Indeed, in this case, following [295a], we have from (53)

$$-\frac{dQ_r}{dt} = \frac{\pi}{2} \frac{\omega_0^3}{k_z^2} f' \frac{\varepsilon_z^2(r)}{4\pi G} \left[\frac{64}{\pi} \int_0^1 d\varkappa \, \frac{2\pi^2 \sin(\pi t/\varkappa K \tau_r)}{\varkappa^5 K^2 (1 + q^2)(1 - q^2)} + O\left(\frac{t}{\tau_0}\right) \right]$$

$$= \frac{\pi}{2} \frac{\omega_0^3}{k_z^2} f' \frac{k_z^2 \Phi_0 J_0^2(k_\perp r)}{4\pi G} \left[1 + O\left(\frac{t}{\tau_0}\right) \right]. \tag{61}$$

Hence, averaging according to (54), we find

$$-\frac{dQ}{dt} = \frac{\pi}{2} \frac{\omega_0^3}{k_z^2} f' \frac{\Phi_0^2 k_z^2}{4\pi G} \left[\frac{2}{R^2} \int_0^R J_0^2(k_\perp r) r \, dr \right]$$

$$= \frac{\pi}{2} \omega_0^3 f' \frac{\Phi_0^2}{4\pi G} J_1^2(k_\perp R). \tag{62}$$

In the derivation of this last equation, we have used the relations (58) and (59).

Finally, by formula (52), making use of (62) and (60), we find

$$\gamma_L = -\left. \frac{dQ}{dt} \right/ 2W = \frac{\pi}{2} \frac{\omega_0^3}{k_z^2} f', \tag{63}$$

which in fact is coincident with the linear growth rate [cf. (23) and (20)].

Make now an estimate of the amplification factor of the wave $\mathcal{K}$.

According to (52),

$$\mathcal{K} = \int_0^\infty \gamma(t) \, dt = \frac{1}{2W} \int_0^\infty \left(-\frac{dQ}{dt} \right) dt = -\frac{\Delta Q}{2W}. \tag{64}$$

Following [295a], we find

$$-\Delta Q = -\int_0^R \Delta Q_r \frac{2\pi r \, dr}{\pi R^2} = O\left[\int_0^R \gamma_L \tau_r k^2 \frac{\Phi_0^2}{4\pi G} J_0^2(k_\perp r) \frac{2r \, dr}{R^2} \right]$$

$$= O\left[\gamma_L k^2 \frac{\Phi_0^2}{4\pi G} \tau_0 \frac{2}{R^2} \int_0^R J_0^{3/2}(k_\perp r) r \, dr \right] = O\left(\gamma_L k^2 \frac{\Phi_0^2}{4\pi G} \tau_0 \right). \tag{65}$$

Since in accordance with (60)

$$2W \sim \frac{k^2 \Phi_0^2}{4\pi G}, \tag{66}$$

then from (64) and (65) we get

$$\int_0^\infty \gamma(t)\, dt = O(\gamma_L \tau_0). \tag{67}$$

From this last equation follows the condition of validity of the approximation of constantness with time of the amplitude of the wave adopted in the
study of the movement of the particles:

$$\gamma_L \tau_0 \ll 1. \tag{68}$$

2.1.5. Range of Applicability of the Theory. The theory constructed above
has its region of applicability at simultaneous fulfilment of all the assumptions
adopted, i.e., (68), the condition of smallness of the "cyclotron" radius (37)
and the condition of domination of the Cherenkov resonance (23)–(25).
These also involve the condition of smallness of the growth rate of the
hydrodynamical beam instability

$$\frac{v_T}{v} > \alpha^{1/2}. \tag{69}$$

Denote $v_T/V = \tilde{v}_T$, $v/V = \tilde{v}$, $V_T/V = \tilde{V}_T$, etc. Take $k_\perp R \sim 1$,

$$\tilde{v}_T \sim 1, \tag{70}$$

$$\frac{1}{\tilde{v}} \sim \frac{\tilde{v}_T}{\tilde{v}} \sim \alpha^{1/2}; \tag{71}$$

then taking into account (8)–(10), and (30), we arrive at the inequalities

$$\frac{\Phi_0}{V^2} \ll 1 \qquad \left(\frac{r_{\text{rot}}}{R} \ll 1\right), \tag{72}$$

$$\frac{\tilde{v}\Phi_0^{1/2}}{V} \gg 1 \qquad (\gamma_L \tau_0 \gg 1), \tag{73}$$

$$1 \gg \tilde{v} e^{-\tilde{v}^2} \qquad (\gamma_b \gg \gamma_c), \tag{74}$$

$$\frac{1}{\tilde{v}^2} \gg \tilde{V}_T e^{-\tilde{V}_T^2} \qquad (\gamma \gg \gamma_J), \tag{75}$$

$$\frac{1}{\tilde{v}^2} \gg \frac{\tilde{v}}{V_T} e^{-\tilde{v}^2/\tilde{V}_T^2} \qquad (\gamma \gg \gamma_M). \tag{76}$$

If now, leaving the quantity $V \sim v_T$ to be fixed, one increases $\tilde{V}_T$ and increases
$\tilde{v}$ according to the law $\tilde{v} \sim \tilde{V}_T^2$, by simultaneously decreasing α according to
the law of (71) and k_z according to the law [cf. (30)]

$$k_z R \sim \frac{\sqrt{2}}{\tilde{v}},$$

then the right-hand sides of (74)–(76) decrease exponentially, while the left sides are not stronger except in a power way. Consequently, beginning with some rather large V_T and v, all the inequalities (74)–(76) will be satisfied. Inequality (73) is also satisfied for a sufficiently high velocity of the beam, v, whatever the amplitude Φ_0 satisfying (72).

2.1.6. Estimations of the Resulting Amplitude of a Monochromatic Wave. From the relations given above it follows that the beam instability of a gravitating cylinder is saturated at

$$\frac{\psi}{\psi_0} \sim \frac{\alpha^2}{(kR)^4} \left(\frac{V_0 v}{v_T^2}\right)^2, \tag{77}$$

where ψ_0 is the equilibrium gravitational potential. It is seen that the ratio ψ/ψ_0, as the function of kR, is maximal at $kR \sim 1$, then

$$\frac{\psi_{max}}{\psi_0} \sim \alpha^2 \left(\frac{V_0 v}{v_T^2}\right)^2. \tag{78}$$

This ratio grows with decreasing thermal scatter of the beam; on the limit of applicability of the concepts of the kinetic instability, i.e., for $(v_T/v) \sim \alpha^{1/3}$ (cf. [86]),

$$\frac{\psi_{max}}{\psi_0} \sim \alpha^{2/3} \left(\frac{V_0}{v}\right)^2. \tag{79}$$

It is interesting to note that under the interaction of a rotating gravitating medium with the beam of a comparable density, $\alpha \sim 1$, and a comparable velocity, $v \sim V_0$, the perturbed gravitational potential ψ turns out to be the same as the equilibrium potential ψ_0 in order of magnitude.

2.2 Nonlinear Saturation of the Instability at the Corotation Radius in the Disk [22[ad]]

This section investigates the Mazitov–O'Neill effect in the stellar disk. Unlike the cylinder, where the wave resonance occurs with a small group (beam) of particles in the velocity space, in the disk with orbits close to circular, the resonance of a spiral wave with stars takes place with almost all the particles of the velocity space located, however, near some definite radii defined by the relation $\Omega(r) - \omega/m = -l\varkappa/m$ (m is the number of the azimuthal mode, l is the "number" of the resonance, $\varkappa$ is the epicyclic frequency, Ω is the angular frequency of rotation of the disk, and ω is the wave frequency). $l = 0$ corresponds to the resonance of "corotation." [11] Lynden-Bell and Kalnajs [289] in the linear approximation show a possibility of amplification of the density wave at the corotation radius. The question of the wave

[11] Refer to Section 2, Chapter XI, for more detail regarding resonances and results of [289].

stabilization level under its nonlinear interaction with stars in the vicinity of the corotation radius is natural. This section is devoted to the clarification of this question. In Section 2.2.1, equations are derived and expressions are obtained for the steady-state distribution function of stars near the corotation radius, and the energy and angular moment transferred to stars are calculated. The estimates of the wave amplitude and the regions of applicability of the results attained are contained in Section 2.2.2.

2.2.1. Stellar Distribution Function Near the Corotation Radius of a Disk.

We turn now to the frame of reference rotating with an angular velocity of a spiral pattern Ω_p. In the epicyclic approximation, the Hamiltonian of the star in this system has the form [60^{ad}]:

$$H = V_0(R) + V_1 + \tfrac{1}{2}\Omega^2(R)R^2 + \varkappa(R)\mathcal{J}_1 - \Omega_p\Omega(R)R^2 + \cdots. \qquad (1)$$

Here V_0 and V_1 are the potential of background and the spiral potential, respectively, R is the radial coordinate of the epicycle center, $\Omega(R)$ is the angular velocity, and $\varkappa(R)$ is the epicyclic frequency; the value H should be considered as a function of the variable angle-action ($\mathcal{J}_1, w_1, \mathcal{J}_2, w_2$): $\mathcal{J}_1 = E/2\varkappa$, E is the energy of the epicyclic movement, $\mathcal{J}_2 = \Omega(R)R^2$ is the angular moment,

$$r - R = (2\mathcal{J}_1/\varkappa)^{1/2} \sin w_1,$$

$$\varphi = w_2 + (2\Omega/\varkappa)(2\mathcal{J}_1/\varkappa R^2)^{1/2} \cos w_1,$$

r, φ are the coordinates of the cylindrical system of coordinates, while the dots in (1) denote the terms of higher order of magnitude in the epicyclic approximation.

Let the radius of corotation be the coordinate $R = R_0$, where $\Omega(R) = \Omega_p$. With interest in the star behavior near the corotation, assume that $x = I_2/(\partial\mathcal{J}_2/\partial R)_{R=R_0} = 2\Omega_p I_2/R_0\varkappa_0^2 \ll R_0$ and $I_2 = \mathcal{J}_2 - \Omega_p R_0^2$, where the index 0 denotes the quantity taken at $R = R_0$. The equations of motion of a star will be derived from the Hamiltonian equation $d\mathcal{J}_i/dt = -\partial H/\partial w_i$ and $dw_i/dt = \partial H/\partial\mathcal{J}_i$. Restricting ourselves to the range of small x, we obtain

$$\frac{dI_1}{dt} = -\frac{\partial V_1}{\partial w_1}, \qquad (2)$$

$$\frac{dI_2}{dt} = \left(\frac{R_0\varkappa_0^2}{2\Omega_p}\right)\frac{dx}{dt} = -\frac{\partial V_1}{\partial w_2}, \qquad (3)$$

$$\frac{dw_1}{dt} = \varkappa_0 + \varkappa_0' x + \frac{\partial V_1}{\partial I_2}, \qquad (4)$$

$$\frac{dw_2}{dt} = \frac{2\Omega_p}{R_0\varkappa_0^2}\left(\varkappa_0' I_1 + 2bx + \frac{\partial V_1}{\partial x}\right), \qquad (5)$$

where $b = \Omega_0' R_0\varkappa_0^2/4\Omega_p$ and $I_1 = \mathcal{J}_1$.

From (3) and (5) it is evident that the equation of corotational resonance in phase space (x, I_1) has the form $2bx + \varkappa_0' I_1 = 0$. In particular, for "cold" particles $(I_1 = 0)$ the condition of resonance $x = 0$ means that the radius of the circular orbit coincides with R_0. In a system with stellar dispersion over I_1, the stars from some vicinity $R = R_0$ prove to be resonant.

Equations (4) and (5) show that $\dot{w}_1 \gg \dot{w}_2$. This allows one to simplify the system of (2)–(5) by making use of the method of averaging over a "fast" phase w_1 (cf., e.g., $[2^{ad}]$). We represent the variables in the form of $I_i = \bar{I}_i + \tilde{I}_i$, $w_i = \bar{w}_i + \tilde{w}_i$. Retaining in the subsequent formulae instead of the mean quantities (with overbar), the previous notations (without overbar), we obtain, instead of (2)–(5),

$$I_1 = \text{const}, \tag{6}$$

$$\left(\frac{R_0 \varkappa_0^2}{2\Omega_p}\right)\dot{x} = \dot{I}_2 = -\frac{\partial \bar{V}_1}{\partial w_2}, \tag{7}$$

$$\dot{w}_2 = \frac{2\Omega_p}{R_0 \varkappa_0^2}\left(\varkappa_0' I_1 + 2bx + \frac{\partial \bar{V}_1}{\partial x}\right). \tag{8}$$

The system of (6)–(8) describes the movement of the epicycle center $(x(t), w_2(t))$. For the spiral potential of the form $V_1 = \psi(r) \cos\{m[\Phi(r) + \varphi]\}$ one easily obtains

$$\bar{V}_1(R, I_1, w_2) = \psi(R) J_0(k'a) \cos\{m[\omega_2 + \Phi(R)]\}, \tag{9}$$

where $k'^2 = k^2 + (2m\Omega/\varkappa R)^2$; $k = m(d\Phi/dr)_{r=R}$, and $a = (2I_1/\varkappa)^{1/2}$ is the size of the epicycle. Near the resonance, one may assume that $k'^2 = k^2(R_0) + (2m\Omega_p/R_0\varkappa_0)^2$, $a = (2I_1/\varkappa_0)^{1/2}$, $\psi(R) \approx \psi(R_0) \equiv \psi_0$, $m\Phi(R) \approx m\Phi(R_0) + kx$; J_0 is the Bessel function of zero order. For a tight spiral, $\tan i = m/kR \ll 1$, and we assume that $k' = k$, by assuming for certainty that $m = 2$. We denote the phase of the spiral wave $\theta = \omega_2 + \Phi(R_0) + kx/2$. Without any restrictions on generality, one may assume that $\Phi(R_0) = 0$. Then we shall obtain finally:

$$\ddot{\theta} = -\frac{4\Omega_0|\Omega_0'|}{R_0\varkappa_0^2}\psi_0 J_0(ka) \sin 2\theta, \tag{10}$$

where x and φ are connected with θ via the relations

$$x = \frac{1}{\Omega_0'}\left(\dot{\theta} - \frac{2\Omega_0\varkappa_0'}{R_0\varkappa_0^2}I_1\right), \qquad \varphi = \theta - \frac{kx}{2}. \tag{11}$$

Thus, the problem of the movement of the epicycle center is reduced to the one-dimensional problem (10) of the nonlinear pendulum. Equation (10) has the integral of "energy" ε:

$$\varepsilon = \frac{\dot{\theta}^2}{2} + \frac{\omega_b^2}{2} \cdot \begin{cases} \sin^2 \theta & \text{for } J_0(ka) > 0, \\ \cos^2 \theta & \text{for } J_0(ka) < 0, \end{cases}$$

where

$$\omega_b^2 = \frac{8\Omega_0 |\Omega_0'|}{R_0 \varkappa_0^2} |J_0(ka)| \psi_0.$$ (12)

The phase plane $(\theta, \dot\theta)$ is separated by the separatrix into the regions of trapped $(q^2 \equiv \omega_b^2/2\varepsilon > 1)$ and transit $(q^2 < 1)$ particles. The parameter q^2 runs through the values from 0 to ∞. The trajectories of trapped particles (epicycles) on the plane R, φ describe a closed curve resembling a banana.[12]

The possibility of reducing the problem to the one-dimensional one allows one to make use of the results concerning the evolution of the distribution functions familiar from plasma physics [78a, 295a]. As a result of stellar intermixing in the phase space, there occurs formation of "plateau" of the distribution function for the characteristic time of the order of

$$\tau_b = \frac{2\pi}{\omega_b} = \frac{2\pi}{\Omega_0} \frac{\varkappa_0}{2\Omega_0} f^{-1/2}(\tan i)^{-1/2} \sim T(f \tan i)^{-1/2}.$$

Here T is the period of rotation of a galaxy on the corotational radius; $f = k\psi_0/\Omega^2 R$ is the amplitude of the gravitational force.

Lynden-Bell and Kalnajs [289] show that the spiral wave, being a wave of negative energy, may amplify at the corotational radius due to the transfer to the stars of the moment and energy. The rate of such transfer, calculated in linear theory, is

$$\dot{\mathscr{L}}_L = -\frac{\pi\psi_0^2}{|\Omega_0'|}\left(\frac{R_0\varkappa_0^2}{2\Omega_0}\right)\frac{\partial\tilde{F}(\mathscr{I}_2 = \tilde{\mathscr{I}}_0)}{\partial\mathscr{I}_2}, \qquad \dot{\mathscr{E}}_L = \Omega_p\dot{\mathscr{L}}_L.$$ (13)

Here F is the distribution function of stars in angular moments,

$$\tilde{F}(\mathscr{I}_2) = (2\pi)^2 \int F_0(\mathscr{I}_1, \mathscr{I}_2)\, d\mathscr{I}_1, \qquad \tilde{\mathscr{I}}_0 = \Omega_p R_0^2,$$

where $F_0(\mathscr{I}_1, \mathscr{I}_2)$ is the initial distribution function. As a result of the exchange of the moment and a production of the "plateau" the amplification of the wave will cease. One may estimate the full moment $\Delta\mathscr{L}$ and the energy $\Delta\mathscr{E}$ transferred by the wave to the resonance stars. For that purpose, let us calculate the finalized distribution function.

Let the initial distribution function be $F_0 = F_0(I_1, I_2(R)) = F_i(I_1, R)$. In the vicinity of the resonance $R = R_0 - (\varkappa_0'/2b)I_1$ we write

$$F_i(I_1, R) \simeq F_0\left(I_1, R_0 - \frac{\varkappa_0'}{2b}I_1\right) + \left(\frac{\partial F_0}{\partial R}\right)_{R=R_0-\varkappa_0'I_1/2b} \cdot \left(x + \frac{\varkappa_0'I_1}{2b}\right).$$ (14)

[12] "Bananas" as considered by Kontopolos [60ad] are drawn by the stars themselves with $I_1 = 0$, while in our case, these are the orbits of the epicycle center.

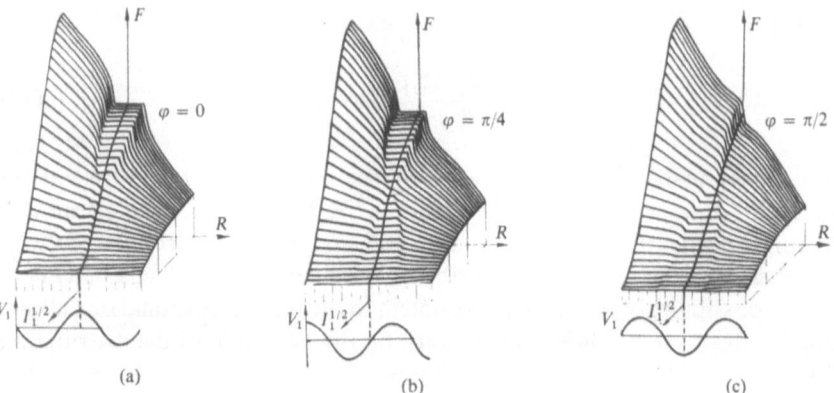

Figure 100. Final star distribution function; (a) $\varphi = 0$, (b) $\varphi = \pi/4$, (c) $\varphi = \pi/2$. The origin of the system of reference $(R, I_1^{1/2})$ is at the corotation radius.

As a result of phase mixing, the distribution function will take on the form

$$
F_f = \begin{cases}
F_{tr} = F_0\left(I_1, R_0 - \dfrac{\varkappa_0' I_1}{2b}\right), & \text{if } q^2 > 1, \\[3mm]
F_c = F_0\left(I_1, R_0 - \varkappa_0' \dfrac{I_1}{2b}\right) - \dfrac{\pi \omega_b \sigma}{2qK(q)|\Omega'|} \dfrac{\partial F_0}{\partial R}, & \text{if } q^2 < 1, \quad (15)
\end{cases}
$$

where F_c and F_{tr} are finalized distribution functions of the transit and trapped particles, respectively, $K(q)$ is the total elliptic integral of the first kind, $\sigma = -\operatorname{sgn}(x + \varkappa_0' I_1/2b)$.

Figure 100 shows the form of the final distribution function $F_f(I_1, R, \varphi)$ near the corotational radius for three different directions: (a) $\varphi = 0$; (b) $\varphi = \pi/4$; (c) $\varphi = \pi/2$. The origin of the angle φ is chosen so that $V_1 \sim \cos[2\varphi + k(R - R_0)]$. Each figure gives the behavior of the spiral potential as a function of r. The figures are made by computer according to (15) for the initial distribution function of the form

$$
F_i(I_1, R) = \frac{2\Omega(R)}{\varkappa(R)} \frac{\sigma_0(R)}{2\pi c^2(R)} \exp\left(-\frac{\varkappa I_1}{c^2(R)}\right),
$$

where $\Omega(R) = \Omega_p R_0/R$, $\sigma_0(R) = \sigma_0 \exp(-R/L)$.

The parameters are chosen so that $f = k\psi_0/R_0\Omega_p^2 = 0.05$, $L/R_0 = 0.5$, $\tan i = 2/kR = 1/7$, $c/R_0\Omega_0 = 0.17$. To estimate the angular moment transfer, one has to calculate the integral

$$
\Delta \mathscr{L} = \int \delta F \, \mathscr{J}_2 \cdot 2\pi \, dw_2 \, dI_0 \, dI_2,
$$

where $\mathscr{J}_2 = I_2 + \hat{\mathscr{J}}_0$ and

$$
\delta F = F_f - F_i = \frac{\omega_b \sigma}{q|\Omega'|} \frac{\partial F}{\partial R} \cdot \begin{cases}
\sqrt{1 - q^2 \sin^2 \theta}, & \text{if } q^2 > 1, \\[3mm]
-\dfrac{\pi}{2K(q)} - \sqrt{1 - q^2 \sin^2 \theta}, & \text{if } q^2 < 1.
\end{cases} \quad (16)
$$

Neglecting the deviation of the resonance line from $R = R_0$, which is valid for a sufficiently small thermal dispersion of stars, we obtain

$$\Delta \mathscr{L} = -\frac{64Q}{|\Omega'|}\frac{\psi_0^2}{\omega_b}\left(\frac{R\varkappa^2}{2\Omega_p}\right)\frac{\partial F_i(\mathscr{J}_2 = \hat{\mathscr{J}}_0)}{\partial \mathscr{J}_2}, \tag{17}$$

where $Q = Q_c + Q_{tr}$ and

$$Q_c = \int_0^1 \frac{dq}{q^4}\left[\frac{E(q)}{\pi} - \frac{\pi}{4K(q)}\right] \simeq 0.028,$$

$$Q_{tr} = \int_0^1 \frac{q\,dq}{\pi}[E(q) + (q^2 - 1)K(q)] \simeq 0.071.$$

The quantities Q_{tr} and Q_c characterize the contributions to the moment transfer and energy transfer of the trapped and transit stars, respectively. It is interesting to note that, although the portion of the untrapped particles in the moment and energy transfer is less than the portion of the trapped particles, nevertheless it is rather significant: $Q_c/Q_{tr} \simeq 0.4$. Comparing (17) and (13), we obtain

$$\Delta \mathscr{L} \simeq 0.32\tau_b \dot{\mathscr{L}}_L. \tag{18}$$

2.2.2. Estimates of the Wave Amplitude and of the Range of Applicability of the Theory. Formula (18) shows that for typical values of the parameters $f \simeq 0.05$, $\tan i \simeq 1/7$, the mechanism of wave amplification, due to the moment and energy transfers to the particles, works approximately during three rotations of the galaxy. For longer time scales, formation of "plateau" leads to the fact that the resonance particles cease to play any role in the energy balance.

Let us estimate the wave amplitude by assuming that this amplification mechanism is the *single* mechanism of generation of the spiral pattern (i.e., we neglect the influence of damping at the inner Lindblad resonance, the influence of the barlike structure in the central region of the galaxy, dissipation in the gas-dynamical shock waves, etc.). Then the upper boundary of the amplitude of resulting spiral pattern can be estimated by equating the moment transferred to the resonance stars and the wave moment. The angular moment of the wave $\mathscr{L}_w$ (cf., e.g., [251]) is $\mathscr{L}_w = 2\pi \int RL(R)\,dR$, where $L(R)$ is the density of the angular moment: $L = (|k|m\psi^2/4G\varkappa)\,dD/dv$ while $D = D(k, v)$ is the "dielectric" permittivity of the stellar disk, $v = m(\Omega_p - \Omega)/\varkappa$. By neglecting the thermal dispersion one may assume that $D \simeq 2\pi G\sigma_0|k|/(1 - v^2)\varkappa^2$. In this case, in order of magnitude

$$\mathscr{L}_w \sim (2\pi)^2\psi_0^2(\tan i)^{-2}\sigma/\varkappa^3.$$

Comparing with (17), we obtain the estimate of the value of the saturation amplitude:

$$f \sim \eta^2 \tan^3 i \sim \tan^3 i \sim 3 \times 10^{-3}, \tag{19}$$

where $\eta = (d \ln \mu/d \ln R)_{R_0} \sim 1$ and $\mu(R)$ is the mass per unit square of the angular moment

$$\mu = \frac{dM}{d\mathcal{J}_2^2} = 2\pi\sigma R \frac{dR}{d\mathcal{J}_2^2} = \frac{\Omega\sigma}{\varkappa^2}.$$

Such a low level of saturation indicates the extreme noneffectiveness of the resonance stars in the dynamics of the spiral pattern.

In conclusion, we make some remarks. As is known (cf. e.g., [39]), the results of the problem of the wave increasing up to the finite amplitude, unlike the results of the problem of the wave damping of finite amplitude, may be considered only as an estimate, because in the calculation of the movements of the particles we neglect the growth of the potential. In the case considered by us, another problem arises—the necessity of a self-consistent consideration. Unlike the plasma, where the resonance involves a small portion of particles in the velocity space, which does not "spoil" the spatial distribution of the potential, here we in fact deal with the resonance in a coordinate space. Therefore, in order to conserve the imposed spiral form of the potential, it is required that the width of the region of plateau in the radius be less than the wavelength of the spiral. The ratio of the width of the region of plateau Δx to the wavelength $\lambda = 2\pi/k$ is of the order $\Delta x/\lambda \simeq (2f/\tan i)^{1/2}/\pi$. The model with the parameters $f = 0.05$, $\tan i = 0.14$ yields $\Delta x/\lambda \lesssim 0.3$.

§ 3 Nonlinear Theory of Gravitational Instability of a Uniform Expanding Medium[13]

We have so far treated the stability of *stationary* systems. This section will deal with an *expanding* uniform medium.

Of most importance in astrophysical applications is the approximate solution of the nonlinear problem of the development of perturbations of arbitrary amplitude in a gravitating uniform (on the average) medium without pressure, as found by Zeldovich [48]. This solution generalizes the results of the perturbation theory and describes the evolution of the increasing mode of potential perturbations at the nonlinear stage.

The solution is constructed on the background of a uniform isotropically expanding medium without pressure, whose evolution is described by hydrodynamical equations with gravity (using the iterating indices—summing):

$$\frac{\partial u_i}{\partial t} + u_r \frac{\partial u_i}{\partial x_r} = -\frac{\partial \Phi}{\partial x_i},$$

$$\frac{\partial \rho}{\partial t} + \frac{\partial}{\partial x_r}(\rho u_r) = 0, \tag{1}$$

$$\frac{\partial^2 \Phi}{\partial x_r^2} = 4\pi G\rho,$$

[13] This section, on request of the authors, is written by A. G. Doroshkevich.

where ρ, Φ, and u are the density, gravitational potential, and the velocity of the medium and G is the gravity constant. The solution corresponding to the isotropic expansion with conservation of uniformity (A. A. Fridman's model) has the form

$$u_i = H(t)r_i, \qquad \rho = \rho(t) = \text{const}/a^3, \qquad a = \exp(-\smallint H \, dt),$$

$$\frac{dH}{dt} + H^2 = -\frac{4\pi}{3} G\rho.$$

In the simplest case [48]

$$\rho = \frac{1}{6\pi Gt^2}, \qquad H = \frac{2}{3t}, \qquad a = \left(\frac{t}{t_0}\right)^{2/3}. \tag{2}$$

This solution is known to be unstable. According to the approximate nonlinear theory suggested by Zeldovich, the movement of an individual element of the matter obeys the relation

$$\mathbf{r}(\mathbf{q}, t) = a(t)[\mathbf{q} - B(t)\mathbf{S}(\mathbf{q})], \tag{3}$$

where $\mathbf{r}$ and $\mathbf{q}$ are the Euler and Lagrange coordinates of the particles, $\mathbf{S}(\mathbf{q})$ is the initial displacement of the particle from the equilibrium location, $a(t)$ describes the general expansion of the medium, and $B(t)$ describes the growth of perturbations. In the simplest case of (2), $a \sim t^{2/3}$ and $B \sim t^{2/3}$. By using these relations, it is easy to find both velocity and density in the particle

$$\mathbf{u} = \frac{d\mathbf{r}}{dt} = H\mathbf{r} - a\dot{B}\mathbf{S}(\mathbf{q}), \qquad \rho = \frac{\rho_0}{D(\mathbf{r})/D(\mathbf{q})} = \frac{\rho_0}{a^3 |\delta_{ik} - B \, \partial S_i/\partial q_k|} \tag{4}$$

where $\dot{B} = dB/dt$, $D(\mathbf{r})/D(\mathbf{q})$ is the Jacobian of transform $\mathbf{r} = \mathbf{r}(\mathbf{q}, t)$. As is well known, in the increasing perturbation mode, the velocities are potential. Therefore, also the approximate theory deals with the potential vector $\mathbf{S}(\mathbf{q})$. Thus, the tensor $\partial S_i/\partial q_k$ is symmetrical and at each point may be reduced to the principal axes. In a corresponding frame of reference

$$\rho = \frac{\rho_0}{a^3(1 - B\lambda_1)(1 - B\lambda_2)(1 - B\lambda_3)}, \tag{5}$$

where $\lambda_1 \geq \lambda_2 \geq \lambda_3$ are the main values of the deformation tensor $\partial S_i/\partial q_k$.

If at the point $\lambda_1 > 0$, then, according to this solution, for a finite time, the density will tend to infinity at the time $B = 1/\lambda_1$. It is typical that the infinity arises due to the vanishing of one factor $(1 - B\lambda_1)$, i.e., due to the compression in the only direction determined by the largest principal value of the deformation tensor $\partial S_i/\partial q_k$. The movement in the plane orthogonal with respect to this direction leads merely to the finite density change. This is the general property of the smooth initial velocity distribution of the particles. In the general, nongenerated case, their intersection occurs on the so-called caustic surface. Only in the degenerated cases ($\lambda_1 = \lambda_2 > \lambda_3$ or $\lambda_1 = \lambda_2 = \lambda_3$) is there focusing, either cylindrical or spherical.

The quantities $\lambda_1, \lambda_2, \lambda_3$ are the functions of coordinates. Infinite density is reached first of all in the particle in which $\lambda_1 > 0$ and is maximal in the region under consideration. Further, infinite density is attained in the neighboring particles located in the plane orthogonal to the direction of the principal axis corresponding to λ_1. A very flattened cloud of compressed gas is thus produced.

In a compressed gas one should not neglect the pressure, which would stop compression at a finite density. Therefore, in the direction of the main axis corresponding to λ_1, the flow running over the compressed matter stops and gets compressed in the shock wave. In more detail, the shape and structure of the compressed gas clouds is studied by Ya. B. Zeldovich with co-workers and in the monograph [48a].

Consider in some detail the question of the region of applicability of the theory advanced. The movement of the matter in accordance with (3) leads to the density distribution described by (4) and (5). On the other hand, in order that the movement might occur in accordance with (3), it is necessary that the acceleration $-\partial\Phi/\partial x_i$ in (1) be generated by the density distribution $\rho^* = -(4\pi G)^{-1}(\partial/\partial x_i)[\partial u_i/\partial t + u_r\, \partial u_i/\partial x_r]$. If $\rho^* = \rho$, then the problem is self-consistent and (3) is an exact solution. Otherwise, the deviation $\rho^* - \rho$ may serve as a measure of approximateness of (3). It is easy to calculate the quantity

$$\Delta = \frac{\rho^*}{\rho} - 1 = [-(3\ddot{a}B + 2\dot{a}\dot{B} + a\bar{B})\mathscr{I}_1 + B(3\ddot{a}B + 4\dot{a}\dot{B} + 2a\bar{B})\mathscr{I}_2$$
$$- 3B^2(\ddot{a}B + 2\dot{a}\dot{B} + a\ddot{B})\mathscr{I}_3]3^{-1}\ddot{a}^{-1}, \quad (6)$$

where the dot denotes time differentiation and

$$\mathscr{I}_1 = \lambda_1 + \lambda_2 + \lambda_3, \qquad \mathscr{I}_2 = \lambda_1\lambda_2 + \lambda_1\lambda_3 + \lambda_2\lambda_3, \qquad \mathscr{I}_3 = \lambda_1\lambda_2\lambda_3$$

are the invariants of the deformation tensor. In expression (3), the dependence $B(t)$ is chosen so that

$$3\ddot{a}B + 2\dot{a}\dot{B} + a\ddot{B} = 0, \quad (7)$$

and is coincident with the expression for the rate of growth of perturbations in linear theory. Therefore, by using (7), we reduce (6) to the form

$$\Delta = B^2\mathscr{I}_2 + 2B^3\mathscr{I}_3. \quad (8)$$

This yields two important results.

(1) The approximate theory is accurate for small perturbations since (8) in this case is quadratic with respect to the amplitude of perturbations.

(2) The theory is accurate for one-dimensional perturbations since for $\lambda_2 = \lambda_3 = 0$, $\mathscr{I}_2 = \mathscr{I}_3 = 0$, and $\Delta \equiv 0$. Substituting $B = 1/\lambda_1$ (the condition of reaching infinite density), we obtain,

$$\Delta = \frac{\mathscr{I}_2}{\lambda_1^2} + 2\mathscr{I}_3/\lambda_1^3, \quad (9)$$

and within the range $\lambda_1 \gg \lambda_2 \geq \lambda_3$ we have again $\Delta \ll 1$. The corrections for the solutions are of the order of the ratio λ_2/λ_1 (or $\lambda_3 \lambda_2/\lambda_1^2$) which can just be considered as a small parameter of the problem. Thus, the approximate theory provides a good accuracy within the ranges $\lambda_1 \gg \lambda_2 \geq \lambda_3$, i.e., in the vicinity of the region of maximum compression.

Verification of the approximate theory by constructing numerical models has confirmed both good accuracy of the approximate theory (not lower than 20%) in the neighborhood of compressions and the one-dimensional character of compression and the production of flat structures [45a].

§ 4 Foundations of Turbulence Theory[14] [53^{ad}]

In this section, we shall treat, in application to gravitating systems, some questions of the theory of weak turbulence. It is well known that, in hydrodynamics, under turbulence there is understood to be a set of a large number of whirls moving a little in space and therefore interacting for a long time (and, consequently, strongly) with each other. The development of the physical theory, in particular, plasma physics, has shown that such an understanding is too narrow. At the present time, under turbulence there is understood to be the movement in which a large number of collective (not necessarily whirls) degrees of freedom are excited, for example, a large number of modes of eigenoscillations of the medium. The study of the latter is just the subject of the weak turbulence theory [39, 15^{ad}].

As far as the wave movement is concerned, individual wave packets are moving in the medium with a group speed and for their lifetime are able to drift apart up to rather great distances. Owing to this, the interaction of each individual pair of the wave packets with each other turns out to be weak, which allows one, in particular, to consider the waves as being nearly linear, i.e., having dispersion properties close to the properties of the linear waves. An essential advantage of the theory of weak turbulence is the possibility of application of the disturbance theory, i.e., expansion of the equations in a small parameter of the ratio of the interaction energy to the total energy, which in many cases is reduced to the expansion with respect to powers of small amplitudes of waves.

The theory of weak turbulence has two aspects: the study of the nonlinear interaction of the waves with each other and the study of the wave–particle interaction. We shall restrict ourselves to the former.

4.1 Hamiltonian Formalism for the Hydrodynamical Model of a Gravitating Medium

It is convenient to place in the basis of our analysis the general method allowing one to at once include the nonlinear wave theory in gravitating systems into the general theory of wave phenomena in nonlinear dispersive

---

[14] This section, on request of the authors, is written by S. M. Churilov and I. G. Shukhman.

media. Such a method is the Hamiltonian formalism developed by Zakharov and his successors (cf. [10^{ad}, 11^{ad}]).

It allows one in a uniform way to describe the nonlinear wave interaction; it also provides a simple algorithm for writing "shortened" equations and calculating the relevant matrix elements which immediately possess the needed symmetry, a fact that with other methods of calculation is achieved only by painstaking efforts and requires great inventiveness. In essence, the Hamiltonian formalism is a method of subsequent expansion of equations with respect to the powers of wave amplitudes.

It is evident that, as in mechanics, the Hamiltonian formalism may be constructed only in neglect of dissipation, i.e., for conservative media possessing, moreover, a translation-invariant Hamiltonian. In addition, the dispersion properties of the waves in the linear approximation must be such that the square of the frequency is $\omega^2(k) > 0$, apart possibly from the value of the wave vector k, where $\omega^2(k) = 0$.

We shall analyze the application of the Hamiltonian formalism in the example of the nonlinear wave processes in an infinitely thin rotating gravitating gaseous layer, restricting ourselves to Jeans oscillations that do not deform the plane of the layer. The case without rotation passes beyond the method since then there is a wide range of wave vectors, for which $\omega^2(k) < 0$ (the region of Jeans instability; cf. Chapter I).

4.1.1. Statement of the Problem and Basic Equations.

Consider an infinitesimally thin gravitating gaseous layer uniformly rotating at an angular velocity Ω and lying in the plane $z = 0$. Assume that the centrifugal force is compensated for by some external force, for example, the gravity force acting from the halo surrounding the layer. The hydrodynamical equations, in the frame of reference rotating at an angular velocity Ω, are of the form

$$\frac{\partial \sigma}{\partial t} + \text{div } \sigma \mathbf{v} = 0, \tag{1}$$

$$\frac{\partial \mathbf{v}}{\partial t} + (\mathbf{v}\nabla)\mathbf{v} + [\varkappa\mathbf{v}] = -\frac{1}{\sigma}\nabla P - \nabla\Phi, \tag{2}$$

$$\Delta\Phi + \frac{\partial^2 \Phi}{\partial z^2} = 4\pi G\sigma\delta(z), \tag{3}$$

where σ is the surface density, $\mathbf{v}$ is the velocity of gas, and Φ is the gravitational potential; all vector values and operators refer to the (x, y) plane, $\varkappa = 2\Omega$.

The solution for the Poisson equation (3) can be written in the form

$$\Phi(r, z) = -G \int \frac{\sigma(\mathbf{r}')\delta(z') \, d\mathbf{r}' \, dz'}{\sqrt{(\mathbf{r} - \mathbf{r}')^2 + (z - z')^2}}.$$

Integrating over z', owing to the δ function, is trivial, while for the potential in the plane of the layer $z = 0$ which is only of interest for us, we obtain

$$\Phi(\mathbf{r}) = -G \int \frac{\sigma(\mathbf{r}') \, d\mathbf{r}'}{|\mathbf{r} - \mathbf{r}'|}. \tag{4}$$

For the barotropic medium, when the pressure is only dependent on the density, $p = p(\sigma)$, one can express the right-hand side of (2) in the form of the gradient from the variational derivative of a certain functional

$$\frac{\partial \mathbf{v}}{\partial t} + (v\nabla)v + [\varkappa v] = -\nabla \frac{\delta E}{\delta \sigma}, \tag{5}$$

where

$$E = \int \varepsilon(\sigma) \, d\mathbf{r} - \frac{G}{2} \int \frac{\sigma(\mathbf{r})\sigma(\mathbf{r}') \, d\mathbf{r} \, d\mathbf{r}'}{|\mathbf{r} - \mathbf{r}'|},$$

$$\frac{d\varepsilon(\sigma)}{d\sigma} = \int \frac{dp}{\sigma} = \int \frac{1}{\sigma} \frac{dp}{d\sigma} \, d\sigma. \tag{6}$$

Multiplying Eq. (1) by $v^2/2$, and Eq. (5) by σv, we obtain, with the aid of continuity Eq. (1),

$$\frac{\partial}{\partial t} \frac{\sigma v^2}{2} + \text{div}\left(\frac{\sigma v^2}{2} \mathbf{v}\right) = -\text{div}\left(\sigma \frac{\delta E}{\delta \sigma} \mathbf{v}\right) - \frac{\delta E}{\delta \sigma} \frac{\partial \sigma}{\partial t}.$$

Integration over the whole area, taking into account the Gaussian theorem, provides the law of conservation of energy:

$$\frac{\partial}{\partial t} \left(\int \frac{\sigma v^2}{2} \, d\mathbf{r} + E \right) = 0. \tag{7}$$

In further development, an essential role will be played by the energy functional, or the Hamiltonian, of the system

$$\mathcal{H} = \int \frac{\sigma v^2}{2} \, d\mathbf{r} + E \equiv \int \left(\frac{\sigma v^2}{2} + \varepsilon\right) d\mathbf{r} - \frac{G}{2} \int \frac{\sigma(\mathbf{r}')\sigma(\mathbf{r}') \, d\mathbf{r} \, d\mathbf{r}'}{|\mathbf{r} - \mathbf{r}'|}. \tag{8}$$

Specify the $\varepsilon(\sigma)$ function. Adopt the polytropic law of the dependence of the pressure on the density $p = A\sigma^\gamma (\gamma > 1)$.[15] Then, from (6), it is easy to get that

$$\varepsilon(\sigma) = \frac{A}{\gamma - 1} \sigma^\gamma. \tag{9}$$

4.1.2. Transition to Canonical Variables. Introduce, in analogy with hydrodynamics, the canonical variables [11[ad]]:

$$V = \frac{\lambda}{\sigma} \nabla \mu + \nabla \varphi - A, \qquad A = \tfrac{1}{2}[\varkappa \mathbf{r}]. \tag{10}$$

The transform (10) resembles very much the expression for the generalized impulse of a charged particle in a magnetic field, where A is similar to the vector potential, and $\varkappa$, to the magnetic field vector. This is still another

---

[15] For the connection of the "surface" and "volume" adiabatic index cf. Section 1.3, Chapter V. For $\gamma = 1$, in (9), instead of the power function, there will be a logarithm; however, one can readily show via direct expansion that all formulae containing γ, beginning with (18), hold true also for $\gamma = 1$.

manifestation of the analogy mentioned in Chapter II between equations describing the rotating gravitating medium and equations describing the plasma in a magnetic field.

Substitution of (10) into (5), taking into account that $(\mathbf{v}\nabla)\mathbf{v} = \nabla(v^2/2) - [\mathbf{v}\,\text{rot}\,\mathbf{v}]$, yields the equation

$$\nabla\left[\frac{\partial\varphi}{\partial t} + \frac{v^2}{2} + \frac{\delta E}{\delta\sigma} - \frac{\lambda}{\sigma}(\mathbf{v}\nabla)\mu\right] + \frac{\lambda}{\sigma}\nabla\left[\frac{\partial\mu}{\partial t} + (\mathbf{v}\nabla)\mu\right] + \frac{\nabla\mu}{\sigma}\left[\frac{\partial\lambda}{\partial t} + \text{div}\,\lambda\mathbf{v}\right] = 0,$$

which is obviously satisfied if φ, λ, and μ satisfy the equations:

$$\frac{\partial\varphi}{\partial t} + \frac{v^2}{2} + \frac{\delta E}{\delta\sigma} - \frac{\lambda}{\sigma}(\mathbf{v}\nabla)\mu = 0,$$

$$\frac{\partial\lambda}{\partial t} + \text{div}\,\lambda\mathbf{v} = 0, \qquad \frac{\partial\mu}{\partial t} + (\mathbf{v}\nabla)\mu = 0. \tag{11}$$

Via direct differentiation, it is easy to show that [cf. (1), (8)]

$$\frac{\partial\sigma}{\partial t} = \frac{\delta\mathcal{H}}{\delta\varphi}, \qquad \frac{\partial\varphi}{\partial t} = -\frac{\delta\mathcal{H}}{\delta\sigma};$$

$$\frac{\partial\lambda}{\partial t} = \frac{\delta\mathcal{H}}{\delta\mu}, \qquad \frac{\partial\mu}{\partial t} = -\frac{\delta\mathcal{H}}{\partial\lambda}. \tag{12}$$

Thus, the pairs of variables σ, φ and λ, μ are canonically conjugate. For each spatial velocity profile $\mathbf{v}$, using formula (10), it is possible (moreover, nonuniquely) to determine the functions λ, μ, and φ, and since Eq. (5) is identically satisfied by Eq. (11), the description via $\mathbf{v}$ and the description via λ, μ, φ are equivalent.

The next step, in analogy with the theory of mechanical oscillations, is the introduction of normal variables, in which, in the linear approximation, each mode of oscillation is described only by its own pair of canonical variables. However, there is a difficulty here similar to that presented in the description of the plasma in a constant magnetic field [cf. remark after formula (10)]: the function A is the linear function of coordinates, and the reduction of the Hamiltonian to the needed form (diagonalization) requires an additional canonical transform, which we shall perform in two stages [10^{ad}]. To begin with, perform "symmetrization" of the variables,

$$\lambda = \sqrt{\frac{\sigma}{2}}(\lambda' + \mu'), \qquad \mu = \frac{1}{\sqrt{2\sigma}}(\mu' - \lambda'), \qquad \varphi = \varphi' + \frac{\lambda'^2 - \mu'^2}{4\sigma}, \tag{13}$$

and then eliminate A:

$$\lambda' = \lambda'' + \sqrt{\varkappa\sigma}\,y, \qquad \mu' = \mu'' - \sqrt{\varkappa\sigma}\,x,$$

$$\varphi' = \varphi'' - \sqrt{\frac{\varkappa}{4\sigma}}(x\lambda'' + y\mu''). \tag{14}$$

It is easily verified that the two sets of variables are canonical. The relevant equations are readily derived from (11) and (10):

$$\frac{\partial \lambda'}{\partial t} + \tfrac{1}{2}(\mathbf{v}\nabla)\lambda' + \tfrac{1}{2}\operatorname{div} \lambda'\mathbf{v} = 0,$$

$$\frac{\partial \mu'}{\partial t} + \tfrac{1}{2}(\mathbf{v}\nabla)\mu' + \tfrac{1}{2}\operatorname{div} \mu'\mathbf{v} = 0,$$

$$\frac{\partial \varphi'}{\partial t} + \frac{v^2}{2} + \frac{\delta E}{\delta \sigma} + \frac{1}{2\sigma}[\mu'(\mathbf{v}\nabla)\lambda' - \lambda'(\mathbf{v}\nabla)\mu'] = 0, \tag{15}$$

$$\mathbf{v} = \frac{\lambda'\nabla\mu' - \mu'\nabla\lambda'}{2\sigma} + \nabla\varphi' - \mathbf{A}$$

and for the second set of variables,

$$\frac{\partial \lambda''}{\partial t} + \sqrt{\varkappa\sigma}\,v_y + \tfrac{1}{2}(\mathbf{v}\nabla)\lambda'' + \tfrac{1}{2}\operatorname{div} \lambda''\mathbf{v} = 0,$$

$$\frac{\partial \mu''}{\partial t} - \sqrt{\varkappa\sigma}\,v_x + \tfrac{1}{2}(\mathbf{v}\nabla)\mu'' + \tfrac{1}{2}\operatorname{div} \mu''\mathbf{v} = 0,$$

$$\frac{\partial \varphi''}{\partial t} + \frac{v^2}{2} + \frac{\delta E}{\delta \sigma} + \frac{1}{2\sigma}[\mu''(\mathbf{v}\nabla)\lambda'' - \lambda''(\mathbf{v}\nabla)\mu''] + \sqrt{\frac{\varkappa}{4\sigma}}(v_x\lambda'' + v_y\mu'') = 0,$$

$$\mathbf{v} = \frac{\lambda''\nabla\mu'' - \mu''\nabla\lambda''}{2\sigma} - \sqrt{\frac{\varkappa}{\sigma}}(\lambda''\mathbf{e}_x + \mu''\mathbf{e}_y) + \nabla\varphi'', \tag{16}$$

where $\mathbf{e}_x$ and $\mathbf{e}_y$ are the unit-vectors on the x and y axes. Further, we utilize the variables λ'', μ'', φ'', omitting, for the sake of brevity, the primes.

Consider the waves on the background of the layer of homogeneous density σ_0. It is convenient to introduce, instead of σ, a new variable τ

$$\sigma = \sigma_0(1 + \tau). \tag{17}$$

The quantity $\tilde{\varphi} = \sigma_0\varphi$ seems obviously to be canonically conjugate to τ. Now, in the unperturbed state, the canonical variables τ, $\tilde{\varphi}$, λ, μ all are equal to zero. We expand the Hamiltonian (8) in powers of canonical variables. We start with the functional E. According to (6) and (9),

$$E = \int \frac{A\sigma_0^\gamma}{\gamma - 1}\left(\frac{\sigma}{\sigma_0}\right)^\gamma d\mathbf{r} - \frac{G}{2}\int \frac{\sigma(\mathbf{r})\sigma(\mathbf{r}')}{|\mathbf{r} - \mathbf{r}'|}\,d\mathbf{r}\,d\mathbf{r}'.$$

Expressing σ by τ and taking into account that

$$(1 + \tau)^\gamma = 1 + \gamma\tau + \gamma(\gamma - 1)\tau^2/2 + \cdots,$$

we obtain

$$E = -\frac{G\sigma_0^2}{2}\int\frac{dr\,dr'}{|r-r'|} - \frac{G\sigma_0^2}{2}\int\frac{\tau(r)+\tau(r')}{|r-r'|}dr\,dr'$$

$$-\frac{G\sigma_0^2}{2}\int\frac{\tau(r)\tau(r')}{|r-r'|}dr'\,dr' + \frac{A\sigma_0^\gamma}{\gamma-1}\int dr + \frac{\sigma_0 c^2}{\gamma-1}\int\tau\,dr$$

$$+ \frac{\sigma_0 c^2}{2}\int dr\left[\tau^2 + \frac{\gamma-2}{3}\tau^3 + \frac{(\gamma-2)(\gamma-3)}{12}\tau^4 + \cdots\right],$$

where $c^2 = \gamma A\sigma_0^{\gamma-1}$ is the square of the sound velocity.

As is known, it is possible to eliminate the constants and subintegral terms (which have a divergent form in Eq. (7)) from the Hamiltonian

$$\frac{d\mathcal{H}}{dt} = 0.$$

Since $\partial\tau/\partial t = -\mathrm{div}(1+\tau)v$ and $s = r - r'$,

$$\frac{\partial}{\partial t}\int\frac{\tau(r)+\tau(r')}{|r-r'|}dr\,dr' = 2\int\frac{ds}{|s|}\int\frac{\partial\tau}{\partial t}dr,$$

then, in E, there remain only the terms in the second and higher powers over τ:

$$E = -\frac{G\sigma_0^2}{2}\int\frac{\tau(r)\tau(r')}{|r-r'|}dr\,dr' + \frac{\sigma_0 c^2}{2}\int dr\left(\tau^2 + \frac{\gamma-2}{3}\tau^3 + \cdots\right). \quad (18)$$

The Hamiltonian appears in the form of a series in powers of canonical variables

$$\mathcal{H} = \mathcal{H}^{(2)} + \mathcal{H}^{(3)} + \mathcal{H}^{(4)} + \cdots, \quad (19)$$

where

$$\mathcal{H}^{(2)} = \frac{\sigma_0}{2}\int(v_1^2 + c^2\tau^2)\,dr - \frac{G\sigma_0^2}{2}\int\frac{\tau(r)\tau(r')\,dr\,dr'}{|r-r'|},$$

$$\mathcal{H}^{(3)} = \frac{\sigma_0}{2}\int\left[\tau v_1^2 + 2(v_1 v_2) + \frac{c^2}{2}\tau^3(\gamma-2)\right]dr,$$

$$\mathcal{H}^{(4)} = \frac{\sigma_0}{2}\int\left[v_2^2 + 2\tau(v_1 v_2) + 2(v_1 v_3) + \frac{(\gamma-2)(\gamma-3)}{12}c^2\tau^4\right]dr,$$

$$v_1 = \nabla\varphi - \sqrt{\frac{\varkappa}{\sigma_0}}(\lambda e_x + \mu e_y),$$

$$v_2 = \frac{\lambda\nabla\mu - \mu\nabla\lambda}{2\sigma_0} + \sqrt{\frac{\varkappa}{4\sigma_0}}\tau(\lambda e_x + \mu e_y),$$

$$v_3 = -\frac{\tau}{2\sigma_0}(\lambda\nabla\mu - \mu\nabla\lambda) - \frac{3}{8}\sqrt{\frac{\varkappa}{\sigma_0}}\tau^2(\lambda e_x + \mu e_y). \quad (20)$$

Consider now the Fourier representation by the formulae

$$A(\mathbf{r}) = \frac{1}{2\pi} \int A_{\mathbf{k}} e^{i\mathbf{kr}}\, d\mathbf{k}, \qquad A_{\mathbf{k}} = \frac{1}{2\pi} \int A(\mathbf{r}) e^{-i\mathbf{kr}}\, d\mathbf{r}.$$

In the expression for $\mathscr{H}^{(2)}$, the transition to the Fourier components in the first integral is easy; let us consider the second integral

$$\int \frac{\tau(\mathbf{r})\tau(\mathbf{r}')}{|\mathbf{r} - \mathbf{r}'|}\, d\mathbf{r}\, d\mathbf{r}' = \frac{1}{(2\pi)^2} \int \frac{d\mathbf{r}\, d\mathbf{r}'\, d\mathbf{k}\, d\mathbf{k}'}{|\mathbf{r} - \mathbf{r}'|} \tau_{\mathbf{k}} \tau_{\mathbf{k}'} e^{i(\mathbf{kr} + \mathbf{k}'\mathbf{r}')}.$$

Since integration over $\mathbf{r}$ and $\mathbf{r}'$ is performed over the whole plane, we may turn to the variables $\mathbf{r}'$ and $s = \mathbf{r} - \mathbf{r}'$. We obtain

$$\frac{1}{(2\pi)^2} \int \frac{d\mathbf{k}\, d\mathbf{k}'\, d\mathbf{s}\, d\mathbf{r}'}{|\mathbf{s}|} \tau_{\mathbf{k}} \tau_{\mathbf{k}'} e^{i(\mathbf{kr}' + \mathbf{k}'\mathbf{r}') + i\mathbf{ks}}$$

$$= \frac{1}{(2\pi)^2} \int d\mathbf{k}\, d\mathbf{k}' \tau_{\mathbf{k}} \tau_{\mathbf{k}'} \int \frac{e^{i\mathbf{ks}}}{|\mathbf{s}|}\, d\mathbf{s} \int e^{i(\mathbf{k} + \mathbf{k}')\mathbf{r}'}\, d\mathbf{r}'.$$

The last integral yields $\delta(\mathbf{k} + \mathbf{k}')$, which eliminates integration over k', while in the integral over s, it is convenient to turn to the polar coordinates s and ψ choosing $\mathbf{k}$ as the polar axis. We obtain, bearing in mind that $\tau_{-k} = \tau_k^*$,

$$\int |\tau_k|^2\, dk \int_0^{2\pi} d\psi \int_0^{\infty} e^{i|k|s\cos\psi}\, ds = 2\pi \int |\tau_k|^2\, dk \int_0^{\infty} \mathscr{J}_0(|k|s)\, ds.$$

The last integral, owing to normalization of the Bessel function, is $1/|k|$ and finally we have

$$\mathscr{H}^{(2)} = \frac{\sigma_0}{2} \int d\mathbf{k} \left[(v_k v_k^*) + \left(c^2 - \frac{2\pi G\sigma_0}{|k|} \right) \tau_k \tau_k^* \right], \tag{21}$$

where v_k is the Fourier component of v_1.

4.1.3. Derivation of the Basic Equation of the Theory in Normal Variables.
Following the general method [10^{ad}], we introduce the wave amplitudes a_k^s, where the index s enumerates the types of oscillations. It is convenient to employ also the negative values of s, by assuming (ω_k^s are frequencies):

$$a_k^{-s} = a_{-k}^{s*}, \qquad \omega_k^{-s} = -\omega_k^s. \tag{22}$$

For the Fourier component of any real quantity A_k, we have

$$A_k = \sum_s A_k^s a_k^s, \qquad A_k^{-s} = A_{-k}^{s*},$$

We linearize Eqs. (1)–(3), taking into account that $a_k^s \sim \exp(-i\omega_k^s t)$:

$$-i\omega_k^s \tau_k^s + i(\mathbf{k} v_k^s) = 0,$$

$$-i\omega_k^s v_k^s = -i\mathbf{k}\left(c^2 - \frac{2\pi G\sigma_0}{|k|} \right)\tau_k^s - [\varkappa v_k^s]. \tag{23}$$

Here we have used the solution for the Poisson equation given in §2, Chapter I:

$$\Phi_k^s = -\frac{2\pi G\sigma_0}{|k|}\tau_k^s.$$

Equating to zero of the determinant of the system (23) provides two branches of oscillations:

1. Jeans branch ($s = 1$)

$$(\omega_k^1)^2 = \varkappa^2 + k^2 c^2 - 2\pi G \sigma_0 |\mathbf{k}|. \tag{24}$$

2. Entropic branch ($s = 2$)

$$\omega_k^2 = 0.$$

The entropic branch dropped out of the region of applicability of the formalism (cf. beginning of the section); therefore, it should be eliminated, i.e., one should assume that $a_k^2 = 0$. For the Jeans branch, one may express the velocity through τ:

$$v_k^1 = \frac{1}{k^2}(\omega_k^1 k - i[\varkappa k])\tau_k^1. \tag{25}$$

Substitution of (25) into Hamiltonian (21) reduces the latter to a diagonal form

$$\mathscr{H}^{(2)} = \frac{\sigma_0}{2} \int dk \, \frac{4(\omega_k^1)^2}{k^2} \, \tau_k^1 \tau_k^1{}^* a_k^1 a_k^1{}^*.$$

We choose the normalization of a_k^1 so that

$$\mathscr{H}^{(2)} = \int \omega_k^1 a_k^1 a_k^1{}^* \, dk, \tag{26}$$

then it is evident that

$$\tau_k^1 = \frac{|k|}{\sqrt{2\sigma_0 \omega_k^1}}.$$

With such normalization, the quantity $a_k^1 a_k^1{}^*$ acquires the meaning of "density" of the number of "quanta" of a given frequency ω_k^1. The quantities a_k^1 are called by the normal variables.

Further, the upper index will be omitted, for the sake of brevity.

We express via (16) and (25) the Fourier components τ_k, φ_k, λ_k, μ_k, and v_k through a_k:

$$\tau_k = \frac{|k|}{\sqrt{2\sigma_0 \omega_k}} (a_k + a_{-k}^*),$$

$$\varphi_k = -i \frac{\omega_k^2 - \varkappa^2}{\sqrt{2\sigma_0} |k| \omega_k^{3/2}} (a_k - a_{-k}^*),$$

$$\mu_k = -\frac{\sqrt{\varkappa}}{\sqrt{2} |k| \omega_k^{3/2}} \{(\varkappa k_y - i\omega_k k_x)a_k - (\varkappa k_y + i\omega_k k_x)a_{-k}^*\}, \tag{27}$$

$$\lambda_k = -\frac{\sqrt{\varkappa}}{\sqrt{2} |k| \omega_k^{3/2}} \{(\varkappa k_x + i\omega_k k_y)a_k - (\varkappa k_x - i\omega_k k_y)a_{-k}^*\}$$

$$v_k = \frac{1}{|k|\sqrt{2\sigma_0 \omega_k}} \{\omega_k k(a_k - a_{-k}^*) - i[\varkappa k](a_k + a_{-k}^*)\}.$$

Write now the equation of motion (16) through a_k. The variables[16] $\tau, \tilde{\varphi}, \lambda, \mu$ are canonical; therefore, in the Fourier components (16) can also be written in the canonical form of (12):

$$\frac{\partial \tau_k}{\partial t} = \frac{\delta \mathcal{H}}{\delta \tilde{\varphi}_k}, \qquad \frac{\partial \tilde{\varphi}_k}{\partial t} = -\frac{\delta \mathcal{H}}{\partial \tau_k};$$

$$\frac{\partial \lambda_k}{\partial t} = \frac{\delta \mathcal{H}}{\partial \mu_k}, \qquad \frac{\partial \mu_k}{\partial t} = -\frac{\delta \mathcal{H}}{\delta \lambda_k}. \tag{28}$$

As is seen from (27), all the canonical variables are the linear functions of a_k and a^*_{-k}, and they can be presented in the form

$$\tau_k = \tau_1 a_k + \tau_1^* a^*_{-k}, \qquad \tilde{\varphi}_k = \tilde{\varphi}_1 a_k + \tilde{\varphi}_1^* a^*_{-k},$$

$$\lambda_k = \lambda_1 a_k - \lambda_1 a^*_{-k}, \qquad \mu_k = \mu_1 a_k - \mu_1^* a^*_{-k}.$$

Multiplying the first equation (28) by $\tilde{\varphi}_1^*$, the second by $(-\tau_1^*)$, the third by $(-\mu_1^*)$, and the fourth by λ_1^*, and summing, we obtain

$$(\tau_1 \tilde{\varphi}_1^* - \tilde{\varphi}_1 \tau_1^* - \lambda_1 \mu_1^* + \mu_1 \lambda_1^*) \frac{\partial a_k}{\partial t}$$

$$= \frac{\delta \mathcal{H}}{\delta \tilde{\varphi}_k} \tilde{\varphi}_1^* + \frac{\delta \mathcal{H}}{\delta \tau_k} \tau_1^* - \frac{\delta \mathcal{H}}{\delta \mu_k} \mu_1^* - \frac{\delta \mathcal{H}}{\delta \lambda_k} \lambda_1^* \equiv \frac{\delta \mathcal{H}}{\delta a^*_{-k}}.$$

By means of direct calculation, it is easy to make sure that the coefficient at $\partial a_k / \partial t$ is i; however, it is seen also from (26). Thus

$$\frac{\partial a_k}{\partial t} = -i \frac{\delta \mathcal{H}}{\delta a^*_{-k}}. \tag{29}$$

This equation is the basic equation of the theory. Using expansion (19) and formulae (20), it is easy to obtain from (29) the so-called shortened equations describing the wave dynamics in any order with respect to the amplitude a_k.

4.2 Three-Wave Interaction

Consider the first nonlinear approximation, i.e., restrict ourselves to the cubic part of the Hamiltonian $\mathcal{H}^{(3)}$. Simple but cumbersome calculations (for more detail, cf. §9, Appendix) yield

$$\mathcal{H}^{(3)} = \int dk\, dk_1\, dk_2 [\tfrac{1}{3} V_{kk_1k_2} a_k a_{k_1} a_{k_2} \delta(k + k_1 + k_2)$$

$$+ V^*_{kk_1k_2} a_k^* a_{k_1} a_{k_2} \delta(k - k_1 - k_2) + \text{c.c.}], \tag{1}$$

---

[16] Recall that $\tilde{\varphi} = \sigma_0 \varphi$.

where c.c. denotes the terms complex conjugate to the written ones and

$$V_{kk_1k_2} = U_{kk_1k_2} + U_{k_1kk_2} + U_{k_2k_1k}$$

$$U_{kk_1k_2} = \frac{1}{16\pi\sqrt{2\sigma_0\,\omega_k^3\omega_{k_1}^3\omega_{k_2}^3}|k||k_1||k_2|}$$

$$\times (2\omega_k\omega_{k_1}\omega_{k_2}\varkappa^2[k_1k_2]^2 + 2\varkappa^4\omega_k([kk_1][kk_2])$$

$$+ 2\omega_k\omega_{k_1}^2\omega_{k_2}^2k^2(k_1k_2) + \tfrac{2}{3}(\gamma - 2)c^2k^2k_1^2k_2^2\,\omega_k\omega_{k_1}\omega_{k_2}$$

$$+ \varkappa^2\{\omega_k^2\omega_{k_1}[(kk_1)(k_1k_2) + \tfrac{1}{2}(kk_2)(k_2^2 - k^2 - k_1^2)]$$

$$+ \omega_k^2\omega_{k_2}[(kk_2)(k_1k_2) + \tfrac{1}{2}(kk_1)(k_1^2 - k^2 - k_2^2)]\}$$

$$- 2i\omega_k^2\omega_{k_1}\omega_{k_2}(\varkappa[k_1k_2])(k_1^2 - k_2^2)$$

$$- i\varkappa^2\{(\omega_{k_1}^2 - \omega_{k_2}^2)(\varkappa[k_1k_2])(k_1k_2)$$

$$+ \omega_{k_1}\omega_{k_2}(\varkappa[k_1k_2])(k_1^2 - k_2^2)\}). \tag{2}$$

The matrix element $V_{kk_1k_2}$ possesses the needed transformation properties: It is symmetrical with respect to kk_1k_2

$$V_{-k-k_1-k_2} = V_{kk_1k_2}.$$

Equation (29) (Section 4.1) in this approximation has the form

$$\frac{\partial a_k}{\partial t} + i\omega_k a_k = -i \int dk_1\, dk_2[V_{kk_1k_2}^* a_{k_1}a_{k_2}\delta(k - k_1 - k_2)$$

$$+ 2V_{kk_1k_2}a_{k_1}^* a_{k_2}\delta(k + k_1 - k_2)$$

$$+ V_{kk_1k_2}^* a_{k_1}^* a_{k_2}^*\delta(k + k_1 + k_2)]. \tag{3}$$

It describes the processes of interaction of three waves with the wave vectors k, k_1, and k_2. We determine the limits of applicability of (3).

For that purpose, we introduce the new amplitudes

$$a_k = A_k e^{-i\omega_k t}. \tag{4}$$

In neglecting the wave interaction, the solution of (3) may be taken in the form

$$A_k^{(0)} = A\delta(\mathbf{k} - \mathbf{k}_0).$$

Allowance for interaction provides corrections. In the first approximation

$$\frac{\partial A_k^{(1)}}{\partial t} = -i\{V_{kk_0k_0}^* A^2 \exp[i(\omega_k - 2\omega_{k_0})t]\delta(k - 2k_0)$$

$$+ 2V_{kk_0k_0} AA^* \exp(i\omega_k t)\delta(k)$$

$$+ V_{kk_0k_0}^* A^{*2} \exp[i(\omega_k + 2\omega_{k_0})t]\delta(k + 2k_0)\},$$

$$A_k^{(1)} = -\frac{V_{2k_0k_0k_0}^*}{\omega_{2k_0} - 2\omega_{k_0}} A^2 \exp[i(\omega_{2k_0} - 2\omega_{k_0})t]\delta(k - 2k_0)$$

$$- \frac{2V_{0k_0k_0}}{\omega_0} |A|^2 \exp(i\omega_0 t)\,\delta(k)$$

$$- \frac{V_{-2k_0k_0k_0}^*}{\omega_{2k_0} + 2\omega_{k_0}} A^{*2} \exp[i(\omega_{2k_0} + 2\omega_{k_0})t]\delta(k + 2k_0).$$

Since all ω_k are positive, the coefficients at the δ functions in the last two terms are small as compared to $|A|$ ($|A|$ itself is also small). Therefore, the condition of applicability of (3) has the form

$$\frac{|V_{2k_0k_0k_0}|}{|\omega_{2k_0} - 2\omega_{k_0}|} \cdot |A| \ll 1.$$

In our problem, the spectrum is far from the linear one [cf. (24), Section 4.1]; therefore, the denominator can be small only for a special choice of k_0. In the general case, the condition of applicability of (3) has, as a rule, the form

$$|V| \cdot |A| \ll 1.$$

4.2.1. Decay Instability. Consider now the problem of the evolution of a small (but finite) perturbation. As seen from (3), the wave with the wave vector **k** can interact with two waves satisfying one of the conditions

$$\mathbf{k} = \mathbf{k}_1 + \mathbf{k}_2, \qquad \mathbf{k} = \mathbf{k}_2 - \mathbf{k}_1, \qquad \mathbf{k} = -\mathbf{k}_1 - \mathbf{k}_2. \qquad (4')$$

For concreteness, we restrict ourselves to one of them. Assuming that

$$A_k = A_0 \delta(k - k_0) + A_1 \delta(k - k_1) + A_2 \delta(k - k_2)$$

and that $|A_0| \gg |A_1|, |A_2|$, we obtain, in the linear approximation with respect to A_1, A_2

$$\frac{dA_1}{dt} = -2iV_{k_0k_1k_2} A_0 A_2^* e^{i\gamma t},$$

$$\frac{dA_2}{dt} = -2iV_{k_0k_1k_2} A_0 A_1^* e^{i\gamma t},$$

$$(5)$$

where

$$\gamma = \omega_{k_1} + \omega_{k_2} - \omega_{k_0}.$$

Equation (5) is easily reduced to the one

$$\frac{d}{dt}\left(e^{-i\gamma t}\frac{dA_1}{dt}\right) = 4|V_{kok_1k_2}|^2|A_0|^2 e^{-i\gamma t}A_1,$$

the solution of which is $A_1 = ce^{qt}$, where

$$q = \frac{i}{2}\gamma \pm \sqrt{4|V_{kok_1k_2}|^2|A_0|^2 - \tfrac{1}{4}\gamma^2}.\tag{6}$$

If the subroot expression (6) is positive, then we are dealing with the so-called decay instability [39], when the wave of "large" amplitude A_0 decays into two other waves.

In order that the decay instability might occur for any A_0, it is necessary that, apart from the triangle condition

$$k_0 = k_1 + k_2,\tag{7a}$$

also the synchronism condition

$$\omega_{k_0} = \omega_{k_1} + \omega_{k_2} \quad \text{or} \quad \gamma = 0\tag{7b}$$

be satisfied.

The spectra, for which these two conditions may be satisfied, are called the decay spectra [39]. It is readily shown, from dispersion equation (24), Section 4 that we are dealing with a decay spectrum. Decay instability map proceed also with violation of (7b), but then a rather large excitation (pumping) wave amplitude A_0 is required. The maximum instability growth rate is

$$\tilde{\gamma}_{max} = 2|V_{kok_1k_2}||A_0|.\tag{8}$$

The decay instability in a gravitating rotating layer for the case of the one-dimensional spectrum in k is dealt with in [70^{ad}]. Our consideration, however, fits also the general case of a two-dimensional spectrum, where the possibilities of decay are much richer: As a rule, the infinite number of pairs (k_2, k_3) corresponds to a given k_1. Let us investigate these possibilities.

The dispersion equation shows that for $k = k_0 = \pi G\sigma_0/c^2$ the frequency has the minimum: $\omega_0^2 = c^2 k_0^2(Q^2 - 1)$, $Q = \varkappa c/\pi G\sigma_0 > 1$. Due to this, k_1 may not be arbitrary: at least the inequality $\omega_1 \geq 2\omega_0$ must be fulfilled. For $\omega_0 \leq \tfrac{1}{2}\varkappa$ $(Q^2 \leq \tfrac{4}{3})$ it provides two regions of variation for k_1:

$$0 \leq k_1 \leq k^- \quad \text{and} \quad k_1 \geq k^+, k^\pm = [1 \pm \sqrt{3(Q^2 - 1)}]k_0.$$

For $\omega_0 > \tfrac{1}{2}\varkappa/(Q^2 > \tfrac{4}{3})$ there is only one region. The condition of a triangle prohibits the decay $\omega_1 \to 2\omega_0$ (since $\omega_{2k_0} = \varkappa < 2\omega_0$), and now the boundary of the region is determined by the inequality $k_1 \geq \tilde{k}^+ = \tfrac{3}{2}Q^2 k_0$.

For decays from the short-wave region ($k_1 \geq k^+$ or $k_1 \geq k^+$) $k_1 > k_2$, k_3, and $\omega_2 = \omega_1 - \omega_3 < c(k_1 - k_3) \leq ck_2$ or $\omega_2/k_2 < c$, then it yields the lower asymptotics of the variation range of $k_2 > k_*^- = k_0 Q^2/2$ and the upper one

$$k_2 < k - k_*, \quad k_* = \begin{cases} \omega_0/c, & Q^2 < 2 \ (k_*^- < k_0), \\ k_*^-, & Q^2 \geq 2 \ (k_*^- \geq k_0). \end{cases}$$

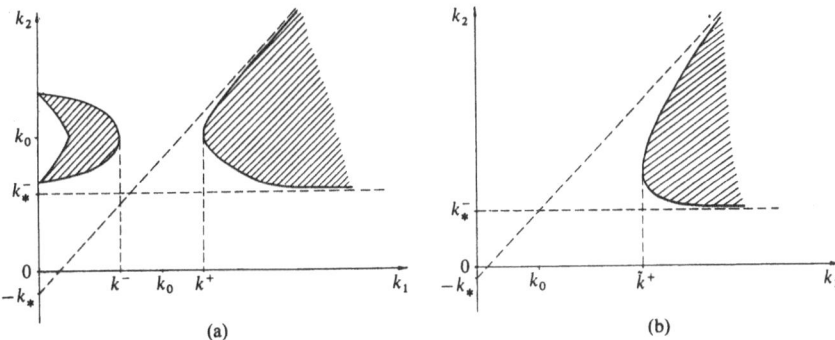

Figure 101. Pairs (k_1, k_2) which may take part in the decay [(a) $0 < \omega_0 < \varkappa/2(1 \leq Q^2 \leq \frac{4}{3})$, (b) $\omega_0 > \varkappa/2(Q^2 > \frac{4}{3})$] for positive energy waves (V. I. Korchagin and P. I. Korchagin have noted that taking part of negative energy waves essentially widens the range of allowed decays).

The variation range of k_3 is the same due to the symmetry. For $\omega_0 > \varkappa/2$ waves with $k \leq k_*^-$ do not take part in decays (dashed regions). Corresponding k_3 and angles between the wave vectors are determined from (23). Formulae describing the boundaries of regions are given in §10, Appendix (see also Fig. 101).

It is evident that Eqs. (5) describe only the initial stage of decay instability. In the course of time, the amplitudes A_1 and A_2 will grow such that they will begin to influence A_0, and there will be the so-called parametric interaction of three waves [39]. In order to describe this process, let us ignore the inequality $|A_0| \gg |A_1|, |A_2|$. By assuming that the conditions (7) are satisfied, we obtain from (3)

$$\frac{dA_0}{dt} = -2iV^*_{kok_1 k_2} A_1 A_2,$$

$$\frac{dA_1}{dt} = -2iV_{kok_1k_2} A_0 A_2^*, \tag{9}$$

$$\frac{dA_2}{dt} = -2iV_{kok_1k_2} A_0 A_1^*.$$

We single out in the quantities in (9) the moduli and phases:

$$A_j = b_j \exp[i\varphi_j(t)], \qquad V_{kok_1k_2} = Ve^{i\psi},$$
$$\theta = \varphi_0 - \varphi_1 - \varphi_2 + \psi.$$

Then (9) will take on the form

$$\frac{db_0}{\partial t} = -2Vb_1b_2 \sin \theta,$$

$$\frac{db_1}{\partial t} = 2Vb_0 b_2 \sin \theta,$$

$$\frac{db_2}{dt} = 2Vb_0 b_1 \sin \theta, \tag{9a}$$

$$\frac{d\theta}{dt} = 2V\left(\frac{b_0 b_2}{b_1} + \frac{b_0 b_1}{b_2} - \frac{b_1 b_2}{b_0}\right) \cos \theta.$$

Multiplying the last equation by $\tan \theta$ and using the first three equations, substituting the fractions on the right-hand side, we obtain

$$b_0 b_1 b_2 \cos \theta = \Gamma = \text{const.} \tag{10}$$

We denote $n_j = b_j^2$. Taking into account the decay condition (7b), we obtain the first integral of the remaining three equations

$$n_0 \omega_{k_0} + n_1 \omega_{k_1} + n_2 \omega_{k_2} = \text{const}, \tag{10a}$$

expressing the law of conservation of energy in the 3-wave interaction (the values n_j have the meaning of the "number of quanta" of a given frequency). In a similar manner, we obtain further three conserving quantities,

$$m_1 = n_0 + n_1 = \text{const},$$
$$m_2 = n_0 + n_2 = \text{const}, \tag{10b}$$
$$m_3 = n_1 - n_2 = \text{const},$$

the meaning of which is easily understood if one notes that, with the vanishing of one quantum with frequency ω_{k_0}, quanta with frequencies ω_{k_1} and ω_{k_2} appear while the quanta ω_{k_1} and ω_{k_2} may vanish only together, producing a quantum ω_{k_0}. Thus, if $\Delta n_0 = \pm 1$, $\Delta n_1 = \Delta n_2 = \mp 1$.

By using the integrals of (10), we obtain from the third equation (9a)

$$\frac{1}{2}\left(\frac{dn_2}{dt}\right)^2 = 8V^2(n_0 n_1 n_2 - \Gamma^2)$$

$$= 8V^2[n_2(m_2 - n_2)(m_3 + n_2) - \Gamma^2]. \tag{11}$$

The solution of this equation is expressed through the elliptical functions (cf., e.g., [39]); however, it would be more obvious if one investigates the solution qualitatively. The equation obtained has the form of the "law of conservation of energy" of a particle moving in the field with the potential

$$U = 8V^2 n_2(n_2 - m_2)(n_2 + m_3) \tag{12}$$

and having negative "energy"

$$E = -8V^2 \Gamma^2,$$

and the role of the "coordinate" is played by n_2. The potential is presented in Fig. 102.

From the figure it is seen that n_2 (and together with it, according to relation (10b), n_0 and n_1) oscillates between the maximum and minimum values. There also occurs periodical pumping of energy from one mode of oscillations to the others. It is interesting that if at the initial time moment one of the modes has not been excited ($\Gamma = 0$), then the process proceeds with alternating disappearance of one of the oscillation modes.

When $n_1 = n_2$ ($m_3 = 0$) and $\Gamma = 0$, Eq. (11) has a simple analytical ("soliton") solution:

$$n_2 = \frac{m_2}{\mathrm{ch}^2\, \alpha(t - t_0)}, \qquad \alpha^2 = 4m_2 V^2. \tag{13}$$

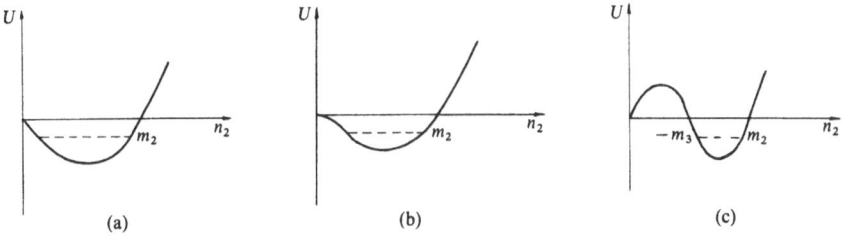

Figure 102. Effective potential; (a) $m_3 > 0$, (b) $m_3 = 0$, (c) $m_3 < 0$.

The process described by the solution of (13) is the following: the wave of a frequency ω_{k_0} with the energy $m_2\omega_{k_0}$ starts to decay into the waves with the frequencies ω_{k_1} and ω_{k_2}, and at the time $t = t_0$ decays completely, and there begins a process of merging that finishes in the complete disappearance of quanta with frequencies ω_{k_1} and ω_{k_2}. The whole cycle is then performed for an infinite time.

We have restricted ourselves to the case, when the first of the conditions in (4′) is satisfied. We write the decay conditions

$$k + k_1 = k_2, \qquad \omega_k + \omega_{k_1} = \omega_{k_2},$$

$$k + k_1 + k_2 = 0, \qquad \omega_k + \omega_{k_1} + \omega_{k_2} = 0, \qquad (7c)$$

corresponding to the two other relations in (4′). The last condition is never satisfied, since we have $\omega_k > 0$; therefore, in the future, under the decay condition either conditions (7a, b) or (7c) are everywhere understood.

4.2.2. Kinetic Equation for Waves.

We have dealt with the three-wave interaction in the case when only a few oscillation frequencies are excited and the interaction has a regular character. If simultaneously a wide frequency spectrum is excited, then since the resonance conditions may be satisfied for a multitude of sets of three waves, their interaction does not usually have a character of a regular process. If the frequencies of different oscillation modes are not comparable, then after some time, even if at the beginning there was a regular spectrum, the phase shifts between them may be considered as occasional. In this case it is convenient to make use of a statistical description with the aid of the so-called kinetic equation for the waves [39]. It is based on the suggestion that the oscillation phases are chaotic i.e., that oscillations with different k do not correlate with each other. The correlation function

$$\langle a_k a_{k_1}^* \rangle = n_k \delta(k - k_1) \qquad (14)$$

is introduced, which is the "number of quanta" with a given wave vector k

[cf. remark after formula (26), Section 4.1)] and obeys the equation (for details of the derivation, cf. §10, Appendix):

$$\frac{dn_k}{dt} = 4\pi \int dk_1 \, dk_2 \, |V_{kk_1k_2}|^2 [(n_{k_1}n_{k_2} - n_k n_{k_1} - n_k n_{k_2})$$

$$\times \delta(\omega_k - \omega_{k_1} - \omega_{k_2})\delta(k - k_1 - k_2) + 2(n_{k_1}n_{k_2} + n_k n_{k_2}$$

$$- n_k n_{k_1})\delta(\omega_k + \omega_{k_1} - \omega_{k_2})\delta(k + k_1 - k_2)], \tag{15}$$

which is just the kinetic equation for waves. The obvious stationary solution for this equation is

$$n_k = N/\omega_k, \tag{16}$$

which corresponds to the equipartition of energy in the "degrees of freedom." Indeed [cf. (26), Section 4.1], the value of energy in the given oscillation mode

$$E_k = \omega_k n_k = N = \text{const},$$

and each wave number k has equal energy.

Comparison of (15) with (18) reveals the connection of the kernel of the kinetic equation for waves with the growth rate of decay instability [note that, in (15), there are only waves satisfying the decay conditions (7)] and allows one to make an estimate of the applicability limits of the kinetic equation [10[ad]].

In a medium let a narrow wave packet propagate, which has a maximum width Δk in the vicinity of k_0. Then

$$\omega_k = \omega_{k_0} + \left(\frac{\partial \omega_k}{\partial k} \Delta k\right),$$

which with the equality $k = k_1 + k_2$ or $k + k_1 = k_2$ accounted for, yields

$$|\omega_k - \omega_{k_1} - \omega_{k_2}| \sim |\omega_k + \omega_{k_1} - \omega_{k_2}| \sim \Delta k \omega',$$

where

$$\omega' = \left|\frac{\partial \omega_k}{\partial k}\right|_{k=k_0}.$$

Replacing in (15) the matrix element by $V_{k_0k_0k_0} = V$ and the δ function of ω by

$$\frac{1}{\omega' \Delta k},$$

we obtain the estimate of the characteristic time of the evolution of the spectrum

$$\frac{1}{\tau} \sim \frac{|V|^2}{\omega' \Delta k} \int n_k \, dk.$$

As $\Delta k \to 0$, $\tau \to 0$. However, the time of the nonlinear interaction must event at all be greater than the inverse maximum growth rate of decay instability (8). For the monocromatic wave, one may formally introduce

$$n_k = |A_0|^2 \delta(k - k_0).$$

Then

$$\tilde{\gamma}_{\max} = 2 \left(|V|^2 \int n_k \, dk \right)^{1/2}.$$

From the requirement $1/\tau \ll \tilde{\gamma}_{\max}$, we obtain the restriction on the packet's width required for the applicability of the kinetic equation (15):

$$\frac{\Delta k}{k} \gg \frac{1}{k\omega'} \left(|V|^2 \int n_k \, dk \right)^{1/2}. \tag{17}$$

Thus, kinetic equation (15) may be applied only to the study of the interaction of wave packets fairly wide in k-space.

4.3 Four-Wave Interaction

If only such waves, for which the decay condition in (7) are not satisfied, are excited, the three-wave interaction is not effective and it is necessary to consider the four-wave interaction. For this purpose, we retain in expansion (19), Section 4.1, the terms up to $\mathcal{H}^{(4)}$. Proceeding in the same way as in the calculation of $\mathcal{H}^{(3)}$, we shall obtain the matrix element of the four-wave interaction

$$\begin{aligned}
\mathcal{H}^{(4)} = \int [& W_{kk_1k_2k_3} a_k a_{k_1} a_{k_2} a_{k_3} \delta(k + k_1 + k_2 + k_3) \\
& + 4W_{kk_1k_2k_3}^* a_k^* a_{k_1} a_{k_2} a_{k_3} \delta(k - k_1 - k_2 - k_3) \\
& + 3W_{kk_1k_2k_3} a_k^* a_{k_1}^* a_{k_2} a_{k_3} \delta(k + k_1 - k_2 - k_3) + \text{c.c.}] \\
& \times dk \, dk_1 \, dk_2 \, dk_3.
\end{aligned} \tag{1}$$

The matrix element W possesses the properties of symmetry similar to the symmetry properties of the matrix element V. We will not give here its explicit expression.

Equation (29), Section 4.1, in this approximation has the form:

$$\begin{aligned}
\frac{\partial a_k}{\partial t} + i\omega_k a_k = & -i \int dk_1 \, dk_2 [V_{kk_1k_2}^* a_{k_1} a_{k_2} \delta(k - k_1 - k_2) \\
& + 2V_{kk_1k_2} a_{k_1}^* a_{k_2} \delta(k + k_1 - k_2) + V_{kk_1k_2}^* a_{k_1}^* a_{k_2}^* \delta(k + k_1 + k_2)] \\
& - i \int dk_1 \, dk_2 \, dk_3 [4W_{kk_1k_2k_3}^* a_{k_1} a_{k_2} a_{k_3} \delta(k - k_1 - k_2 - k_3) \\
& + 6(W_{kk_1k_2k_3} + W_{kk_1k_2k_3}^*) a_{k_1}^* a_{k_2} a_{k_3} \delta(k + k_1 - k_2 - k_3) \\
& + 12W_{kk_1k_2k_3} a_{k_1} a_{k_2}^* a_{k_3}^* \delta(k - k_1 + k_2 + k_3) \\
& + 4W_{kk_1k_2k_3}^* a_{k_1}^* a_{k_2}^* a_{k_3}^* \delta(k + k_1 + k_2 + k_3)]. \tag{2}
\end{aligned}$$

4.3.1. Nonlinear Parabolic Equation. We consider with the aid of (2) the evolution of waves close to linear ones [13^{ad}]. We divide a_k into fast oscillating and slowly varying parts:

$$a_k = (A_k + f_k)e^{-i\omega_k t},$$

where the fast oscillating part is much less than the slowly varying one

$$|f_k| \ll |A_k|.$$

Restricting ourselves, in the equation for f_k, only to the terms quadratic with respect to A_k, we obtain

$$\frac{\partial f_k}{\partial t} = -i \int dk_1\, dk_2 \{ V^*_{kk_1k_2} A_{k_1} A_{k_2} \exp[i(\omega_k - \omega_{k_1} - \omega_{k_2})t]$$

$$\times\, \delta(k - k_1 - k_2) + 2V_{kk_1k_2} A^*_{k_1} A_{k_2} \exp[i(\omega_k - \omega_{k_1} - \omega_{k_2})t]$$

$$\times\, \delta(k + k_1 - k_2) + V^*_{kk_1k_2} A^*_{k_1} A^*_{k_2} \exp[i(\omega_k + \omega_{k_1} + \omega_{k_2})t]$$

$$\delta(k + k_1 + k_2)\}.$$

We integrate this equation by assuming that, for the time of the variation of f_k, A_k does not change:

$$f_k = - \int dk_1\, dk_2 \left\{ \frac{V^*_{kk_1k_2} A_{k_1} A_{k_2}}{\omega_k - \omega_{k_1} - \omega_{k_2}} \exp[i(\omega_k - \omega_{k_1} - \omega_{k_2})t] \right.$$

$$\times\, \delta(k - k_1 - k_2) + \frac{2V_{kk_1k_2} A^*_{k_1} A_{k_2}}{\omega_k + \omega_{k_1} - \omega_{k_2}} \exp[i(\omega_k + \omega_{k_1} - \omega_{k_2})t]$$

$$\times\, \delta(k + k_1 - k_2) + \frac{V^*_{kk_1k_2} A^*_{k_1} A^*_{k_2}}{\omega_k + \omega_{k_1} + \omega_{k_2}} \exp[i(\omega_k + \omega_{k_1} + \omega_{k_2})t]$$

$$\left. \times\, \delta(k + k_1 + k_2)\right\}. \tag{3}$$

Let us, in the equation for A_k, take into account the slowest exponents, i.e., the terms of the form A^*AA. Taking into account (3), we have

$$\frac{\partial A_k}{\partial t} = -i \int T_{kk_1k_2k_3} A^*_{k_1} A_{k_2} A_{k_3} \exp[i(\omega_k + \omega_{k_1} - \omega_{k_2} - \omega_{k_3})t]$$

$$\times\, \delta(k + k_1 - k_2 - k_3)\, dk_1\, dk_2\, dk_3, \tag{4}$$

where

$$T_{kk_1k_2k_3} = -2 \frac{V^*_{kk_1(-k-k_1)} V_{(-k_2-k_3)k_2k_3}}{\omega_{k_2+k_3} + \omega_{k_2} + \omega_{k_3}}$$

$$- 2 \frac{V_{kk_1 k + k_1} V^*_{k_2 + k_3 k_2 k_3}}{\omega_{k_2+k_3} - \omega_{k_2} - \omega_{k_3}} - 4 \frac{V^*_{kk_2 k - k_2} V_{k_3 - k_1 k_1 k_3}}{\omega_{k_1-k_3} + \omega_{k_1} - \omega_{k_3}}$$

$$- 4 \frac{V_{kk_3 k_3 - k} V^*_{k_1 - k_2 k_1 k_3}}{\omega_{k_1-k_2} + \omega_{k_2} - \omega_{k_1}} + 6(W_{kk_1k_2k_3} + W^*_{kk_1k_2k_3}).$$

If the spectrum contains waves for which the decay conditions are satisfied, then singularities will appear in T. Such a difficulty is not encountered for the wave packets narrow in k-space. Consider such a "packet" with the "carrier" wave number k_0 and with a width Δk, such that

$$|k - k_0| \ll k_0.$$

We expand the frequency ω_k with respect to $q = k - k_0$:

$$\omega_k = \omega_{k_0} + (qv) + \tfrac{1}{2} D_{ij} q_i q_j + \cdots,$$

where

$$v = \left. \frac{\partial \omega_k}{\partial k} \right|_{k=k_0}, \qquad D_{ij} = \left. \frac{\partial \omega_k}{\partial k_i \partial k_j} \right|_{k=k_0}.$$

Replace $T_{kk_1 k_2 k_3}$ by $w = T_{k_0 k_0 k_0 k_0}$ in (4) and put

$$b_k = A_k \exp\{-i[(qv) - \tfrac{1}{2} D_{ij} q_i q_j]t\} \equiv A_k e^{i(\omega_{k_0} - \omega_k)t}.$$

Then we obtain from (4)

$$\frac{\partial b_k}{\partial t} + i[(qv) + \tfrac{1}{2} D_{ij} q_i q_j] b_k$$

$$= -2w \int b_{k_1}^* b_{k_2} b_{k_3} \delta(k + k_1 - k_2 - k_3) \, dk_1 \, dk_2 \, dk_3$$

(the temporal exponents are subtracted).

We perform the inverse Fourier transform:

$$b(\mathbf{r}, t) = \frac{1}{2\pi} \int b_q e^{-iqr} \, dq.$$

The quantity $b(\mathbf{r}, t)$ has the meaning of the envelope of the wave packet. We obtain

$$\frac{\partial b}{\partial t} + (v\nabla)b - \frac{i}{2} D_{ik} \frac{\partial^2 b}{\partial x_i \partial x_k} = -\frac{iw}{2\pi} \int e^{-iqr} b_{q_1}^* b_{q_2} b_{q_3}$$

$$\times \delta(q + q_1 - q_2 - q_3) \, dq_1 \, dq_2 \, dq_3 \, dq$$

$$= -\frac{i\omega}{2\pi} \int dq_1 \, dq_2 \, dq_3 \, b_{q_1}^* \, b_{q_2} b_{q_3} e^{i(q_1 - q_2 - q_3)r}$$

$$= -(2\pi)^2 iw |b|^2 b.$$

We have obtained the nonlinear parabolic equation

$$\frac{\partial b}{\partial t} + (v\nabla)b - \frac{i}{2} D_{ij} \frac{\partial^2 b}{\partial x_i \partial x_k} + (2\pi)^2 iw |b|^2 b = 0, \qquad (5)$$

which describes such important effects as the nonlinear correction for the monochromatic wave frequency

$$\Delta\omega = (2\pi)^2 w |b|^2,$$

and self-focusing and self-compression (modulational instability) of the wave packets, leading to the production of envelope solitons and "waveguides" and other phenomena. Some of them have already been considered in the previous sections. We refer the reader for a detailed discussion of these effects to the literature [15[ad]].

§ 5   Concluding Remarks

5.1   When Can an Unstable Gravitating Disk be Regarded as an Infinitesimally Thin One? [39[ad]]

5.1.1. Formulation of the Problem. The disk is incorporated as the main part in many models of astrophysical objects. Indeed, flat subsystems of spiral galaxies, a later stage of evolution of the protoplanetary cloud, the rings of Saturn, and, finally, pancakes and accreting disks around compact masses— this is an incomplete list of objects represented in the form of gravitating disks. Definition of the stability of such objects, as the simplest investigation of one of the possible ways of their evolution, is inevitably associated with further simplifications of the model. The simplest and, therefore, of course, the most popular model was found to be that of an infinitesimally thin disk, i.e., a disk, the thickness h of which is many times less than the perturbation wavelengths λ, $\lambda \ll h$ in question.

However, already in [209] it was shown[17] that the maximum of instability of a gravitating disk (with a not very large temperature anisotropy $\alpha = T_\perp/T_\parallel$, where $T_\perp$ and $T_\parallel$ are the temperatures across and along the rotation axis, respectively) lies in the range of wavelengths comparable to the disk thickness, $\lambda \sim h$. The last condition means that the collisionless disk model in the form being used is found to be inapplicable for the most unstable wavelengths, i.e., just for those developing with a maximum growth rate so that with necessity we must use only a disk model of finite thickness, the stability study of which is an essentially more labor consuming task [31[ad]].

The aim of this section is to point out conditions under which in the infinitesimally thin disk model one can investigate correctly the stability and related fundamental questions: nonlinear density wave evolution [90a, 90, 20[ad], 32[ad]] and weak turbulence [53[ad]]. The study of these problems, as we shall make sure, is possible in the approximation of an infinitesimally thin disk if the latter is immersed in a massive halo. Thus, at what ratios between the basic parameters of these two subsystems can disk stability be investigated by assuming it to be infinitesimally thin?

---

[17] Goldreigh and Lynden-Bell in [209] have given the proof for the case when the disk is near the stability boundary, $\omega^2 \simeq 0$. In the present example, this statement is proved in the general case [see formula (20)].

5.1.2. Vertical Density Distribution of a Light Gaseous Component and a Massive Stellar Halo.

For the sake of concreteness let us consider a gaseous disk immersed in a stellar halo. In addition, we shall denote stellar values by the asterisk subscript and gaseous values by the "g" index. Let the stellar halo density be many times greater than the gas density. Thus, the first condition

$$\frac{\rho_{0*}}{\rho_{0g}} \gg 1 \tag{1}$$

allows one in the Poisson equation for steady-state values to neglect the gaseous component density. Assume the system to be so much extended along the $z = 0$ plane that the gravitational potential along z changes far more abruptly than along r so that $|\partial^2\Phi_0/\partial z^2| \gg |\partial^2\Phi_0/\partial r^2|$. Due to the last remarks, the Poisson equation takes on the form:

$$\frac{\partial^2\Phi_0}{\partial z^2} = 4\pi G\rho_{0*}. \tag{2}$$

Write the equilibrium equation of the stellar component along the z-axis as

$$\frac{\partial\Phi_0}{\partial z} = -\frac{1}{\rho_{0*}}\frac{\partial P_{0*}}{\partial z}. \tag{3}$$

For the "barotropic" stellar component

$$P_{0*} = P_{0*}(\rho_{0*}). \tag{4}$$

Rewrite condition (3) in the form

$$\frac{\partial\Phi_0}{\partial z} = -\frac{c_{\|*}^2}{\rho_{0*}}\frac{\partial\rho_{0*}}{\partial z}, \tag{5}$$

where $c_{\|*}^2 = \partial P_{0*}/\partial\rho_{0*}$ is the square of stellar velocity dispersion along the z-axis. We assume further that $c_{\|*}^2$ is little dependent on z (as compared to density ρ_{0*}) so that

$$\frac{1}{c_{\|*}^2}\frac{\partial c_{\|*}^2}{\partial z}\bigg/\frac{1}{\rho_{0*}}\frac{\partial\rho_{0*}}{\partial z} \ll 1. \tag{6}$$

Differentiating (5) over z and making use of condition (6) and Poisson equation (2), we obtain the Emden equation

$$u'' + be^u = 0. \tag{7}$$

Here, the primes denote differentiation in z,

$$u(z) = \ln\rho_{0*}, \qquad b = \frac{4\pi G}{c_{\|*}^2}. \tag{8}$$

We seek solution of Eq. (7) in the form

$$u(z) = \ln\frac{A}{\mathrm{ch}^2\,az}, \tag{9}$$

where A, a are two constants [Eq. (7)—of second order] which we find by substituting (9) into (7):

$$A = \rho_{0*}(0), \qquad a = \frac{1}{h_*} = \frac{\sqrt{2\pi G \rho_{0*}(0)}}{c_{\|*}} \tag{10}$$

Ultimately from (8), (10) we have [209]

$$\rho_{0*}(z) = \frac{\rho_{0*}(0)}{\cosh^2(z/h_*)}. \tag{11}$$

For the barotropic gas component

$$P_{0g} = P_{0g}(\rho_{0g})$$

the equilibrium condition along z has the form

$$\frac{\partial \Phi_0}{\partial z} = -\frac{c_g^2}{\rho_{0g}} \frac{\partial \rho_{0g}}{\partial z}, \tag{12}$$

where $c_g^2 = \partial P_{0g}/\partial \rho_{0g}$ is the square of sound velocity in gas.

The left-hand sides of Eqs. (5) and (12) are equal; by making the right-hand sides equal, we obtain

$$\frac{\rho_{0g}(z)}{\rho_{0g}(0)} = \left[\frac{\rho_{0*}(z)}{\rho_{0*}(0)} \right]^{c_{\|*}^2/c_g^2}, \tag{13}$$

or, taking into account (11),

$$\rho_{0g}(z) = \frac{\rho_{0g}(0)}{[\cosh^2(z/h_*)]^{c_{\|*}^2/c_g^2}}. \tag{14}$$

In a partial case, $c_{\|*}^2 = c_g^2$, the gas density distribution precisely repeats the star density distribution. Usually, the stellar halo has a temperature greater than the gas temperature, i.e., $c_{\|*}^2 > c_g^2$. In this case, the characteristic thickness of the gaseous disk h_g is found to be less than the stellar one, $h_g < h^*$ [see formula (23) below].

5.1.3. Why One Gaseous Disk Cannot Be Regarded as an Infinitesimally Thin One.
Let us now make sure that in the absence of a massive stellar halo the most unstable modes in a gaseous disk (or in a stellar disk with not very large temperature anisotropy) corresponds to wavelengths comparable to the disk thickness, $\lambda \sim h_g$. The most unstable mode k_0 is obtained from the dispersion curve minimum condition [333] $\partial \omega^2/\partial k = 0$, where $\omega^2 = \varkappa^2 + k^2 c_g^2 - 2\pi G \sigma_{0g} k$,

$$k_0 = \frac{\pi G \sigma_{0g}}{c_g^2}, \tag{15}$$

where

$$\sigma_{0g} = \int_{-\infty}^{\infty} \rho_{0g}(z) \, dz. \tag{16}$$

In the case wherein there is no stellar halo, the potential Φ_0 is determined only by the gas component, and hence, similarly to (11)

$$\rho_{0g}(z) = \frac{\rho_{0g}(0)}{\cosh^2(z/h_g)}, \tag{17}$$

where

$$h_g = \frac{c_g}{\sqrt{2\pi G \rho_{0g}(0)}}. \tag{18}$$

Substituting (17) into (16), we obtain

$$\sigma_{0g} = 2h_g \rho_{0g}(0). \tag{19}$$

Using (15), (18), and (19), we find [209]

$$k_0 h_g = 1. \tag{20}$$

From the last relation it is evident that the infinitesimally thin disk representation turns out to be inapplicable in the vicinity of the wave vector $k = k_0$, corresponding to the most unstable mode.

5.1.4. What Does the Presence of the Stellar Halo Change? If the stellar halo surrounding the stellar disk is taken into account, the situation changes. We determine in this case the surface density of the gaseous disk, for which purpose let us make use of formula (14):

$$\sigma_{0g} = \rho_{0g}(0) h_* \int_{-\infty}^{\infty} \frac{dx}{\cosh^{2v} x}. \tag{21}$$

Here we have introduced the dimensionless variables $x = z/h_*$; $v \equiv c_{\|*}^2/c_g^2$. The integral in (21) is easily calculated by using the relation of [42]

$$\int_{-\infty}^{\infty} \frac{dx}{\cosh^{2v} x} = B(\tfrac{1}{2}, v) = \frac{\Gamma(\tfrac{1}{2})\Gamma(v)}{\Gamma(v + \tfrac{1}{2})}.$$

Next, at $v \gg 1$ the asymptotic formula of [42] can be used

$$\Gamma(az + p) = \sqrt{2\pi} e^{-az}(az)^{az + p - 1/2}.$$

Finally we obtain

$$\sigma_g = \sqrt{\pi} \rho_{0g}(0) h_* \frac{c_g}{c_{\|*}} \tag{22}$$

(at $c_{\|*}/c_g \gg 1$). From the last formula it is seen that the characteristic thickness of the gaseous disk[18] is

$$h_g = h_* \frac{c_g}{c_{\|*}}. \tag{23}$$

With due regard for (10), (15), (23) we obtain [31[ad]]

$$k_0 h_g = \sqrt{\frac{\pi}{2} \frac{\rho_{0g}(0)}{\rho_{0*}(0)}} \ll 1. \tag{24}$$

Thus, if the stellar density exceeds the density of the gas component, for the stability study of the gaseous disk it will be correct to regard the latter as infinitesimally thin, provided some additional condition is satisfied, the derivation of which we are now attempting to determine.

5.1.5. The Basic Theorem. In calculating inequality (24) we have made use of expression (15) which has been derived from the dispersion equation describing small oscillations of a gaseous disk (see Section 2.2, Chapter V) in the absence of the influence of the stellar component. Consequently, we have to obtain the condition of negligible contribution by the stellar component to the perturbed gravitational potential Φ_1.

Let us write the ratios between the perturbed surface density and the unperturbed one for the gaseous and stellar disks:

$$\left(\frac{\sigma_1}{\sigma_0}\right)_g = \frac{k^2 \Phi_1}{\omega^2 - \varkappa^2 - k^2 c_g^2}, \tag{25}$$

$$\left(\frac{\sigma_1}{\sigma_0}\right)_* = \frac{k^2 \Phi_1 I_*}{\omega^2 - \varkappa^2 - k^2 c_{\perp*}^2}. \tag{26}$$

According to Eq. (24), we consider the gaseous disk as thin. Therefore, the reduced factor I_* that provides the correction for thickness is taken into account only for the stellar disk.

It is appropriate to note here that, although in the works of Shu [323] and Toomre [333] the reduced factors had been obtained on the assumption $kh \ll 1$, one can, by direct calculation, make sure that they provide a true

---

[18] Note that the estimate in (23) is valid at the arbitrary value of the parameter $\nu \equiv c_{\|*}^2/c_g^2$. Let us obtain (23) directly from (14) [formulae (14) and (21) are true at any ν!]

$$\frac{1}{\mathrm{ch}^{2\nu} x} = \left(\frac{2}{e^x + e^{-x}}\right)^{2\nu}$$

$$\approx \left(\frac{1}{1 + x^2/2}\right)^{2\nu} \approx \left(\frac{1}{e^{x^2/2}}\right)^{2\nu} = e^{-\nu x^2}.$$

With $1 > x \gtrsim 1/2\nu$ by using (21') we obtain (23).

asymptotic also in the opposite limiting case $kh \gg 1$. Indeed, let us make use of the coupling between Φ_1 and σ_1 (Chapter V)

$$\Phi_1 = -\frac{2\pi G\sigma_1}{|k|}. \tag{27}$$

At $kh \gg 1$

$$I = \frac{2}{kh}. \tag{28}$$

Substituting (27) and (28) into (26) and using the fact that $\rho_0 = \sigma_0/h$, we obtain

$$\omega^2 = \varkappa^2 + k^2 c_g^2 - \omega_0^2, \tag{29}$$

where $\omega_0^2 = 4\pi G\rho_0$. Equation (29) is the dispersion equation describing small oscillations in a rotating gravitating cylinder in the plane perpendicular to the generatrix (Chapter II).

In connection with the above, the reduction factor in (26) can be used in two opposite cases: (1) $kh_* \ll 1$; (2) $kh_* \gg 1$. We examine the first case. In the lowest (zero) order in kh_*, the reduction factors of Shu and Toomre $I_* = 1$.[19]

For a two-component medium of gas and stars, the relationship between the perturbed potential and density has the form (in accord with (27))

$$\Phi_1 = -\frac{2\pi G}{k}(\sigma_{1*} + \sigma_{1g}). \tag{30}$$

From (30) it is evident that the contribution to the perturbed potential of the star component can be neglected, provided that

$$\sigma_{1g} \gg \sigma_{1*}, \tag{31}$$

or, using formulae (25) and (26) and by taking into account $I_* = 1$ [84],

$$\frac{\sigma_{0g}}{c_g^2} \gg \frac{\sigma_{0*}}{c_{\perp*}^2}. \tag{32}$$

By introducing the coefficient of stellar disk anisotropy

$$\alpha = \frac{c_{\perp*}^2}{c_{\parallel*}^2}, \tag{33}$$

using (23), we shall ultimately obtain the following conditions of negligible contribution by the stellar component to the perturbed gravitational potential:

$$1 \gg \frac{\rho_{0g}(0)}{\rho_{0*}(0)} \gg \frac{h_g}{\alpha h_*} \qquad \text{(for } kh_* \ll 1\text{)}. \tag{34}$$

---

[19] In the following (first) order of expansion of I_* over $kh_* \ll 1$ the reduction factors of [323] and [333] differ by a factor of 2.

As with the case of (2) $(kh_* \gg 1)$, by restricting ourselves to the first term of expansion over $1/kh_*$, being equal for reduction factors of the two types of [323] and [333], we have

$$I_* = \frac{2}{kh_*}, \tag{35}$$

and the condition in (31) takes the form

$$\frac{\rho_{0g}^2(0)}{\rho_{0*}^2(0)} \gg \frac{h_g^2}{\alpha h_*^2} \qquad \text{(for } kh_* \gg 1). \tag{36}$$

From the above, there follows:

Theorem. *The necessary and sufficient condition for the existence of an infinitesimally thin disk approximation for the region of unstable wave vectors is the presence of a stellar halo with parameters satisfying inequalities (34) and (36).*

5.1.6. Constraints of the Model. It is obvious that if $\alpha \sim 1$, inequality (36) follows automatically from the right-hand inequality in (34.) Therefore, for models with stellar halo anisotropy not very much different from 1, in terms of the theorem, the requirement of satisfying inequality (36) can be omitted.

Despite the fact that this section is devoted to the correctness of an infinitesimally thin gaseous disk approximation the same problem exists also for a stellar disk with not very great anisotropy α_1[20] of star velocity dispersions. Of course, it is solved by a similar requirement of a massive stellar halo, and conditions (34) and (36) will further incorporate the parameter α_1 that facilitates their fulfillment at $\alpha_1 > 1$.

We propose the reader makes sure that the fulfillment of conditions (34) and (36) for many models is problematic.

5.2 On Future Soliton Theory of Spiral Structure

From applications of the above developed theory of nonlinear density waves in gravitating disks to galaxies two possibilities can be discussed. The former consists in the representation of the flat component of a spiral galaxy in the form of the envelope soliton, the filling of which are the spiral arms. The latter is to represent a separate spiral arm as a soliton if one assumes the existence of inner structure of the arm.[21]

---

[20] It is easy to make sure that for the stellar disk $k_0 h_{*1} \sim 1/\alpha_1 \ll 1$ at $\alpha_1 \gg 1$ (do not confuse α_1 and h_{*1} with α and h_* of stellar halo).

[21] Inner structure of the arm of the Galaxy was investigated, for example, in works by I. V. Gosachinsky [41a].

Let us examine the first possibility. By representing the flat component of the spiral galaxy in the form of a soliton we thereby assume the absence in the soliton of the group velocity, $c_g = d\omega/dk = 0$ (for certainty, we shall bear in mind the stable case—see Fig. 93). In this case there appears an interesting relationship between the radial extent of the spiral structure of the galaxy (which is then simply just a size of the soliton) and the equilibrium parameters of the disk (in particular, the stability reserve Q). Indeed, the characteristic spatial scale of the soliton is, according to (29), of Section 5.1 equal to

$$\Delta = \frac{c_s}{\gamma_0}. \tag{1}$$

But from the dispersion equation, provided that $c_g = d\omega/dk = 0$, it follows that

$$\gamma_0^2 = \varkappa_0^2\left(1 - \frac{1}{Q^2}\right), \tag{2}$$

where $Q = \varkappa_0 c_s/\pi G \sigma_0$ is the Toomre stability reserve. Therefore, instead of (1) we get for the characteristic size of the global picture of the spiral structure

$$\Delta = \frac{c_s}{\varkappa_0\sqrt{1 - 1/Q^2}}. \tag{3}$$

The fundamental question for the possibility of applications to real systems is that of how the various inhomogeneities (density and velocity dispersion inhomogeneities, differentiality of rotation) will affect even the very existence of the soliton structure at rest.[22] The point is that if the condition $c_g = 0$ is satisfied at some one point of the inhomogeneous system, then at another point it generally will not be satisfied. One may, however, suggest that disturbances that possess a required property are automatically chosen from initial disturbances by the system itself (the galaxy): the wave groups having a nonzero group velocity leave the system[23]; at the same time disturbances with $c_g = 0$ remain in it for ever.

The above remark follows from linear theory. If, however, a wave packet having a group velocity $c_g(r) = 0$ is provided (in the framework of linear theory), then the effects of nonlinearity in turn will ensure (for corresponding values of the effective adiabatic exponent γ) nonspreading of the wave packet: the influences of the dispersion and nonlinearity will compensate for each other.

In reality, however, one has to require that simultaneously two conditions, rather than one should be satisfied: (1) the above condition of the

---

[22] The applicability of the theory that deals, for example, with a single soliton moving along the radius, is limited by the passage time to the nearest point of reflection. Indeed, in an inhomogeneous system there occur reflections, refractions, and transformations of (long-wave into short-wave and inversely) waves when these arrive at some special circumferences: resonances, nontransparency region boundaries, etc.

[23] Of course, in the future full theory one will have to take into consideration both the possibility of wave reflection and the effects of wave amplification or damping.

absence of radial drift of the wave packet and (2) weakness (or still better the absence) of angular twisting of the disturbance by differential rotation of the galaxy. The latter condition means

$$\Omega_p = \frac{\omega}{m} \simeq \text{const} \neq f(r). \tag{4}$$

If the two conditions are satisfied, we shall have a packet of spiral waves not subjected to either drift or twisting and therefore not needing any generator for its regeneration. The nonlinearity will stabilize this packet also from dispersion spreading.

Condition (4) leads to a definite dependence of the spiral tilt angle to the circumferences $r = \text{const}$:

$$\theta(r) \simeq \frac{\pi G \sigma_0 Q^2 \sqrt{1 - Q^2}}{|\Omega - \Omega_p| r \varkappa_0}, \qquad \theta \ll 1. \tag{5}$$

Another possibility—the representation of a separate spiral arm or ring in the form of a soliton—was discussed in [90a, 89a, 50^{ad}, 52^{ad}]. In contrast to the picture stated above (soliton—the entire flat subsystem) this model suggests the motion of the soliton arm (or ring) either at subsonic or supersonic velocity.

The spiral structure theory should account for both the formation of ring galaxies and the existence of ring structures in normal galaxies. The theoretical and observational aspects of this problem, as well as the necessary references can be found in [50^{ad}] and [47^{ad}].

Problems

1. Derive the nonlinear equation for density waves in a rotating infinitesimally thin disk [cf. (30), Section 1.1] by using the Lagrange description.

Solution. By introducing the local Cartesian coordinates $x = r, y = r\varphi$, and assuming for the sake of simplicity the rotation to be uniform ($\varkappa_0 = 2\Omega_0$), write the equation of motion

$$\ddot{x} - 2\Omega_0 \dot{y} = -\frac{\partial}{\partial x}\left[\Phi_1 + \frac{c_s^2}{\varkappa - 1}\left(\frac{\sigma}{\sigma_0}\right)^{\varkappa - 1}\right], \tag{1}$$

$$\ddot{y} + 2\Omega_0 \dot{x} = 0, \tag{2}$$

where Φ_1 is the perturbed potential and σ is the full surface density that satisfies the continuity equation

$$\sigma(x)\, dx = \sigma_0\, dx_0. \tag{3}$$

Introducing the displacement ξ through the relation $\xi = x - x_0$, rewrite the system of (1) and (2) in the form of one equation:

$$\ddot{\xi} + 4\Omega_0^2 \xi = -\frac{\partial}{\partial x}\left[\Phi_1 + \frac{c_s^2}{\varkappa - 1}\left(\frac{\sigma}{\sigma_0}\right)^{\varkappa - 1}\right]. \tag{4}$$

Use the Fourier expansions

$$\xi = \sum_k \xi_k e^{ikx_0}, \qquad \xi_k = \frac{1}{2L} \int_{-L}^{L} e^{-ikx_0} \xi \, dx_0, \tag{5}$$

$$\sigma = \sum_k \sigma_k e^{ikx}, \qquad \sigma_k = \frac{1}{2L} \int_{-L}^{L} e^{-ikx} \sigma(x) \, dx = \frac{\sigma_0}{2L} \int_{-L}^{L} e^{-ikx} \, dx_0, \tag{6}$$

with the help of which the relation between σ_k and ξ_k is established (the relation between ψ_k and σ_k is known from the Poisson equation: $\psi_k = -2\pi G \sigma_k / |k|$). Calculating then the Fourier components from the right-hand side of (4) up to the third order of magnitude over ξ_k inclusively, write (4) for the kth harmonic:

$$\frac{1}{c_s^2}\left(\frac{d^2}{dt^2} + \hat\omega_k^2\right)\xi_{\tilde k} = i \sum_{k_1,k_2} \xi_{k_1} \xi_{k_2} (\tilde k Q_{\tilde k, k_1, k_2} - k_1^2 P_{k_1})$$

$$+ \sum_{k_1,k_2,k_3}\left[\tilde k R_{k,k_1,k_2,k_3} - (\tilde k - k_3)^2 Q_{\tilde k - k_3, k_1, k_2} + \frac{k_1^3}{2} P_{k_1}\right]\xi_{k_1}\xi_{k_2}\xi_{k_3}. \tag{7}$$

Here the notations

$$\hat\omega_k^2 = 4\Omega_0^2 + k^2 c_s^2\left(1 - \frac{2|k_0|}{|k|}\right), \qquad |k_0| = \frac{\pi G \sigma_0}{c_s^2}, \tag{8}$$

$$P_k = k\left(1 - 2\frac{|k_0|}{|k|}\right), \tag{9}$$

$$Q_{k,k_1+k_2} = \delta_{k,k_1+k_2}\left[\frac{k}{2}P_k + \frac{(\varkappa - 2)}{2}k_1 k_2\right], \tag{10}$$

$$R_{k,k_1,k_2,k_3} = \delta_{k,k_1+k_2+k_3}\left[\frac{k^2}{6}P_k + \frac{(\varkappa - 2)}{2}k_1(k_2 + k_3)^2 + \frac{(\varkappa - 2)(\varkappa - 3)}{6}k_1 k_2 k_3\right] \tag{11}$$

are introduced.

As in Section 1.1, we assume that only the harmonics $k = \pm k_0$ are excited and $\hat\omega_{k_0}^2 \ll \Omega_0^2$ (then $\hat\omega_{2k_0}^2 \simeq k_0^2 c_s^2 \simeq 4\Omega_0^2$). Then calculating the amplitudes of excited overtones,

$$\xi_{2k_0} \simeq ik_0 \xi_{k_0}^2 \cdot (\varkappa - 1),$$

$$\xi_0 = \xi_{k=0} = 0, \tag{12}$$

we obtain the nonlinear equation of oscillations for perturbations

$$\left(\frac{d^2}{dt^2} + \hat\omega_{k_0}^2\right)\xi_{k_0} = k_0^2 c_s^2 \cdot A |k_0 \xi_{k_0}|^2 \xi_{k_0}, \tag{13}$$

where

$$A = \tfrac{1}{2}(3\varkappa^2 - 11\varkappa + 10). \tag{14}$$

Equation (13) is the sought-for oscillation equation that takes into account nonlinear terms up to the third order of magnitude in amplitude, inclusively.

2. Determine for the nonrotating isothermal layer the nonlinear correction for the critical wavelength that corresponds to the stability boundary [31^{ad}].

Solution. The equilibrium density ρ_0, pressure P_0, and gravity force F_0 in this case are

$$\rho_0 = \frac{1}{\cosh^2 z}, \qquad P_0 = \frac{1}{2\cosh^2 z}, \qquad F_0 = -\frac{\partial \Phi_0}{\partial z} = -\tanh z, \qquad (1)$$

where units are assumed, in which $\rho_0(0) = 1$, $4\pi G\rho_0(0) = 1$, and the "semithickness" of the layer $a = 1$ (the usual dimensional vertical coordinate $z' = za$). The system of equations describing the isothermic layer on the stability boundary is the following:

$$\frac{1}{\rho}\nabla P + \nabla\Phi = 0, \qquad (2)$$

$$\Delta\Phi = \rho, \qquad (3)$$

$$P = \frac{\rho}{2}. \qquad (4)$$

Equations (2)–(4) are readily reduced to a single equation for $\xi = \delta\rho/\rho_0$ $[\rho = \rho_0(1 + \xi)]$; within the values of third order with respect to ξ it may be written in the form

$$\Delta\left(\xi - \frac{\xi^2}{2} + \frac{\xi^3}{3}\right) = 2\frac{\xi}{\cosh^2 z}. \qquad (5)$$

Substituting the variable $z \to \mu = \tanh z$, we obtain

$$\left[\frac{\partial}{\partial\mu}(1 - \mu^2)\frac{\partial}{\partial\mu} + \frac{1}{1-\mu^2}\frac{\partial^2}{\partial x^2}\right]\left(\xi - \frac{\xi^2}{2} + \frac{\xi^3}{3}\right) + 2\xi = 0. \qquad (6)$$

By representing ξ as

$$\xi = \xi_0 + (\xi_1 e^{ikx} + \text{c.c.}) + (\xi_2 e^{2ikx} + \text{c.c.}) + \cdots, \qquad (7)$$

where c.c. denoted a complex conjugate quantity, then in the linear approximation from (6) we find

$$\left[\frac{\partial}{\partial\mu}(1 - \mu^2)\frac{\partial}{\partial\mu} + \left(2 - \frac{k^2}{1-\mu^2}\right)\right]\xi_1^{(1)} = 0. \qquad (8)$$

Hence it follows that

$$\xi_1^{(1)} = A\sqrt{1 - \mu^2}, \qquad k_0^2 = 1 \qquad (A = \text{const}). \qquad (9)$$

In second order (7) yields two equations. One of them, for ξ_2, is

$$\frac{\partial}{\partial\mu}(1 - \mu^2)\frac{\partial}{\partial\mu}\left(\xi_2 - \frac{\xi_1^2}{2}\right) + \left(2 - \frac{4}{1-\mu^2}\right)\left(\xi_2^2 - \frac{\xi_1^2}{2}\right) = -A^2(1 - \mu^2), \qquad (10)$$

while the second equation is for ξ_0

$$\left[\frac{\partial}{\partial\mu}(1 - \mu^2)\frac{\partial}{\partial\mu} + 2\right]\xi_0 = -2(1 - 3\mu^2)|A|^2. \qquad (11)$$

From (10) we find for the combination $\varphi_2 = \xi_2 - \xi_1^2/2$:

$$\varphi_2 = A^2(1 - \mu^2)/4;$$

therefore

$$\xi_2 = \tfrac{3}{4}A^2(1 - \mu^2). \qquad (12)$$

The solution of Eq. (11) is

$$\xi_0 = -\tfrac{1}{2}|A|^2(1 - 3\mu^2). \tag{13}$$

The equation for the "full" $\xi_1 = \xi_1^{(1)} + \xi_1^{(3)}$ is as follows:

$$\left[\frac{\partial}{\partial\mu}(1 - \mu^2)\frac{\partial}{\partial\mu} - \frac{k^2}{1 - \mu^2}\right](\xi_1 - \xi_0\xi_1^{(1)} - \xi_2\bar\xi_1^{(1)} + \bar\xi_1^{(1)}\xi_1^2) + 2\xi_1 = 0. \tag{14}$$

Represent it in the symbolical form

$$\hat L(k^2)\xi_1 = -\hat L(k_0^2)(\bar\xi_1^{(1)}\xi_1^2 - \xi_0\xi_1^{(1)} - \xi_2\bar\xi_1^{(1)}) \equiv -\hat L R(\mu), \tag{15}$$

where

$$L(k^2) = \frac{\partial}{\partial\mu}(1 - \mu^2)\frac{\partial}{\partial\mu} - \frac{k^2}{1 - \mu^2}, \tag{16}$$

$$R(\mu) = \bar\xi_1^{(1)}\xi_1^2 - \xi_0\xi_1^{(1)} - \xi_2\bar\xi_1^{(1)} = \frac{A|A|^2}{4}\sqrt{1 - \mu^2}(5\mu^2 - 1), \tag{17}$$

$$-\hat L(k_c^2)R(\mu) = 3A|A|^2\sqrt{1 - \mu^2}(5\mu^2 - 1). \tag{18}$$

Therefore, we get

$$\hat L(k^2)\xi_1 = 3A|A|^2\sqrt{1 - \mu^2}(5\mu^2 - 1) \equiv \hat\varepsilon(\mu)\xi_1, \tag{19}$$

where

$$\hat\varepsilon(\mu) \equiv 3|A|^2(5\mu^2 - 1). \tag{20}$$

Find now from (19) and (20) the nonlinear correction δk^2 to $k_0^2 = 1$ from perturbation theory. We have

$$\hat L(k_0^2)\delta\xi + \frac{\partial\hat L}{\partial k^2}\delta k^2\xi^{(0)} = \hat\varepsilon(\mu)\xi^{(0)}, \qquad \xi^{(0)} \equiv A\sqrt{1 - \mu^2}. \tag{21}$$

Multiply (21) on the left by $\bar\xi^{(0)}$ and integrate over μ from (-1) to $(+1)$:

$$\int_{-1}^{1} d\mu\,\bar\xi^{(0)}\hat L(k_0^2)\delta\xi + \delta k^2\int_{-1}^{1} d\mu\,\bar\xi^{(0)}\frac{\partial\hat L}{\partial k^2}\xi^{(0)}$$

$$= \int_{-1}^{1} d\mu\,\bar\xi^{(0)}\hat\varepsilon(\mu)\xi^{(0)}. \tag{22}$$

But the first of the integrals in (22) is zero due to self-conjugateness of the operator $\hat L(k_0^2)$; therefore,

$$\delta k^2 = \frac{\int_{-1}^{1}\hat\varepsilon(\mu)\xi^{(0)2}\,d\mu}{\int_{-1}^{1}\xi^{(0)}(\partial\hat L/\partial k^2)\xi^{(0)}\,d\mu}$$

$$= -\frac{1}{2|A|^2}\int_{-1}^{1} 3|A|^4(1 - \mu^2)(5\mu^2 - 1)\,d\mu = 0. \tag{23}$$

Consequently, in lower order with respect to the amplitude of perturbation the nonlinear correction for the critical wavelength of the nonrotating isothermic layer is absent.

3. Same as in Problem 2, for the rotating isothermic layer [31^{ad}].

Solution. The dependence of equilibrium values ρ_0, P_0, F_0 on the vertical coordinate

z remains the same [cf. formulae (1) in the previous problem]. For the ordering usually taken by us, $\partial/\partial t \sim \varepsilon^2$, $u \sim w \sim \varepsilon^3$, $v \sim \varepsilon$, $\delta\rho \sim \varepsilon$, $\delta\Phi \sim \varepsilon$ (cf. §1.2 of main text), we get the following system of equations of motion ($\varkappa$ is the epicyclic frequency) :

$$\varkappa v = \frac{1}{\rho}\frac{\partial P}{\partial x} + \frac{\partial \Phi}{\partial x}, \tag{1}$$

$$\varkappa u = -\left(\frac{\partial v}{\partial t} + u\frac{\partial v}{\partial x} + w\frac{\partial v}{\partial z}\right), \tag{2}$$

$$\frac{1}{\rho}\frac{\partial P}{\partial z} + \frac{\partial \Phi}{\partial z} = 0, \tag{3}$$

$$\frac{\partial \rho}{\partial t} + \frac{\partial}{\partial x}(\rho u) + \frac{\partial}{\partial z}(\rho w) = 0, \tag{4}$$

$$P = \frac{\rho}{2}, \tag{5}$$

$$\Delta\Phi = 4\pi G\rho. \tag{6}$$

Assuming that $\rho = \rho_0 + \delta\rho$ and $\Phi = \Phi_0 + \delta\Phi$, we find from (3)–(5)

$$\frac{1}{2}\frac{\partial}{\partial z}\left[\frac{\delta\rho}{\rho_0} - \frac{1}{2}\left(\frac{\delta\rho}{\rho_0}\right)^2 + \frac{1}{3}\left(\frac{\delta\rho}{\rho_0}\right)^3\right] + \frac{\partial\delta\Phi}{\partial z} = 0. \tag{7}$$

Denote $\theta = \delta\rho/\rho_0$; then from (7) it follows that

$$\frac{1}{2}\left(\theta - \frac{\theta^2}{2} + \frac{\theta^3}{3}\right) + \delta\Phi = -\lambda(x). \tag{8}$$

Equation (1) becomes

$$v = -\frac{1}{\varkappa}\frac{\partial\lambda}{\partial x}, \tag{9}$$

while Eq. (2) yields

$$\varkappa^2 u = \frac{\partial}{\partial t}\frac{\partial\lambda}{\partial x} + u\frac{\partial^2\lambda}{\partial x^2}. \tag{10}$$

We have further the continuity equation (4), which is written in the form

$$\rho w|_{-\infty}^{\infty} = 0,$$

or

$$\int_{-\infty}^{\infty}\left[\rho_0\frac{\partial\theta}{\partial t} + \frac{\partial}{\partial x}(\rho_0 u + \rho_0\theta u)\right]dz = 0, \tag{11}$$

and will be used below as the boundary condition. By adding to the written equations the Poisson equation and turning from z to the new variable $\mu = \tanh z$, we finally have the following system:

$$\varkappa^2 u = \frac{\partial}{\partial t}\frac{\partial\lambda}{\partial x} + u\frac{\partial^2\lambda}{\partial x^2}, \tag{12}$$

$$\Delta\delta\Phi = \rho_0\theta = \theta(1 - \mu^2), \tag{13}$$

$$\frac{1}{2}\left(\theta - \frac{\theta^2}{2} + \frac{\theta^3}{3}\right) + \delta\Phi = -\lambda(x), \tag{14}$$

$$\int_{-1}^{1}d\mu\left[\frac{\partial\theta}{\partial t} + \frac{\partial}{\partial x}(u + \theta u)\right] = 0. \tag{15}$$

Let us solve the system of equations (12)–(15) by successive approximations.

1. *Linear approximation* was investigated [209] by Goldreich and Lynden-Bell, and we first follow their work. We have, instead of (12)–(15),

$$\left[\frac{\partial}{\partial\mu}(1-\mu^2)\frac{\partial}{\partial\mu} + \left(2 - \frac{k^2}{1-\mu^2}\right)\right]\theta_1 \equiv \hat{L}_k\theta_1 = \frac{2k^2}{1-\mu^2}\lambda_1; \tag{16}$$

therefore [209],

$$\theta_1 = \lambda_1\psi_k(\mu), \tag{17}$$

where

$$\psi_k(\mu) = \frac{k}{1-k^2}\left[(k+\mu)\left(\frac{1-\mu}{1+\mu}\right)^{k/2}\int_{-1}^{\mu}\frac{(k-v)}{(1-v^2)}\left(\frac{1+v}{1-v}\right)^{k/2}dv\right.$$

$$\left. + (k-\mu)\left(\frac{1+\mu}{1-\mu}\right)^{k/2}\int_{\mu}^{1}\frac{(k+v)}{(1-v^2)}\left(\frac{1-v}{1+v}\right)^{k/2}dv\right] \equiv \hat{L}_k^{-1}\left(\frac{2k^2}{1-\mu^2}\right). \tag{18}$$

The boundary condition (15) yields the equation

$$\frac{2k^2}{\varkappa^2} = \int_{-1}^{1}\psi_k(\mu)\,d\mu \tag{19}$$

for the determination of the critical wave number k_0 (and the corresponding parameter $\varkappa^2$). Write (19) in the form

$$F(k) = A(k), \tag{20}$$

where the notations

$$F(k) = \frac{2k^2}{\varkappa^2}, \qquad A(k) = \int_{-1}^{1}\psi_k(\mu)\,d\mu \tag{21}$$

are introduced. The behavior of the curve, determined by Eq. (19), is qualitatively the same as that in Fig. 94. At the point of contact $k = k_0 = 0.47$, and $1/\varkappa^2 \simeq 4.38$.

2. *Second order.* For the second harmonic, from (12)–(14) one can obtain the equation

$$\hat{L}_{2k}\theta_2 = \frac{2(2k^2)^2\lambda_2}{1-\mu^2} + (\hat{L}_{2k} - 2)\frac{\theta_1^2}{2}; \tag{22}$$

therefore,

$$\theta_2 = \lambda_2\psi_{2k} + (1 - 2\hat{L}_{2k}^{-1})\psi_k^2\frac{\lambda_1^2}{2}. \tag{23}$$

Substituting (23) into the boundary condition of (15), we find

$$\lambda_2 = -\lambda_1^2\frac{B(k)}{A(2k) - 4A(k)}, \tag{24}$$

where

$$B(2k) = \frac{1}{2}\int_{-1}^{1}(1 - 2\hat{L}_{2k}^{-1})\psi_k^2\,d\mu. \tag{25}$$

In the second order, however, we also get the equation for the zero harmonic θ_0:

$$\hat{L}_0\theta_0 = (\hat{L}_0 - 2)|\theta_1|^2 \qquad \left(\hat{L}_0 \equiv \frac{\partial}{\partial\mu}(1-\mu^2)\frac{\partial}{\partial\mu} + 2\right), \tag{26}$$

with the solution

$$\theta_0 = (1 - 2\hat{L}_0^{-2})|\theta_1|^2 = |\lambda_1|^2(1 - 2\hat{L}_0^{-1})\psi_k^2. \tag{27}$$

Here the boundary condition

$$\int_{-1}^{1} \theta_0 \, d\mu = 0$$

is satisfied automatically. The operator $\hat{L}_0^{-1}$ acts in the following manner:

$$\hat{L}_0^{-1}X(\mu) = \tfrac{1}{2}\left[\mu \int_{\mu}^{1} y(v)X(v) \, dv + y(\mu) \int_{-1}^{\mu} vX(v) \, dv \right.$$

$$\left. - \mu \int_{-1}^{\mu} y(v)X(v) \, dv - y(\mu) \int_{\mu}^{1} vX(v) \, dv \right], \tag{28}$$

where

$$y(\mu) = \frac{\mu}{2} \ln \frac{1 + \mu}{1 - \mu} - 1. \tag{29}$$

It is easy to see that the solution of (27) is even and has no singularities at $\mu = \pm 1$.

3. *Third order.* Somewhat more cumbersome than those written above but in principle similar to them, calculations finally lead to the following equation for the determination of the nonlinear correction for the critical wave number k_0:

$$A(k) - \frac{2k^2}{\varkappa^2} = |\lambda_1|^2\left[C(k) - \frac{B(k)D(k)}{A(2k) - 4A(k)}\right], \tag{30}$$

where

$$C(k) = -\int_{-1}^{1} d\mu\left[\frac{k^2}{\varkappa^2}\left[(-\tfrac{3}{2} + 2L_0^{-1} + \hat{L}_{2k}^{-1})\psi_k^2\right.\right.$$

$$\left.\left. + (1 - 2\hat{L}_k^{-1})\psi_k(\tfrac{1}{2} - \hat{L}_{2k}^{-1} - 2\hat{L}_0^{-1})\psi_k^2\right], \tag{31}$$

$$D(k) = \int_{-1}^{1}\left[\frac{k^2}{\varkappa^2}\left(\psi_{2k} - \frac{8k^2}{\varkappa^2}\right) - (1 - 2\hat{L}_k^{-1})\psi_{2k}\psi_k\right] d\mu. \tag{32}$$

The computation made by these formulae has shown that the nonlinearity in this case plays a destabilizing role: it somewhat broadens the region of unstable wave numbers ($\delta k^2 > 0$).

4. Same as in Problem 2, for the rotating incompressible layer[24] (the adiabatic index $\gamma = \infty$) and for the rotating layer with $\gamma = 2$.

Solution. In both cases, the solutions of equations ensuing in the scheme of successive approximations, are expressed in elementary functions. Since the schemes of calculations largely repeat the ones already repeatedly used by us earlier (for example, in Problem 2), we shall give only the final form of nonlinear dispersion equations (on the stability boundary) and determine the sign δk^2.

---

[24] The solution is obtained by S. M. Churilov.

1. *Incompressible layer.* To determine δk^2 in this case, we obtain a system of two equations:

$$\left[\frac{1}{k(1-2k+e^{-2k})}-\frac{1}{x^2}\right]\Lambda^{(1)}=\frac{2k^3}{x^4}\frac{(1+e^{-2k})^2}{(1-2k+e^{-2k})}\Lambda^{(1)}\Lambda^{(2)}$$

$$+\frac{k^8}{2x^6}\cdot\frac{3(k^2+x^2)-k(1-e^{-4k})}{(1-2k+e^{-2k})}\Lambda^{(1)^3}, \qquad (1)$$

$$\left[\frac{1}{2k(1-4k+e^{-4k})}-\frac{1}{x^2}\right]\Lambda^{(2)}=\frac{k^3}{8x^4}\frac{(1+e^{-2k})^2}{1-4k+e^{-4k}}\Lambda^{(1)^2}. \qquad (2)$$

Here $\Lambda^{(1)}$ and $\Lambda^{(2)}$ are the expansion coefficients of the value $\chi=P/\rho_0-\Phi=\Lambda(x,t)$ in the series

$$\Lambda=\Lambda^{(0)}+(\Lambda^{(1)}e^{ikx-i\omega t}+\text{c.c})+(\Lambda^{(2)}e^{2ikx-2i\omega t}+\text{c.c.})+\cdots; \qquad (3)$$

units are used, in which the half-thickness of the layer $c=1$, $4\pi G=1$, and the density $\rho_0=1$.

The minimum of the function

$$F_k=\frac{1}{k(1-2k+e^{-2k})}$$

is reached at $2k=2k_0=0.607$.[25] Here it is easy to see that $1-2k_0+e^{-2k_0}>0$; the calculation shows, in addition, that also $1-4k_0+e^{-4k_0}\simeq0.05>0$. Since

$$\frac{1}{k(1-2k+e^{-2k})}\simeq\frac{1}{x^2}$$

and

$$\frac{1}{2k(1-4k+e^{-4k})}>\frac{1}{k(1-2k+e^{-2k})},$$

then, according to (2), $\Lambda^{(2)}>0$. The equation for $(\delta k)^2$ will be reduced to the form

$$\frac{1}{2}(\delta k)^2F_{k_0}''=\frac{2k_0^3}{x^4}\frac{(1+e^{-2k_0})^2}{1-2k_0+e^{-2k_0}}\Lambda^{(2)}$$

$$+\frac{k_0^8}{2x^6}\frac{3(k_0^2+x^2)-k_0(1-e^{-4k_0})}{1-2k_0+e^{-2k_0}}\Lambda^{(1)^2}. \qquad (4)$$

The first term on the right-hand side of (4) is positive owing to $\Lambda^{(2)}>0$, just proved. Since $x^2=\frac{1}{4}(\frac{1}{4}x^2=1.75)$, therefore,

$$3(k_0^2+x^2)-k_0(1-e^{-4k_0})>3(k_0^2+x^2)-k_0\simeq0.15,$$

and the second term is also positive. If one takes into account that $F_{k_0}''>0$ (cf. Fig. 94), we arrive at the conclusion about the destabilizing character of the nonlinear correction: $(\delta k)^2>0$.

In the limiting case of a nonrotating layer from (1) and (2), one may obtain

$$[1-k(1+\tanh k)]=\left[\frac{2k(1-\tanh 2k)(2k-\frac{1}{2})}{1-2k(1+\tanh 2k)}-k(\tfrac{7}{2}k-1)\right]|A^{(1)}|^2, \qquad (5)$$

[25] This and subsequent values of the required quantities are taken from a paper of Goldreich and Lynden-Bell [209].

where $A^{(1)}$ is the amplitude of the perturbed potential

$$(\Phi^{(1)} = A^{(1)} \cosh kz / \cosh k).$$

Here also $\delta k > 0$, i.e., the nonlinearity effect is destabilizing.

2. *Polytropic layer for* $\gamma = 2$. The equation for the nonlinear correction $(\delta k)^2$ has the form:

$$\tfrac{1}{2} F''_k (\delta k)^2 = D \Lambda^{(1)^2}, \tag{6}$$

where

$$D = \frac{C^2}{2(F_{2k} - F_k)} - A,$$

$$A = \frac{k^2}{\rho_c^2 l_k^4} \left[\frac{k \tan(\pi l_k/2) + l_k}{l_k \tan(\pi l_k/2) - k} \right]^4 \left\{ \frac{1}{2[l_{2k} \tan(\pi l_{2k}/2) - 2k]} + \frac{2 \tan(\pi l_k/2)}{k \tan(\pi l_k/2) + l_k} \right\}, \tag{7}$$

$$C = \frac{k}{\rho_c l_{2k} l_k^2} \left(\frac{k \tan(\pi l_k/2) + l_k}{l_k \tan(\pi l_k/2) - k} \right)^2 \frac{2k \tan(\pi l_{2k}/2) + l_{2k}}{l_{2k} \tan(\pi l_{2k}/2) - 2k}, \tag{8}$$

$$F_k = \frac{\pi}{2l_k} + \frac{\tan(\pi l_k/2)}{k l_k^3 [l_k \tan(\pi l_k/2) - k]}, \qquad l_k^2 = 1 - k^2.$$

In the unperturbed state, the density $\rho_0 = \rho_c \cos z$; the pressure is $P = \rho^2/2$; the angular velocity is $\Omega = \tfrac{1}{2}$, $4\pi G = 1$; the half-thickness of the layer is $a = \pi/2$. The numerical calculation at the stability boundary yields

$$k_0 = 0.39,$$

$$F_{k_0} = 5.62, \qquad F_{2k_0} = 53.79, \tag{9}$$

$$A = 0.0054, \qquad C = 0.500, \qquad D = -0.0028.$$

Since $D < 0$ and $F''_{k_0} > 0$, therefore, $(\delta k)^2 < 0$, i.e., stabilization takes place.

On the stability boundary of the nonrotating layer with $\gamma = 2$ we get the following equation replacing (6):

$$k \left(l_k - k \cot \frac{\pi l_k}{2} \right) = \frac{l_k^2 A^{(1)^2}}{2\rho_c^2} \cdot \frac{(8k^2 - 1) \cosh(\pi l_k/2) + 4 l_k k \sinh(\pi l_k/2)}{l_k \sinh(\pi l_k/2) + 2k \cosh(\pi l_k/2)}, \tag{10}$$

where $l_k^2 = 4k^2 - 1$, $l_k^2 = 1 - k^2$, and $l_k^2 > 0$; since the equation for l_{k_0}, $l_{k_0} = k_0 \cot(\pi l_{k_0}/2)$ (or, what is the same, $l = \cos(\pi l/2)$), has the solution $l = 0.6$ ($k \simeq 0.8$). On the right-hand side of (10), a positive value stands which we shall denote by B^2. Expanding the left-hand side of (10) near $k = k_0$, we obtain

$$-\left(\frac{1}{2l} + \frac{1}{kl} \right) \delta k = B^2; \tag{11}$$

therefore, $\delta k < 0$. This means that the region of unstable k is somewhat reduced with due regard for nonlinearity (stabilization).

5. Derive the nonlinear equation for disturbances conserving the surface density in the fastly rotating homogeneous gas layer (the density ρ_0, the thickness $2c$, the angular velocity of rotation Ω; $\Omega^2 \gg 4\pi G \rho_0$). Obtain the solutions of the soliton-like type. Investigate the modulation instability of the layer and the collapse of two-dimensional nonlinear waves [30[ad]].

Solution. This problem in terms of one and the same equilibrium model (item 1) similar to the Goldreich and Lynden-Bell model [209] will deal with: solitary waves (solitons), in item 2; modulational instability of nonlinear monochromatic waves (leading to their division into individual packets), in item 3; and finally, collapse of two-dimensional nonlinear waves which is formally manifested in the appearance of a singularity in the solution of the basic equation after some marginal time (item 4).

1. *Equilibrium model and derivation of the basic equation.* Consider a homogeneous gaseous layer of density ρ_0 and thickness $2c$, maintained in equilibrium along the z-axis as a result of the balance of forces of gravitation and pressure $P_0 = \frac{1}{2}\Omega_0^2\rho_0(c^2 - z^2)$. In order to avoid Jeans instability in the (x, y) plane, let us assume the layer to be rotating with an angular velocity Ω in an external field with the potential $\Phi_0 = \Omega^2(x^2 + y^2)/2$.[26] The maximum growth rate of Jeans instability for $\Omega = 0$ is of the order Ω_0. For $\Omega^2 \gtrsim \Omega_0^2$, this instability stabilizes. Further, we are interested in oscillations of the layer with a frequency of the order $\omega_0 = \sqrt{4\pi G\rho_0} \sim \Omega_0$. For the sake of simplicity, let us assume that $\Omega^2 \gg \Omega_0^2 \sim \omega^2$ and make use of the dimensionless variables, in which $\rho_0 = 1$, $c = 1$, and $4\pi G\rho_0 = 1$.

Find now the spectrum of small, long-wave oscillations of the layer in the plane (x, y).

In the frame of reference rotating with an angular velocity Ω, the linearized equations have the form

$$\frac{\partial \rho_1}{\partial t} + \frac{\partial v_{z1}}{\partial z} + i\mathbf{k}_\perp \mathbf{v}_{\perp 1} = 0, \tag{1}$$

$$\frac{\partial \mathbf{v}_{\perp 1}}{\partial t} = 2[\mathbf{\Omega}\mathbf{v}_{\perp 1}] - i\mathbf{k}_\perp(P_1 + \Phi_1), \tag{2}$$

$$\frac{\partial v_{z1}}{\partial t} = -\frac{\partial}{\partial z}(P_1 + \Phi_1) + \rho_1 P_0', \tag{3}$$

$$\frac{\partial}{\partial t}(P_1 - \gamma P_0\rho_1) + v_{z1}P_0' = 0, \tag{4}$$

$$\frac{\partial^2\Phi_1}{\partial z^2} - k_\perp^2\Phi_1 = \rho_1, \tag{5}$$

where $\mathbf{k}_\perp \equiv (k_x, k_y)$, γ is the adiabatic index, and the prime denotes the derivative with respect to z. From Eq. (2), one can find $v_{\perp 1}$. Omitting $\partial/\partial t$ in comparison with $\Omega(\Omega \gg \omega)$ and multiplying this equation vectorially by $\mathbf{\Omega}$, we obtain

$$\mathbf{v}_{\perp 1} = -\frac{i}{2}\frac{[\mathbf{\Omega}\mathbf{k}_\perp]}{\Omega^2}(P_1 + \Phi_1). \tag{6}$$

Substituting (6) into (1), we see that the term with $k_\perp v_\perp$ is eliminated from the latter equation. Introducing the quantity ξ (vertical displacement along the z-axis), via the relation $v_{z1} = -i\omega\xi$, we obtain the following simplified system of equations:

$$\rho_1 + \xi' = 0, \tag{7}$$

$$-\omega^2\xi = -\Phi_1' - P_1' - \rho_1 z, \tag{8}$$

$$P_1 = \frac{\gamma}{2}(1 - z^2)\rho_1 + z\xi, \tag{9}$$

$$\Phi_1'' - k^2\Phi_1 = \rho_1. \tag{10}$$

---

[26] A similar model was applied by Goldreich and Lynden-Bell [210] in the construction of the regenerative theory of spiral arms in galaxies.

Differentiating (8) over z and substituting into the equations thus obtained (7), (9), (10), we arrive at the following equation for the perturbed potential Φ_1:

$$\left\{\omega^2 \frac{d^2}{dz^2} + \frac{d^2}{dz^2}\left[\frac{\gamma}{2}(1-z^2)\frac{d^2}{dz^2}\right]\right\}\Phi_1 = (\omega^2 - 1)k^2\Phi_1 + k^2 \frac{d^2}{dz^2}\frac{\gamma}{2}(1-z^2)\Phi_1. \quad (11)$$

In the case $k_\perp = 0$, the order of Eq. (11) is lowered:

$$\omega^2\Phi_1 + \frac{\gamma}{2}(1-z^2)\frac{d^2\Phi_1}{dz^2} = 0. \quad (12)$$

It may be easily tested that this equation has eigenfunctions

$$\Phi_1^{(n)} = (1-z^2)P'_{n+1}(z) \sim P_{n+2}(z) - P_n(z) \quad (13)$$

and eigenfrequencies

$$\omega_n^2 = \tfrac{1}{2}\gamma(n+1)(n+2), \quad (14)$$

where $n = 0, 1, 2, \cdots$; $P_n(z)$ are the Legendre polynomials. In particular, for the mode $n = 0$, corresponding to uniform extensions–contractions of the layer along the z-axis, $\omega^2 = \gamma$. We shall be interested just in this mode since it is most convenient to obtain the nonlinear dispersion equation for it.

Calculate now the corrections of the order of k^2 for the eigenfrequencies, assuming that $k^2 \ll 1$.

Represent Eq. (11) in the form

$$\hat{R}(\omega^2)\Phi_1 = k^2\hat{P}\Phi_1, \quad (15)$$

where

$$\hat{R}(\omega^2) = \omega^2 \frac{d^2}{dz^2} + \frac{d^2}{dz^2}\left[\frac{\gamma}{2}(1-z^2)\frac{d^2}{dz^2}\right], \quad (16)$$

$$\hat{P} = -(\gamma + 1) - 2\gamma z \frac{d}{dz}. \quad (17)$$

In the zero order of perturbation theory

$$\hat{R}(\omega_n^{(0)^2})\Phi_{10}^{(n)} = 0, \quad (18)$$

where $\Phi_{10}^{(n)}$ and $\omega_n^{(0)^2}$ are defined by expressions (13) and (14). Assume that $\Phi_1^{(n)} = \Phi_{10}^{(n)} + \varphi_n$; then

$$\hat{R}(\omega_n^{(0)^2})\varphi_n + \frac{\partial \hat{R}\Phi_{10}^{(n)}}{\partial \omega^2}\delta\omega^2 = k^2\hat{P}\Phi_{10}^{(n)}. \quad (19)$$

Define the scalar product

$$\langle \Phi, \Psi \rangle \equiv \int_{-1}^{1} \Phi\Psi \, dz. \quad (20)$$

Multiply (39) on the left scalarly by $\Phi_{10}^{(n)}$:

$$\langle \Phi_{10}^{(n)}, \hat{R}(\omega_n^{(0)^2})\varphi_n \rangle + \delta\omega^2 \left\langle \Phi_{10}^{(n)}, \frac{\partial \hat{R}}{\partial \omega^2}\Phi_{10}^{(n)} \right\rangle = k^2\langle \Phi_{10}^{(n)}, \hat{P}\Phi_{10}^{(n)} \rangle. \quad (21)$$

Hence we find [taking into account (18) and also the self-conjugateness of the operator $\hat{R}, \langle \Phi, \hat{R}\Psi \rangle = \langle \Psi, \hat{R}\Phi \rangle$]:

$$\delta\omega^2 = k^2 \frac{\langle \Phi_{10}^{(n)}, \hat{P}\Phi_{10}^{(n)} \rangle}{\langle \Phi_{10}^{(n)}, (\partial \hat{R}/\partial \omega^2)\Phi_{10}^{(n)} \rangle}. \tag{22}$$

Since

$$\frac{\partial \hat{R}}{\partial \omega^2} = \frac{d^2}{dz^2}, \quad \left\langle \Phi, \frac{d^2\Phi}{dz^2} \right\rangle = -\langle \Phi', \Phi' \rangle, \quad \langle \Phi, \hat{P}\Phi \rangle = -(\gamma + 1)\langle \Phi, \Phi \rangle$$

$$= -2\gamma \left\langle \Phi, z\frac{d}{dz}\Phi \right\rangle = -(\gamma + 1)\langle \Phi, \Phi \rangle + \gamma\langle \Phi, \Phi \rangle = -\langle \Phi, \Phi \rangle,$$

we obtain

$$\delta\omega_n^2 = k^2 \frac{\langle \Phi_{10}^{(n)}, \Phi_{10}^{(n)} \rangle}{\langle \Phi_{10}^{(n)'}, \Phi_{10}^{(n)'} \rangle} > 0. \tag{23}$$

Finally, since

$$\langle \Phi_{10}^{(n)}, \Phi_{10}^{(n)} \rangle = 4\frac{(n + 1)^2(n + 2)^2}{(2n + 1)(2n + 3)(2n + 5)},$$

$$\langle \Phi_{10}^{(n)'}, \Phi_{10}^{(n)'} \rangle = 2\frac{(n + 1)^2(n + 2)^2}{2n + 3},$$

we finally find the following expression for the correction to the frequency:

$$\delta\omega_n^2 = \frac{2k^2}{(2n + 1)(2n + 5)}. \tag{24}$$

In particular, for $n = 0$, the correction is $\delta\omega^2 = 2k^2/5$.

Calculate now the nonlinear correction for the frequency $\omega^2 = \gamma$. Assuming that the nonlinearity is sufficiently small, the nonlinear correction may be calculated at $k_\perp = 0$. But such oscillations, as one may easily ascertain, are described by the equation

$$\ddot{c} = -1 + c^{-\gamma}, \tag{25}$$

which remains true at arbitrary amplitudes. Assuming that $c = 1 + h$, $h \ll 1$, we find from (25)

$$\ddot{h} + \omega_0^2 h = -\alpha h^2 - \beta h^3, \tag{26}$$

where

$$\alpha = -\tfrac{1}{2}\gamma(\gamma + 1), \quad \beta = \tfrac{1}{6}\gamma(\gamma + 1)(\gamma + 2). \tag{27}$$

Equation (26) has a standard form [69] of the equation for oscillations of the anharmonic oscillator. Therefore, one can immediately write the expression for the nonlinear correction to the frequency

$$\delta\omega = \left(\frac{3}{8}\frac{\beta}{\omega_0} - \frac{5\alpha^2}{12\omega_0^3}\right)|h|^2 \tag{28}$$

or

$$\delta\omega = -\lambda^2\omega_0|h|^2, \quad \lambda^2 \equiv \tfrac{1}{48}(2\gamma - 1)(\gamma + 1). \tag{29}$$

Combining (24) and (29), we obtain the sought-for nonlinear dispersion equation

$$\omega^2 = \omega_0^2(1 + 3k_\perp^2/5\gamma - 2\lambda^2|h|^2). \tag{30}$$

2. *Solitons.* Assuming now in (30) $k_\perp^2 = k_y^2 - \partial^2/\partial x^2$, we arrive at the differential equation

$$\frac{2}{5\gamma}\frac{\partial^2 h}{\partial x^2} = \frac{\omega_0^2[1 + (2/5\gamma)k_y^2] - \omega^2}{\omega_0^2}h - 2\lambda^2|h^2|h. \tag{31}$$

Multiplying (31) by $\partial h^*/\partial x$ and adding with the complex conjugate equation, we obtain

$$\frac{2}{5\gamma}\frac{\partial}{\partial x}\left|\frac{\partial h}{\partial x}\right|^2 = \frac{\omega_0^2[1 + (2/5\gamma)k_y^2] - \omega^2}{\omega_0^2}\frac{\partial}{\partial x}|h|^2 - \lambda^2\frac{\partial}{\partial x}|h|^4.$$

Integrating over x, we find

$$\frac{2}{5\gamma}\left(\frac{\partial|h|}{\partial x}\right)^2 = \left\{1 + \frac{2}{5\gamma}k_y^2 - \frac{\omega^2}{\omega_0^2}\right\}|h|^2 - \lambda^2|h|^4 + \text{const}. \tag{32}$$

The soliton solution corresponds to the case: const $= 0$. Denote

$$q^2 = \frac{5\gamma\lambda^2}{2}, \qquad A^2 = \frac{1}{\lambda^2}\left(1 + \frac{2}{5\gamma}k_y^2 - \frac{\omega^2}{\omega_0^2}\right), \tag{33}$$

then

$$\frac{1}{q}\frac{\partial|h|}{\partial x} = \pm\sqrt{A^2 - h^2}. \tag{34}$$

The solution for this equation is

$$|h| = \frac{A}{\cosh(Aqx)}, \tag{35}$$

the width of the soliton

$$\Delta \approx \frac{1}{Aq} = \left[\frac{2}{5\gamma}\frac{1}{1 + (2/5\gamma)k_y^2 - \omega^2/\omega_0^2}\right]^{1/2}, \tag{36}$$

and the amplitude $|h|_{max} = A$. The minimum size of the soliton (on the limit of applicability of the theory) has the order of the layer thickness.

Note that the soliton solutions, similar to that obtained above, are easily found also in the model of a uniform gaseous cylinder.

It is, however, interesting that for the *collisionless* layer and cylinder, there are no similar solitons (the nonlinear correction for the frequency ω_0 and the correction for $k_\perp^2$ have the *same* sign).

The solution of (35) describes a soliton being at rest. However, it is easy to construct a solution also for a running soliton.

Let $h = \frac{1}{2}(ae^{-i\omega_0 t} + a^*e^{i\omega_0 t})$, and $b \equiv ae^{-i\omega_0 t}$; then for b, we have the equation

$$i\frac{\partial b}{\partial t} - \omega_0 b + \alpha\Delta b + \beta b|b|^2 = 0. \tag{37}$$

We seek the solution in the form $b = C(x, t)e^{-i\omega t + i\mathbf{kr}}$, and for C we obtain the equation

$$i\left(\frac{\partial C}{\partial t} + 2\alpha k_x\frac{\partial C}{\partial x}\right) + (\omega - \omega_0 + \alpha k^2)C + \alpha\Delta C + \beta C|C|^2 = 0. \tag{38}$$

One may assume C to be a real function, $C = C(\xi) \equiv C(x - v_g t)$, where $v_g = 2\alpha k_x$ is the group velocity. Then from (38), for $C(\xi)$, we obtain

$$\alpha \frac{\partial^2 C}{\partial \xi^2} + (\omega - \omega_0 - \alpha k^2)C + \beta C^3 = 0; \qquad (39)$$

therefore,

$$C = \frac{C_{max}}{\cosh k_0 \xi}, \qquad (40)$$

where

$$k_0 = \sqrt{\frac{\beta}{2\alpha}} C_{max}, \qquad \omega = \omega_0 + \alpha k^2 - \frac{\beta}{2} C_{max}^2 = \omega_0 + \alpha(k^2 - k_0^2).$$

This solution is coincident with (35) if in (40) it is assumed that $k_x = 0$. The perturbation h has the form

$$h = \cos(\omega t - \mathbf{kr}) \frac{C_{max}}{\cosh k_0 (x - v_g t)}. \qquad (41)$$

3. *Modulation instability.* We are investigating the problem of the stability of nonlinear monochromatic waves in the above adopted model of a uniform gaseous layer. Write the dispersion equation (30) in the form

$$\omega = \omega_0 + \alpha k_\perp^2 - \beta|h|^2, \qquad (42)$$

where $\alpha = \omega_0/5\gamma$, $\beta = \omega_0 \lambda^2$. Let $h = \frac{1}{2}(ae^{-i\omega_0 t} + a^* e^{i\omega_0 t})$, where $a = a(\mathbf{r}_\perp, t)$ is the envelope. For it we have the equation

$$i\frac{\partial a}{\partial t} + \alpha \Delta a + \beta|a|^2 a = 0. \qquad (43)$$

This equation admits a solution in the form of a plane monochromatic wave

$$a = b_0 e^{-i\omega_k t + i\mathbf{k}_\perp \mathbf{r}}, \qquad (44)$$

where

$$\omega_k = \alpha k_\perp^2 - \beta|b_0|^2. \qquad (45)$$

We investigate the stability of this solution. For this purpose write $a = be^{i\varphi}$, $b = b_0 + b_1$, $\varphi = \varphi_0 + \varphi_1$, $\varphi_0 \equiv -\omega_k t + \mathbf{k}_\perp \mathbf{r}_\perp$. Then

$$a_1 = (b_1 + b_0 i\varphi_1)e^{i\varphi_0}, \qquad a_1^* = (b_1 - b_0 i\varphi_1)e^{-i\varphi_0}. \qquad (46)$$

We linearize Eq. (43):

$$i\frac{\partial a_1}{\partial t} + \alpha \Delta a_1 + 2\beta|a_0|^2 a_1 + \beta a_0^2 a_1^* = 0. \qquad (47)$$

Substituting into (47) expressions (46), we obtain

$$i\frac{\partial b_1}{\partial t} - b_0 \frac{\partial \varphi_1}{\partial t} + \omega_k(b_1 + i\varphi_1 b_0) + \alpha(\Delta b_1 + ib_0 \Delta \varphi_1) + 2\alpha(\nabla b_1 - ib_0 \nabla \varphi_1)i\nabla \varphi_0$$

$$+ \alpha(b_1 + ib_0 \varphi_1)i\Delta \varphi_0 - \alpha(b_1 + ib_0 \varphi_1)(\nabla \varphi_0)^2$$

$$+ 2\beta b_0^2(b_1 + ib_0 \varphi_1) + \beta(b_1 - ib_0 \varphi_1)b_0^2 = 0. \qquad (48)$$

Separate the real and imaginary parts

$$\frac{\partial b_1}{\partial t} + \omega_k b_0 \varphi_1 + \alpha b_0 \Delta\varphi_1 + 2\alpha k \nabla b_1 - \alpha k^2 b_0 \varphi_1 + \beta b_0^3 \varphi_1 = 0, \tag{49}$$

$$-b_0 \frac{\partial\varphi_1}{\partial t} + \omega_k b_1 + \alpha\Delta b_1 - 2\alpha b_0 k\nabla\varphi_1 - \alpha k^2 b_1 + 3\beta b_0 b_1^2 = 0. \tag{50}$$

Taking into account (45), eqs. (49) and (50) will be rewritten in the form

$$\left(\frac{\partial}{\partial t} + 2\alpha k\nabla\right)b_1 + \alpha b_0 \Delta\varphi_1 = 0,$$

$$\left(\frac{\partial}{\partial t} + 2\alpha k\nabla\right)\varphi_1 - \frac{\alpha}{b_0}\Delta b_1 - 2\beta b_0 b_1 = 0,$$

or

$$\left(\frac{\partial}{\partial t} + 2\alpha k\nabla\right)^2 b_1 + \alpha^2 \Delta^2 b_1 + 2\alpha\beta b_0^2 \Delta b_1 = 0. \tag{51}$$

Substituting b_1 in the form of $b_1 \sim e^{-i\Omega t + i\varkappa r}$ into (51), we obtain[27]

$$(\Omega - 2\alpha k\varkappa)^2 = \alpha\varkappa^2(\alpha\varkappa^2 - 2\beta b_0^2). \tag{52}$$

This yields the instability condition

$$2\beta b_0^2 > \alpha\varkappa^2. \tag{53}$$

Thus, for sufficiently large b_0^2, there will be instability with respect to long-wave modulations.

Note in conclusion that the self-modulation leading to the division of the wave into individual packets, as is well known, always is the case if the general Lighthill criterion [265a]

$$\frac{\partial^2\omega}{\partial k^2}\bigg|_{h=0} \frac{\partial\omega}{\partial h^2} < 0 \tag{54}$$

is satisfied.

In our case, this criterion is just identical to the requirement that the correction for the frequency of linear oscillations $\sim k_1^2 \neq 0$ and those proportional to the square of the finite amplitude h^2 have different signs.

4. *Collapse of nonlinear waves.* Consider the case of two-dimensional waves, whose amplitude is dependent only on $\sqrt{x^2 + y^2} = r$. Equation (37) will take the form

$$i\frac{\partial b}{\partial t} + \frac{\mu}{r}\frac{\partial}{\partial r}r\frac{\partial b}{\partial r} + vb|b|^2 = 0. \tag{55}$$

Equation (65), as may be shown, has the integrals of motion

$$I_1 = \int_0^\infty r\,dr|b|^2, \tag{56}$$

$$I_2 = \int_0^\infty r\,dr\left(\tfrac{1}{2}v|b|^4 - \mu\left|\frac{\partial b}{\partial r}\right|^2\right). \tag{57}$$

---

[27] Note that (52) coincides with Eq. (27.13) in Karpman's book [55a].

Let us show that under definite conditions, in the solution of (55) for a marginal time, singularity appears. For that purpose, let us introduce the quantity

$$A = \int_0^\infty r^3 \, dr \, |b|^2 > 0. \tag{58}$$

It may be shown that

$$\frac{d^2 A}{dt^2} = -8\mu I_2 \tag{59}$$

so that

$$A = -4\mu I_2 t^2 + C_1 t + A(0). \tag{60}$$

Since $A > 0$, then under condition $I_2 > 0$, for a marginal time, singularity arises. The condition $I_2 > 0$ qualitatively coincides with the condition of the modulational instability in (53) [30[ad]].

6. Show that instability criterion of a uniform-density circular cylinder and Maclaurin disk are universal with respect to the amplitude of nonradial oscillations (V. A. Antonov and S. N. Nuritdinov) [12a].

Solution. Let us show that the instability criterion obtained earlier for perturbations of small amplitude of a circular cylinder (2), §1, Chapter II, and of the "Maclaurin disk" (6), §1, Chapter V, remain unchanged in the case of the finite amplitude. We shall perform the proof with the aid of the energy principle, following [12a].

Consider only oscillations of a special kind which keep the system spatially homogeneous ($\rho = \text{const}$) and also in the perturbed state.

The connection of the perturbed system with the unperturbed one is expressed via a certain affine transformation of the phase coordinates. For the velocities one may write (cf. similar formulae in Problem 3, Chapter I, etc.)

$$v_x - v'_x = \alpha(t)x + \beta(t)y, \tag{1}$$

$$v_y - v'_y = \delta(t)x + d(t)y, \tag{2}$$

where v'_x, v'_y are the components of the peculiar velocity, α, β, δ, d are some unknown time functions.

One may make sure that the quantities

$$C_1 = \overline{x^2 v'^2_x} + \overline{y^2 v'^2_y} + (\beta - \delta)^2 \overline{x^2} \, \overline{y^2}, \tag{3}$$

$$C_2 = \overline{x^2 y^2} [\overline{v'^2_x v'^2_y} - (\overline{v'_x v'_y})^2], \tag{4}$$

$$C_3 = \delta \overline{x^2} - \beta \overline{y^2} \tag{5}$$

are invariant. The overbar denotes the averaging over the phase density of the perturbed system. The quantity C_2 is proportional to the square of the phase volume of the system, $C_3 = L/M$, where $L = M(\overline{xv_y - yv_x})$ is the total angular moment, and M is the mass of the system. It is more difficult to ascribe a definite physical meaning to the quantity C_1.

Calculate the values of C_1, C_2, C_3 for the cylinder and the disk. For the cylinder, we first of all have

$$\overline{x^2} = \tfrac{1}{4}a^2, \qquad \overline{y^2} = \tfrac{1}{4}b^2, \tag{6}$$

where a and b are, respectively, the large and small semiaxes of the elliptical cross section of the perturbed cylinder. From the form of the argument of the δ-functional distribution in this case it is easy to notice that

$$\overline{v_x'^2} = \tfrac{1}{4}a^2(1 - \gamma^2), \qquad \overline{v_y'^2} = \tfrac{1}{4}b^2(1 - \gamma^2). \tag{7}$$

In the stationary state $a = b = 1, \delta = \gamma = -\beta, \overline{v_x'v_y'} = 0$. Therefore, for the case of the cylinder we obtain

$$C_1 = \tfrac{1}{8}(1 + \gamma^2), \qquad C_2 = \tfrac{1}{256}(1 - \gamma^2)^2, \qquad C_3 = \tfrac{1}{2}\gamma. \tag{8}$$

For the disk

$$\overline{x^2} = \tfrac{1}{5}a^2, \qquad \overline{y^2} = \tfrac{1}{5}b^2, \qquad \overline{v_x'^2} = \tfrac{1}{5}a^2(1 - \gamma^2), \qquad \overline{v_y'^2} = \tfrac{1}{5}b^2(1 - \gamma^2). \tag{9}$$

Taking into account the stationary values of a, b, δ, β and $\overline{v_x'v_y'}$, we find the following values of the invariants for the Maclaurin disk:

$$C_1 = \tfrac{2}{25}(1 + \gamma^2), \qquad C_2 = \tfrac{1}{625}(1 - \gamma^2)^2, \qquad C_3 = \tfrac{2}{5}\gamma. \tag{10}$$

Note that in both cases, $C_3^2 = C_1 - 2\sqrt{C_2}$.

Turn now to the study of the stability of the models in question with respect to non-radial oscillations of the "affine" type. The task is to minimize the total energy under condition of conservation of the invariants C_1, C_2, C_3.

The kinetic energy of a two-dimensional system is

$$T = \tfrac{1}{2}M(\overline{v_x^2} + \overline{v_y^2}) = \tfrac{1}{2}M[(\alpha^2 + \delta^2)\overline{x^2} + (\beta^2 + d^2)\overline{y^2} + \overline{v_x'^2} + \overline{v_y'^2}]. \tag{11}$$

From (3) and (5), it follows

$$\beta = \frac{\overline{x^2}}{\overline{x^2} - \overline{y^2}} \left(\frac{C_3}{\overline{x^2}} - \sqrt{\frac{C_1 - \xi}{\overline{x^2}\,\overline{y^2}}} \right),$$

$$\delta = \frac{\overline{y^2}}{\overline{x^2} - \overline{y^2}} \left(\frac{C_3}{\overline{y^2}} - \sqrt{\frac{C_1 - \xi}{\overline{x^2}\,\overline{y^2}}} \right), \tag{12}$$

where the quantity $\xi = \overline{x^2 v_x'^2} + \overline{y^2 v_y'^2}$ is introduced, the values of which in the stationary state for the cylinder and the disk are coincident: $\xi \to \xi_0 = 2\sqrt{C_2}$. Let $\eta = \overline{x^2 v_x'^2} - \overline{y^2 v_y'^2}$. From (4), it follows that

$$\xi^2 - \eta^2 = 4\overline{x^2 y^2}\,\overline{v_x'^2 v_y'^2} \geq 4C_2, \qquad \eta \leq \sqrt{\xi^2 - 4C_2}. \tag{13}$$

Then the total energy satisfies the inequality

$$E \geq \frac{M}{2(\overline{x^2} - \overline{y^2})} \left\{ (C_3^2 + C_1)(\overline{x^2} + \overline{y^2}) + \frac{\xi}{2} \left[\frac{(\overline{x^2})^2}{\overline{y^2}} + \frac{(\overline{y^2})^2}{\overline{x^2}} - 3(\overline{x^2} + \overline{y^2}) \right] \right.$$

$$\left. - \frac{(\overline{x^2} + \overline{y^2})^3}{2\overline{x^2}\,\overline{y^2}} \sqrt{\xi^2 - 4C_2} - 4C_3\sqrt{\overline{x^2}\,\overline{y^2}(C_1 - \xi)} \right\} + W, \tag{14}$$

where W is the potential energy. In the derivation of (14), it is assumed that $\alpha = d = 0$, for $\alpha^2 \overline{x^2} + d^2 \overline{y^2} \geq 0$ and α and d themselves are not connected with the invariants at all.

Thus, we have taken into account all the invariants, and the function being minimized depends on the three remaining independent variables: $\overline{x^2}, \overline{y^2}$ and ξ. Further, for the sake of certainty, we assume that $\overline{x^2} \geq \overline{y^2}$, i.e., $a \geq b$. Note also that $2\sqrt{C_2} \leq \xi \leq C_1$.

1. *Cylinder.* The potential energy in this case is

$$W = \tfrac{1}{2}M \ln(a + b) + \text{const} = \tfrac{1}{2}M \ln(\sqrt{\overline{x^2}} + \sqrt{\overline{y^2}}) + \text{const.} \qquad (15)$$

Make use of the following relations:

$$\sqrt{C_1 - \xi} \le C_3 - \frac{\xi - 2\sqrt{C_2}}{2C_3} \qquad (C_3 = \sqrt{C_1 - 2\sqrt{C_2}}), \qquad (16)$$

$$\min_{\xi}(\mu_1\xi - \mu_2\sqrt{\xi^2 - 4C_2}) = 2\sqrt{C_2}\sqrt{\mu_1^2 - \mu_2^2} \qquad (\mu_1 > \mu_2 > 0). \qquad (17)$$

Then, introducing the notations

$$m = \sqrt{\overline{x^2}} + \sqrt{\overline{y^2}}, \qquad n = \sqrt{\overline{x^2}} - \sqrt{\overline{y^2}} \qquad (m \ge n), \qquad (18)$$

we obtain

$$\frac{E}{M} \ge \frac{1 + 3\gamma^2}{16m^2} + \frac{1 - \gamma^2}{16m^2}\sqrt{9 + \frac{8n^2}{m^2 - n^2}} + \frac{1}{2}\ln m \ge \frac{1}{4m^2} + \frac{1}{2}\ln m \equiv E_1(m), \qquad (19)$$

if $\gamma < 1$. The stationary values $m_0 = 1$ and $n_0 = 0$. As is seen, $E_2(m) \ge E_1(m_0)$. Therefore, for $\gamma < 1$, there is nonlinear stability of the cylinder.

2. *Disk.* Rewrite inequality (14) in the form

$$\frac{E}{M} \ge \frac{(C_3 + \sqrt{C_1 - \xi})^2}{4m^2} + \frac{(C_3 - \sqrt{C_1 - \xi})^2}{4n^2}$$

$$+ \frac{2(m^2 + n^2)}{(m^2 - n^2)^2}\xi - \frac{4mn}{(m^2 - n^2)^2}\sqrt{\xi^2 - 4C_2} + \frac{W}{M} = F. \qquad (20)$$

Introduce a new notation of l in the following manner:

$$C_3 - \sqrt{C_1 - \xi} = l^2n^2. \qquad (21)$$

With the aim to find the *local* minimum, expand the function F over n^2 within an accuracy of n^2:

$$F(m, n, l) = F_1(m) + n^2F_2(m, l) + O(n^4), \qquad (22)$$

where

$$F_1(m) = \frac{4}{25m^2} - \frac{4}{5\sqrt{5}m} \ge F_1\left(m_0 = \frac{2}{\sqrt{5}}\right),$$

$$F_2(m, l) = \frac{l^4}{4} + \frac{6\gamma}{5m^2}l^2 - \frac{16\sqrt{\gamma(1 - \gamma^2)}}{5\sqrt{5}m^3}l + \frac{12(1 - \gamma^2)}{25m^4} - \frac{1}{5\sqrt{5}m^3}. \qquad (23)$$

It is evident that regarding m, there is a minimum for the arbitrary γ, and in F_2, instead of m, one may substitute its stationary value $m_0 = 2/\sqrt{5}$.

Either stability or instability of the model depends on the sign F_2. The critical point corresponds to $F_2(m_0, l) = 0$ for a certain l. At this point, there appear multiple roots, and $(\partial F_2/\partial l)_{m=m_0} = 0$. An investigation combining the last two equations gives the critical value $\gamma_c = \sqrt{125/486}$, the same as in linear theory.

In the paper [12a], a proof is given of the *absolute* character of the minimum thus found. We do not consider this point here.

In conclusion we note that the analytical consideration given above may be easily generalized on the case of a disk immersed into the massive halo if we suppose that the latter has also an uniform volume density and, consequently, creates the quadratic potential

$$\Phi_h = \Omega^2 r^2/2 + \text{const.}$$

The expansion of function $F(n, m, l)$ with an accuracy up to n^4 has the following form:

$$F = F_1 + n^2 F_2 + n^4 F_3,$$

$$F_1 = \frac{4}{25m^2} - \frac{4(1 - \Omega^2)}{5\sqrt{5}} + \frac{\Omega^2 m^2}{4}, \qquad F_2 = \frac{l^4}{4} + \frac{6\gamma}{5m^2} l^2$$

$$+ \frac{12(1 - \gamma^2)}{25\sqrt{5}} + \frac{\Omega^2 m^2}{4} - \frac{16\sqrt{\gamma(1 - \gamma^2)}}{5\sqrt{5}m^3} l - \frac{(1 - \Omega^2)}{5\sqrt{5}m^3}, \qquad F_3 = \frac{5l^4}{16}$$

$$+ \frac{5}{2}\left[\frac{5(1 - \gamma^2)}{8} + 3\gamma l^2 - l^4\right] + \frac{15l^3(1 + 5\gamma^2)}{8\sqrt{\gamma(1 - \gamma^2)}} - \frac{45(1 - \Omega^2)}{512}. \tag{24}$$

The investigation above says nothing about nonlinear stage of the barlike instability in the disk systems unstable according to linear theory. This question is considered (by the numerical method) in the next problem. In the same place we investigate nonlinear evolution of disturbances in the *elliptical* disks of Freeman.

7. Investigate the *nonlinear evolution* of the barlike perturbations on the example of Freeman's disk models.

Solution. Write the connection between the current x, y, v_x, v_y and initial $x_0, y_0, v_{x_0}, v_{y_0}$ coordinates and velocities of the particle in the form

$$x = u_1 x_0 + u_2 y_0 + u_3 v'_{x_0} + u_4 v'_{y_0},$$

$$y = v_1 x_0 + v_2 y_0 + v_3 v'_{x_0} + v_4 v'_{y_0},$$

$$v_x = \dot{x} = \dot{u}_1 x_0 + \dot{u}_2 y_0 + \dot{u}_3 v'_{x_0} + \dot{u}_4 v'_{y_0}, \tag{1}$$

$$v_y = \dot{y} = \dot{v}_1 x_0 + \dot{v}_2 y_0 + \dot{v}_3 v'_{x_0} + \dot{v}_4 v'_{y_0},$$

where $v'_{x_0} = v_{x_0} + \gamma y_0, v'_{y_0} = v_{y_0} - \gamma x_0, \gamma$ is the angular velocity of a disk, and u_i, v_i are unknown time functions, $i = 1, 2, 3, 4$. The total kinetic energy of a disk is

$$T = \tfrac{1}{2} \int dx_0 \, dy_0 \, dv'_{x_0} \, dv'_{y_0} \; F_0(x_0, y_0, v'_{x_0}, v'_{y_0})(\dot{x}^2 + \dot{y}^2), \tag{2}$$

where the equilibrium distribution function

$$F_0 = \frac{\sigma_0}{2\pi\sqrt{1 - \gamma^2}} \left[(1 - \gamma^2)(1 - x_0^2 - y_0^2) - v'^2_{x_0} - v'^2_{y_0}\right]^{-1/2}, \tag{3}$$

σ_0 being the surface density at the center of a disk $[\sigma_0(r) = \sigma_0\sqrt{1 - r^2/R^2}]$. Assuming the angular velocity of a particle in the circular orbit and the radius of a disk to be unit, $\Omega = 1$ and $R = 1$, we obtain, after calculations of simple integrals in (2),

$$T = \frac{\pi\sigma_0}{15} [\dot{u}_1^2 + \dot{v}_1^2 + \dot{u}_2^2 + \dot{v}_2^2 + (1 - \gamma^2)(\dot{u}_3^2 + \dot{v}_3^2 + \dot{u}_4^2 + \dot{v}_4^2)]. \tag{4}$$

The potential energy W of the elliptical disk may be expressed through the values of semi-axes a, b: $W = W(a, b)$.[28] The Lagrangian of the system $L = T - W$, and the Lagrange equations are

$$C_i \ddot{u}_i = -\frac{\partial W}{\partial u_i} = -\frac{\partial W}{\partial a}\frac{\partial a}{\partial u_i} - \frac{\partial W}{\partial b}\frac{\partial b}{\partial u_i}, \tag{5}$$

$$D_i \ddot{v}_i = -\frac{\partial W}{\partial v_i} = -\frac{\partial W}{\partial a}\frac{\partial a}{\partial v_i} - \frac{\partial W}{\partial b}\frac{\partial b}{\partial v_i}, \tag{6}$$

where $i = 1, 2, 3, 4$; $C_i = 2\pi\sigma_0/15$, $D_i = C_i(1 - \gamma^2)$. Now we describe how one may express a and b by u_i and v_i. Determining $x_0, y_0, v'_{x_0}, v'_{y_0}$ from Eq. (1), we obtain

$$x_0 = \alpha_1 x + \alpha_2 y + \alpha_3 v_x + \alpha_4 v_y,$$
$$y_0 = \beta_1 x + \beta_2 y + \beta_3 v_x + \beta_4 v_y,$$
$$v'_{x_0} = \gamma_1 x + \gamma_2 y + \gamma_3 v_x + \gamma_4 v_y, \tag{7}$$
$$v'_{y_0} = \delta_1 x + \delta_2 y + \delta_3 v_x + \delta_4 v_y,$$

where $\alpha_i, \beta_i, \gamma_i, \delta_i$ denote the corresponding coefficients which are the time functions. Now, substituting (7) into the equation for the boundary of the system's phase region

$$(1 - \gamma^2) = (1 - \gamma^2)(x_0^2 + y_0^2) + v'^2_{x_0} + v'^2_{y_0}, \tag{8}$$

we find

$$(1 - \gamma^2) = A_1 x^2 + A_2 y^2 + 2A_3 xy + 2D_1 xv_x + 2D_2 xv_y + 2D_3 yv_x$$
$$+ 2D_4 yv_y + B_1 v_x^2 + B_2 v_y^2 + 2B_3 v_x v_y, \tag{9}$$

where the coefficients A_i, D_i, and B_i may be defined through $\alpha_i, \beta_i, \gamma_i, \delta_i$ by the symmetrical formulae

$$A_1 = (1 - \gamma^2)(\alpha_1^2 + \beta_1^2) + (\gamma_1^2 + d_1^2)$$
$$A_2 = (1 - \gamma^2)(\alpha_2^2 + \beta_2^2) + (\gamma_2^2 + d_2^2) \tag{10}$$
$$A_3 = (1 - \gamma^2)(\alpha_1\alpha_2 + \beta_1\beta_3) + (\gamma_1\gamma_3 + d_1 d_3) \text{ and so on.}$$

From Eq. (9) one may find, for the boundary of the elliptical disk, the following equation:

$$(1 - \gamma^2) = A_1 x^2 + A_2 y^2 + 2A_3 xy - \frac{1}{B_1}(D_1 x + D_3 y)$$

$$- \frac{[(D_2 B_1 - B_3 D_1)x + (D_4 B_1 - B_3 D_3)y]^2}{B_1(B_2 B_1 - B_3^2)}. \tag{11}$$

Now the semiaxes a, b may be easily determined from Eq. (11).

[28] Note the conventional representation of W, suggested by Antonov and Nuritdinov (see Problem 6) in a form of the seria

$$W = -\frac{4}{5\sqrt{5m}}\sum_{i=0}^{\infty} C_{i+1}\left(\frac{n}{m}\right)^{2i}, \tag{4'}$$

where $n = (a - b)/\sqrt{5}$, $m = (a + b)/\sqrt{5}$, $C_1 = 1$, $C_{i+1} = C_i[(2i - 1)/2i]^2$.

The set of eight two-order equations (5), (6) can be simply solved with the help of a computer. In the equilibrium state

$$u_1 = \cos t, \qquad u_2 = -\gamma \sin t, \qquad u_3 = \sin t, \qquad u_4 = 0,$$
$$v_1 = \gamma \sin t, \qquad v_2 = \cos t, \qquad v_3 = 0, \qquad v_4 = \sin t. \tag{12}$$

Accordingly, if we substitute the following, as the initial values, into the equation of motion,

$$u_1 = 1, \qquad u_2 = 0, \qquad u_3 = 0, \qquad u_4 = 0,$$
$$v_1 = 0, \qquad v_2 = 1, \qquad v_3 = 0, \qquad v_4 = 0,$$
$$\dot{u}_1 = 0, \qquad \dot{u}_2 = -\gamma, \qquad \dot{u}_3 = 1, \qquad \dot{u}_4 = 0, \tag{13}$$
$$\dot{v}_1 = \gamma, \qquad \dot{v}_2 = 0, \qquad \dot{v}_3 = 0, \qquad \dot{v}_4 = 1,$$

then, as it must be, the solution shows that a and b do not depend on the time (remaining as const $= 1$). Now, we shall use the values u_i, v_i, $\dot{u}_i$, $\dot{v}_i$ corresponding to the linear barlike disturbances as the initial values, and then we shall follow the evolution of these disturbances in the nonlinear regime. Corresponding corrections Δu_i, Δv_i, $\Delta \dot{u}_i$, $\Delta \dot{v}_i$ to the equilibrium values (13) may easily be found from the solution already known to us (see, for example, the end of Section 4.4, Chapter V).

Let us describe the obtained results. As was to be expected, the manner of the evolution of the initial disturbance depends essentially on whether the concrete model under consideration was stable or unstable (according to the linear theory). Let us recall that the models are stable for $\gamma < 0.507$ and unstable for $\gamma > 0.507$. For the stable models there are the oscillations with the amplitudes of the order of initial disturbance amplitude. In the unstable (according to the linear theory) region of γ, for the given initial amplitude, the oscillation amplitudes in the nonlinear regime are greater for larger γ. This is seen from Fig. 103, where we present typical graphs $a(t)$ and $b(t)$ for $\gamma = 0.6$ and $\gamma = 0.7$. From the figure one may see also that the major a and minor b semiaxes oscillate (with different frequencies) near some mean values ($\bar{a} \approx 1.4$, $\bar{b} \approx 0.68$ for $\gamma = 0.6$ and $\bar{a} \approx 1.65$, $\bar{b} = 0.55$ for $\gamma = 0.7$). The picture of evolution for γ sufficiently far from $\gamma = \gamma_c$ is weakly dependent on a value of the initial disturbance ε. For γ near $\gamma_c (\gamma \gtrsim \gamma_c)$ the oscillation amplitude remains small (for small ε), but quickly increases with growth of γ.

Thus, it is seen that, as a result of development of the barlike instability, we obtain the elliptical disk with a greater degree of the mean flatness the more quickly the system rotates in the initial state. With the aim of controling the accuracy of counting, we checked, at each integration step, conservation of the total energy and total angular moment

$$E = T + W, \tag{14}$$

$$L_z = x\bar{v}_y - y\bar{v}_x \equiv C_3, \tag{15}$$

and also of the following values which must be the integrals of the movement:

$$C_1 = \bar{x}^2\bar{v}_x^2 + \bar{y}^2\bar{v}_y^2 + 2\overline{xy}\,\overline{v_x v_y} - \overline{xv_x^2} - \overline{yv_y^2} - 2\overline{xv_y}\,\overline{yv_x}, \tag{16}$$

$$\begin{aligned}
C_2 = \;& (\bar{x}^2\bar{y}^2 - \overline{xy}^2)(\bar{v}_x^2\bar{v}_y^2 - \overline{v_x v_y}^2) - \bar{x}^2\bar{v}_y^2\,\overline{yv_x^2} - \bar{x}^2\bar{v}_x^2\,\overline{yv_y^2} \\
& + 2\bar{x}^2\overline{yv_x}\,\overline{yv_y}\,\overline{v_x v_y} + 2\bar{y}^2\overline{xv_x}\,\overline{xv_y}\,\overline{v_x v_y} - \bar{y}^2\bar{v}_x^2\overline{xv_y^2} - \bar{y}^2\bar{v}_y^2\overline{xv_x^2} \\
& + \overline{yv_y^2}\,\overline{xv_x^2} + \overline{xv_y^2}\,\overline{yv_y^2} - 2\overline{xv_y}\,\overline{yv_x}\,\overline{xv_x}\,\overline{yv_y} \\
& + 2\overline{xy}(\overline{yv_x}\,v_y^2\overline{xv_x} - \overline{yv_x}\cdot\overline{xv_y}\,\overline{v_x v_y} - \overline{yv_y}\,\overline{xv_x}\,\overline{v_x v_y} + \overline{yv_y}\,v_x^2\overline{xv_y}). \tag{17}
\end{aligned}$$

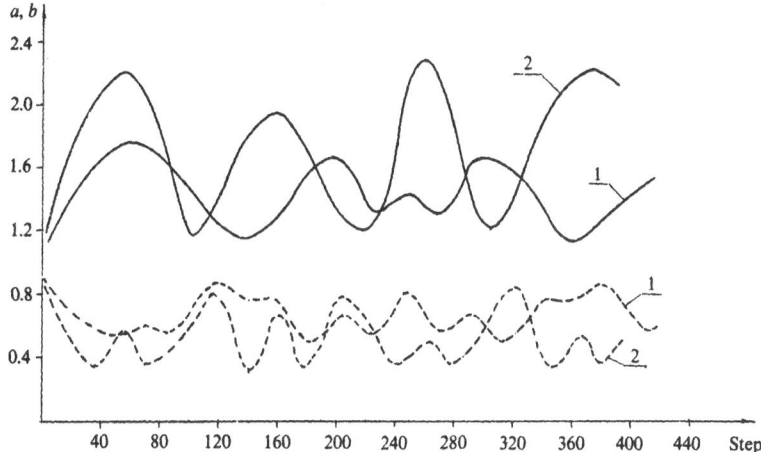

Figure 103. Dynamics of the bar-like mode for Freeman's circular disks with $\gamma = 0.6$ (1) and $\gamma = 0.7$ (2), in the nonlinear regime; solid line—$a(t)$, dotted line—$b(t)$.

The second momenta involving (15)–(17) may be easily expressed through $u_i, v_i, \dot{u}_i, \dot{v}_i$; for example,

$$\bar{x}^2 = \tfrac{1}{5}[u_1^2 + u_2^2 + (1 - \gamma^2)(u_3^2 + u_4^2)],$$

$$\overline{v_x v_y} = \tfrac{1}{5}[\dot{u}_1\dot{v}_1 + \dot{u}_2\dot{v}_2 + (1 - \gamma^2)(\dot{u}_3\dot{v}_3 + \dot{u}_4\dot{v}_4)], \quad \text{and so on.} \tag{18}$$

If the initial small disturbance can be obtained by a continuous deformation of the Freeman's circular disk, then the corresponding values of integrals C_i proved to be the following (see previous problem):

$$C_1 = \tfrac{2}{25}(1 + \gamma^2), \qquad C_2 = \tfrac{1}{625}(1 - \gamma^2)^2, \qquad C_3 = \tfrac{2}{5}\gamma$$

In fact, for all the cases investigated, these values practically do not change (for example, the accuracy of the conservation of total energy was better than 0.1%).

Elliptical disks. Recall that, according to the linear theory, elliptical disks are unstable in the triangular region of parameters $(b/a, \Omega^2/A^2)$—see Fig. 49. The nonlinear theory in this case is constructed approximately in the same manner as for the circular disks (item 1). Let us write by analogy with (1)

$$x(t) = u_1 x_0 + u_2 y_0 + u_3 \tilde{c}_{x_0} + u_4 \tilde{c}_{y_0},$$

$$y(t) = v_1 x_0 + v_2 y_0 + v_3 \tilde{c}_{x_0} + v_4 \tilde{c}_{y_0}, \tag{19}$$

where u_i and v_i are the time functions, $\tilde{c}_{x_0} = v_{x_0} + \theta y_0/b_2$, $\tilde{c}_{y_0} = v_{y_0} - \theta x_0/a^2$, and the expressions for θ and other values we meet with below (see in §1, Chapter IV). The calculation of kinetic energy T gives

$$T = \frac{M}{2}\left[\frac{a^5}{5}(\dot{u}_1^2 + \dot{v}_1^2) + \frac{b^2}{5}(\dot{u}_2 + \dot{v}_2^2) + \frac{k_\alpha^2 k_\beta^2}{5\Lambda^2 b^2}(\dot{u}_3^2 + \dot{v}_3^2) + \frac{1}{5\Lambda^2 a^2}(\dot{u}_4^2 + \dot{v}_4^2)\right], \tag{20}$$

where M is a disk mass. For the potential energy $W(a, b)$ one may use, for example, a previous expression (4′). The initial values of the invariants $C_1, C_2, C_3 = L_z$ are now the following:

$$C_1 = \tfrac{1}{25}\left[\frac{k_\alpha^2 k_\beta^2 a^2}{\Lambda^2 b^2} + \frac{b^2}{\Lambda^2 a^2} + \left(2\Omega - \frac{\theta}{b^2} - \frac{\theta}{a^2}\right)^2 a^2 b^2\right], \tag{21}$$

$$C_2 = \frac{k_\alpha^2 k_\beta^2}{625\Lambda^4}, \qquad C_3 = \tfrac{1}{5}(2\theta - \Omega a^2 - \Omega b^2); \tag{22}$$

at the same time the expressions (15)–(17) for its current values remain without changes. Similarly, mainly "the equations of motion" do not change—they have the form (5), (6), with slightly different values of the effective masses C_i and D_i which are determined according to (20). The equilibrium state of the elliptical disk corresponds to the following initial values of $u_i, v_i, \dot{u}_i, \dot{v}_i$:

$$u_1 = 1, \qquad v_1 = 0, \qquad \dot{u}_1 = 0, \qquad\qquad \dot{v}_1 = -\Omega + \frac{\theta}{a^2};$$

$$u_2 = 0, \qquad v_2 = 1, \qquad \dot{u}_2 = -\frac{\theta}{b^2} + \Omega, \qquad \dot{v}_2 = 0; \tag{23}$$

$$u_3 = 0, \qquad v_3 = 0, \qquad \dot{u}_3 = 1, \qquad\qquad \dot{v}_3 = 0;$$

$$u_4 = 0, \qquad v_4 = 0, \qquad \dot{u}_4 = 0, \qquad\qquad \dot{v}_4 = 1.$$

In the initial moment ($t = 0$) we assumed the certain small perturbations from the equilibrium values (23); moreover, we considered the perturbations of two types. For the perturbations of the first type the initial u_i and v_i ("velocities") were supposed to be equal to their equilibrium values, and small corrections to "coordinates" δu_i and δv_i were determined so that the initial values of invariants C_1, C_2, C_3 calculated according to (15)–(17), were equal to their equilibrium values (21) and (22). Then one might assume $\delta u_2 = \delta u_3 = \delta v_1 = \delta v_4 = 0$, and, putting the amplitude of disturbance $\delta v_3 = \varepsilon$,

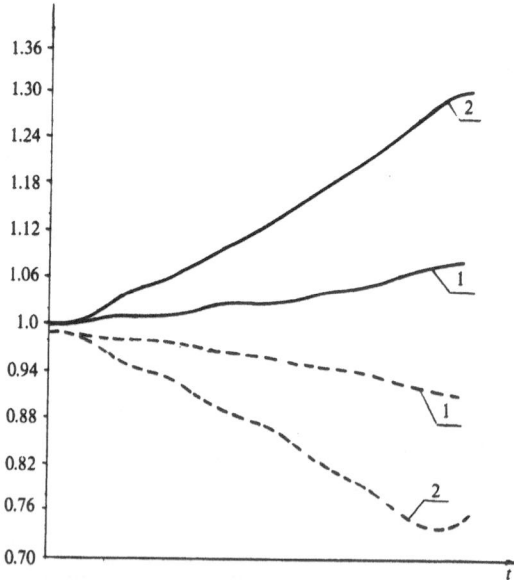

Figure 104. Initial parts of evolution of the major $a(t)$ (solid line) and the minor $b(t)$ (dotted line) for perturbations of first type (1) and second type (2).

determine the remaining disturbances δu_1, δv_2, δu_4 from the conditions $\delta C_1 = \delta C_2 = \delta C_3 = 0$ (in the linear approximation). For the disturbances of the second type (where at the moment $t = 0$ we introduced small corrections only to the "velocities" $\delta \dot{u}_i$ and $\delta \dot{v}_i$) one might assume $\delta \dot{u}_1 = \delta \dot{u}_4 = \delta \dot{v}_2 = \delta \dot{v}_3 = 0$, put $\delta \dot{v}_4 = \varepsilon$, and determine $\delta \dot{u}_2$, $\delta \dot{u}_3$, $\delta \dot{v}_1$ from the same conditions $\delta C_1 = \delta C_2 = \delta C_3 = 0$.

The character of the perturbation evolution is essentially dependent on whether the parameters of a model b/a and Ω^2/A^2 correspond to stable or unstable solution (in the linear theory, see again Fig. 49). Here, as in the case of the circular disks, oscillations with the amplitude $\sim \varepsilon$, but with deepening into the region of the unstable models (within the triangle in Fig. 49), the amplitude of oscillations increases. In Fig. 104 we present, for illustration, the typical initial parts of graphs $a(t)$ and $b(t)$ for the elliptical disk with $\Omega^2/A^2 = 0.85$ and $b/a = 0.99$ for the disturbances of first and second types with $\varepsilon = 0.01$ (we note that, according to the linear theory, this model is slightly unstable).

The accuracy control of the evolution computation we perform in the same manner as for the circular disks, i.e., we check the conservation of the total energy, angular moment, and invariants C_1 and C_2 (for the example in Fig. 104 the exactness of conservation of the energy was $\sim 10^{-5}$).

PART II
ASTROPHYSICAL APPLICATIONS

CHAPTER VIII

General Remarks

A gentleman went out for
a walk along a street,
he was struck on the skull:
a flower-pot had dropped on him,
thrown down by some hooligans!

But, to tell the truth there were no
hooligans anywhere around,
as today all hooligans
were sitting in Prevel's Hall
listening to Mozart.

Therefore it was the wind.
But that again is not true,
as the wind is abominably inert
and calm today and could not
topple the flower-pot.

But nevertheless
the gentleman going for a walk
has been struck on the skull.
How! Apropos of nothing?
How! Without the slightest cause?

It was incomprehensible, and the
profundity of this problem no one
could measure.

The gentleman applies
an antiseptic plaster
to his bruised skull
and does not believe
in anything since. . .

REMON CENO

Translated by T. G. Galenpolskii
with M. E. Fridman and F. Ya. Shanebaum

§ 1 Oort's Antievolutionary Hypothesis

The study of equilibrium and stability of different gravitating systems must evidently play an important role in the construction of the picture of their evolution. The instants when the system loses its stability are defined by the critical points of the line of evolutionary development, when the smooth evolution of the system should be replaced by a rapid reconstruction. Some authors [195] indeed observe some traces of jumplike transformations which have occurred in our Galaxy. On the other hand, the smooth evolution must evidently follow the sequence of quasistationary states. On the line of evolutionary development, the quasistationary states present such points, to which all the stellar systems observed in sufficient abundance must correspond. This is similar to the situation in the world of stars [150], where the spectrum-luminosity diagram shows the presence, in appreciable quantities, of only stars in the regions of the diagram familiar in its stationarity (above all, on the Main Sequence): the regions corresponding to strongly nonstationary states are very rapidly "rushed" by the evolving system.

As is well known, the theory of stellar evolution was constructed on the basis of numerous calculations of their equilibrium configurations with different values of parameters (mass, chemical composition, etc., see, e.g., [150]). The major role in the stellar evolution is played, as is now obvious, by burning out the nuclear fuel which produces a slow increase in molecular weight of the stellar matter. It prescribes the direction of the evolution. The rate of this process is determined by the luminosity of the star, which may be calculated at each given instant from its quasiequilibrium configuration. Thus one may trace the evolution of the star along the sequence of the quasi-equilibrium states (with consistently changing chemical composition) up to the exhaustion time of nuclear fuel, after which must follow a "disruption" with the rapid transition to the next quasiequilibrium phase, etc. At the present time, the evolutionary "tracks" of stars with different masses have been restored up to very advanced stages of evolution.

Such a way of constructing the stellar system evolutionary theory is, in principle, possible. However, in this case, we do not yet know the real *cause* of evolution, which would be, in its degree of reliability, to some extent similar to the burning out of hydrogen in stars of the Main Sequence. For example, evolution has been studied in great detail, which is due to stellar dissipation from galaxies. It is not quite obvious, however, that this dissipation does indeed occur at a sufficient rate. A similar mechanism suggests a constant restoration of the high-energetic "tail" of the distribution of particles in velocities—a process which is quite natural only in a collisional system.

More frequently, when speaking about galactic evolution, the possibility of evolution along the "tuning-fork" Hubble sequence (Fig. 105) was implied, with conversion of elliptical galaxies to spiral galaxies, or vice versa. Recently, the question of such an evolution of the present-day shapes of galaxies (of its real effectiveness) is again under consideration. It should be noted, however,

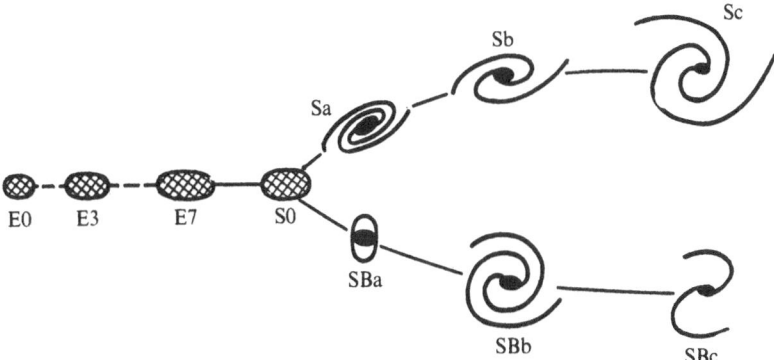

Figure 105. "Tuning-fork" diagram of galaxies by Hubble.

that as a result of prolonged discussions about the evolution of present-day galactic shapes, most investigators were inclined to a quite unexpected, "antievolutionary" point of view, in all likelihood first suggested by Oort [297]. According to Oort, the mean characteristics, say, of spherical and elliptical galaxies (such as mass, specific angular momentum, etc.) are so different that their mutual conversion is practically impossible. This is again quite similar to the causes of the rejection of old ideas of stellar evolution *along* the Main Sequence, which would require too great a mass loss. The above, of course, does not prohibit evolution inside each of the shapes. It is also not excluded that the collisionless evolution (relaxation) of the system may, at certain times, lead to instability—"disruption".

§ 2 Is There a Relationship Between the Rotational Momentum of an Elliptical Galaxy and the Degree of Oblateness?

The antievolutionary point of view of Oort agrees well with the current view of the nature of elliptical galaxies and, primarily, with the reason for their oblateness. Until recently (about 1975) there have been no *systematic* measurements of the rotation velocity of elliptical galaxies. Nevertheless, the observed oblateness of each E-galaxy was attributed to the quantity of its rotation momentum. In this we see the traditional relationship between oblateness and rotation in liquid or gaseous configurations with isotropic pressure. In addition, the rotation-produced dynamical form of self-consistent models (Gott [95[ad]], Larson [96[ad]], and Wilson [351]) matched well the surface brightness distribution in E-galaxies and it was believed that the models correctly predicted the oblateness of the galactic figure, depending on the value of the ratio of the maximum rotation velocity v_{rot} to $\langle v \rangle$—the velocity dispersion of chaotic motions of stars. However, when it became possible to measure rotation in a great number of E-galaxies (see Illingworth [97[ad]] and the references therein) a surprising fact was found—the value $(v_{rot})_{max}/\langle v \rangle$

was, on average, one-third of that predicted by the models. For example, for the galaxies NGC 4406, 4621 and 4697, the theory predicted this ratio to be 0.5, 0.8, and 0.8. Observations, however, give 0.12, 0.20 ÷ 0.36, and 0.30 ÷ 0.45, respectively.

A question arises: if not rotation, what then plays the determining role in the formation of the figures of elliptical galaxies? Perhaps the oblateness is due to sufficiently strong anisotropy of stellar "pressures": the velocity dispersion in the equatorial plane must be more than the velocity dispersion in the direction of the minor axis. The first phase models of the simplest (homogeneous) ellipsoidal systems, the shape of which is determined primarily by pressure anisotropy (so that, for example, the stellar ellipsoid at rest may have an arbitrary degree of oblateness) were constructed [117]. As far as the question of the (*evolutionary*) *origin* of elliptical galaxies (and, in particular, their oblateness) is concerned, there may presently be several different points of view, but we shall deal with only one here.

As will be shown below (in §5); the fundamental difference of elliptical (and other) galaxies from liquid and gaseous gravitating configurations consists in the fact that the former (with a good approximation) are collisionless. The originally available anisotropy in the velocity dispersion of stars (at the time of birth of such systems) cannot be canceled completely as a result of their evolution (by collective processes, see Chapter X). Therefore, it is natural that Binney's theory follows, advanced by him in 1976 [58[ad]], which states that the reason for compression of elliptical galaxies is due to residual anisotropy of velocity dispersion of stars.[1] By applying the tensor virial theorem to systems, the density of matter is constant on ellipsoidal surfaces similar to each other, i.e., has the form $\rho(m^2)$, where $m^2 = x^2/a^2 + y^2/b^2 + z^2/c^2 \leq 1$, Binney [98[ad]] studied the influence of the anisotropy on the oblateness of the models of the oblate and prolate spheroids. Comparing the conclusions of the theory with observational results, he concluded that the oblateness of elliptical galaxies is not directly associated with their rotation, but is due to some anisotropy of velocity dispersion. Schechter and Gunn [94[ad]] measured rotation of another twelve elliptical galaxies and, comparing their observational data with Binney's models, arrived at similar conclusions: observations rule out all the models with isotropic pressure (both axisymmetric—oblate spheroid, and nonaxisymmetric—prolate spheroid or the three-axis ellipsoid).

Despite the above serious arguments we are not inclined to think that the problem of rotation, in view of the question of the value of oblateness of elliptical galaxies, should be buried in oblivion.[2] The models of Binney

---

[1] We have already pointed out that such a possibility is not the only one. For example, to some extent a contrary point of view [118[ad]] is, generally speaking, not prohibited, which assumes expansion out of a strongly oblate stellar "pancake". In this case, the evolution also ceases at a finite value of velocity anisotropy of stars. Below, in §5, Chapter IX, a possibility is pointed out of forming the elliptical galaxies due to large-scale instability in a spherical stellar cluster, which is compressed out of a state being far from equilibrium.

[2] Due to the dependence of the value $v_{rot}/\langle v \rangle$ on the form of the models; for example, for models with incompressible density $v_{rot}/\langle v \rangle = 0$ at any v_{rot}; see also below.

[98[ad]] have one difference from real galaxies which looks insignificant on the face of it but, however, provided accurate calculation, can give an additional numerical factor of the order of 3. The point is that Binney [98[ad]] restricted himself to considering a special case, where the layers of equal density are ellipsoids similar to, and concentric with, the boundary ellipsoid. It may be shown that in models with such layers of equal density, the ratio of the rotational energy and the gravitational energy is independent of the law of matter density distribution. For example, for axisymmetric models with similar layers and an arbitrary density distribution, this ratio coincides precisely with a similar ratio for the classical uniform spheroid of Maclaurin. If, however, the layer oblateness is changed, this ratio depends on concentration of matter density toward the center. In E-galaxies, the oblatenesses of isophotes, as shown by observations, are not constant but vary with distance from the center of the systems.[3] The observed change of the isophotes' oblatenesses means that layers of equal density in E-galaxies are not similar to each other. Recognizing the fact that taking into account this latter fact should lead to some numerical result, which however may turn out to be qualitatively decisive, Binney in his next paper [99[ad]] applied the tensor virial theorem to the subsystem of elliptical configuration for taking into account the oblateness profile effect on the rotation curves of the *axisymmetric* model. Without going into details of calculations which were later made by B. Kondrat'ev note also that the influence of the deviation of layers of equal density from the similar and concentric ellipsoids on the values of mass and gravitational energy proves to be more effective in the *prolate* spheroid model than in the *axisymmetric* model. Further detailed numerical analysis will probably allow one to come to know the particulars of the question of primary importance for the understanding of the whole picture of the evolution of elliptical galaxies.

§ 3 General Principles of the Construction of Models of Spherically Symmetric Systems

The models of collisionless systems get, with time, still more complicated since finer details are taken into account and attempts are made to provide an explanation of the new observational data. In particular, a very large number of papers is devoted to the construction of models of spherically-symmetrical stellar systems. Therefore, quite a few models of such systems are known at present (cf. review [32]).

It is clear that in the future most of the models will be of academic interest only. Direct observations do not yet give unambiguous information regarding the velocity distribution of particles. On the other hand, as we have already noted, the problem of the construction of the distribution functions from the given density distribution is ambiguous. In reality, the problem

---

[3] The same effect is observed in globular clusters [73].

under consideration must be solved in parallel with the problem of the evolution and origin of these systems. This may lead to an essential restriction of numerous stationary possibilities or even to an unambiguous solution. In the case of spherical star clusters such work as was performed by Michie [293], King [260], Henon [214] and others, has already led to essential progress. The difficulty is, however, that we have not yet obtained a detailed solution of the fundamental problem of evolution.

Therefore the real settlement of the problem of the theoretical description of quasistationary systems admits at present a very significant arbitrariness. In the construction of the models, only some "evolutionary considerations" of the qualitative character are employed. It may be said that now nearly any distribution function satisfying only some natural "criteria of reasonableness" is suitable. Among them, the stability condition is of course necessary.

Very frequently, the observed data on the distribution of brightness and the number of "particles" in different spherically-symmetrical systems are compared with simple theoretical models of the type considered in §1, Chapter III. For example, Kamm [180] obtained quite a good coincidence of some generalized polytropes of his (25), §1, with the observed data for the globular clusters M5, M15, M92. The isothermic model, somewhat corrected on the system periphery, describes well, according to Zwicky, some clusters of galaxies. These questions are dealt with in more detail in Veltmann's review [32].

The "most probable" distributions are also constructed [101], often without indicating a specific statistical mechanism, which must lead to their formation in a real collisionless system.

§ 4 Lynden-Bell's Collisionless Relaxation

In this respect, the paper by Lynden-Bell [286], who had considered the problem of the collisionless relaxation in a vigorously nonstationary process of formation of the equilibrium state, differs advantageously from other papers. Qualitatively, the course of Lynden-Bell's considerations is as follows. Assume that initially we have a very nonequilibrium configuration, i.e., the system "starts" sufficiently far from equilibrium (for example, the virial theorem is strongly violated). In the phase space, some regions are occupied by particles while some are vacant. As a result of the interaction via the self-consistent field, a chaotic intermixing of the elements in the phase space will occur. The suggestion about the violent character of relaxation is necessary to Lynden-Bell in order that he might assume that the system, by reaching the final equilibrium, has indeed been well intermixed so that any typical element is, with the same probability, to be found in any place of the phase space (with some general limitations). In this case, one may perform a corresponding statistical calculation (such calculations are always based on the assumption about the equal probability of certain states): the equilibrium state must be the most probable with the limitations mentioned.

Any final equilibrium state, attainable in principle by the system, must have the same total energy E, mass M (or the total number of particles N), the total angular momentum L and the total impulse P. If only these limitations are imposed, the statistical mechanics will give the Maxwellian distribution (in a respectively moving and rotating coordinate system).

But the collisionless character of the systems under consideration imposes an additional constraint, since it means the conservation of one more value — the phase density. The flow of the phase "fluid" is incompressible, and therefore must occasion the "exclusion principle": the distribution function in the given element of the phase space is either zero, if a cell without particles has arrived there, or is equal to the original value in that cell, which after intermixing has arrived at a given point. If at the initial time the distribution function was unity throughout the region, where the particles were present, then we obtain Pauli's exclusion principle. In this case, the final equilibrium function must coincide exactly with the Fermi–Dirac distribution:

$$\bar{f} = \frac{1}{e^{\beta(E-\mu)} + 1}.$$

In this formula, β and μ are the constants determined by the total number of particles N and the energy E of the system (here they play the same role as the reverse temperature $1/T$ and the chemical potential μ in a "usual" Fermi distribution), while the bar over f means that this form must not indeed be tended to the exact distribution function (which in the course of time becomes still more "cut") but the one averaged over the small energy intervals. Lynden-Bell believes that his new statistics must explain the remarkable regularity which we observe in the light distribution of spherical and elliptical galaxies. Note, however, that Lynden-Bell's mixing mechanism itself is effective only provided that instabilities are absent in the evolving system (for details, see §7, Chapter IX).

§ 5 Estimates of "Collisionlessness" of Particles in Different Real Systems

By using the observed data, consider, first of all, in what degree all the systems described below are really collisionless.

The problem of collisions in stellar systems was studied in detail by many authors, beginning with Chandrasekhar [147]. The time of establishment of the quasi-Maxwellian distribution due to collisions of particles of one sort (the time of collisonal relaxation) may be written in the form (see, e.g., [101])

$$\tau \sim \tau_0/\lambda \tag{1}$$

The value τ_0 in this formula has the meaning of a mean time between close collisions of particles, leading to an essential change in the direction of motion (scattering by an angle $\theta \sim 1$):

$$\tau_0 \sim \frac{v^3}{G^2 n_0 m^2},\tag{2}$$

where v is the mean velocity of "thermal" motion of particles, n_0 is their mean density, and m is the mass of one particle.[4] The second value in (1), λ, is the so-called "Coulomb logarithm", taking into account the fact that in systems with far-acting forces (gravitational and electrical) the main role is played by far passages of the particles. To make estimates in application to such objects as galaxies, it is generally assumed that $\lambda \sim 20$. Since the Coulomb logarithm λ is a slow function of v, n_0, the quantity $\lambda \sim 20$ also changes little in the transition to other objects which are considered here.

We have already given the parameters typical for galaxies: the number of stars $N \sim 10^{11}$, the radius $R \sim 10$ kps, therefore the density of the number of stars must be $n_0 \sim 10^{-57}$ cm^{-3}. Since according to the virial theorem it would be $v^2 \sim GM/R$, thus we obtain $v \sim 200$ km/s. Since finally $m \sim M_\odot = 2 \cdot 10^{33}$ g, then by substituting all these values into (2), we shall find $\tau \sim 10^{15}$ years. Accordingly, $\tau_0 \sim 10^{17}$ years.

The estimates obtained for the time between the pair collisions of stars may be compared with the lifetime of the Universe $T \sim 10^{10}$ years, and we arrive at the conclusion that for each star, during the lifetime of the Universe, none of the collisions are likely to have occurred, so that the relaxation of the distribution of stars in velocities could not be collisional. In this they differ from *globular clusters*, which are essentially denser formations. For a typical spherical cluster, the number of stars $N \sim 10^5$, while the radius $R \sim 10$ ps, so that by formula (1) for τ we have the value of only an order of a billion years. Therefore for the lifetime of spherical clusters, $(4 \div 8) \cdot 10^9$ years, a sufficient number of collisions might have occurred, and as a result the phase distribution of these systems could have relaxed toward the most probable (quasi-Maxwellian) distribution in the "usual way". In reality, however, the situation here is not so simple, due to a strong non-homogeneity of spherical clusters (the other complicating factor is the possibility of stellar dissipation from the system). Therefore it would be more correct to determine the *local* times between the collisions dependent on the radius. Then the system turns out to be collisional (and consequently, Maxwellian) in its central part, but collisionless (and therefore with an *a priori* unknown distribution function) on its periphery. The distribution function on the periphery must transform smoothly to the Maxwellian function at the center, which in this case is the boundary condition lacking in the case of galaxies, which are "purely" collisionless systems.

---

[4] Similar expressions in plasma physics ensue by substituting the effective gravitational "charge" $m\sqrt{G}$ by the electric charge e.

It should be stressed here that the "collisionness" (at least partial) of globular clusters manifest itself only on long time intervals of the order of the lifetime of this system. For example, as far as small *oscillations* of these systems are concerned, they are characterized by quite other times: by the periods of orbital revolutions (or radial oscillations) of the stars, which have the order $t \sim R/v \sim 10^6$ years. This value, on the contrary, is much less than the time between the collisions, so that in the study of collective oscillations or stability (and this is just the problem of most interest to us) the spherical clusters may be considered as *collisionless*: no collisions occur in the course of many periods of oscillations.

The requirement of the dependence of the stationary collisionless distribution function on the single-valued integrals of motion, together with the above boundary condition of matching with the Maxwellian distribution at the center, in the case of spherical clusters, greatly restricts the class of admissible distribution functions. In any event, we do not have such an *a priori* arbitrariness as in the case of galaxies.

However, for the problem of interest, the difference is probably not so deep. Indeed, observed distribution functions of even collisionless systems, like galaxies, are by no means arbitrary functions of integrals of motion. The mechanisms, which have not yet been established unambiguously, lead to the form which closely resembles the same Maxwellian distribution function (or Schwarzschield's one, i.e., anisotropic Maxwellian as in our Galaxy). However, in principle, the number of possibilities here remains large. In addition, in different cases, various mechanisms of establishment of an equilibrium distribution (of collisionless relaxation) may operate.

In case of a *system of globular clusters*, similar estimates of the average time of pair collisions yield $\tau \sim 10^{11} \div 10^{13}$ years. This time, in any event, significantly exceeds the average time of oscillations T_0, which equals a value of the order of 10^8 years.

For *compact clusters of galaxies*, these times are respectively $\tau \sim 10^{10} \div 10^{12}$ years, $T_0 \sim 10^9$ years. Therefore, such systems are also described by the collisionless kinetic equation and only near the center will this approximation become inapplicable. But the latter is already evident from the description of such clusters [146], as formations, at whose center the galaxies are in contact with each other.

Thus, all the systems described may with sufficient accuracy be assumed to be collisionless and to be described in the framework of the proper mathematical formalism of the physical kinetics.

CHAPTER IX

Spherical Systems

§ 1 A Brief Description of Observational Data

As already noted (see beginning of Chapter III), under spherical collisionless gravitating systems we understand the following objects: (1) globular clusters of stars; (2) spherical galaxies (or, roughly, elliptical galaxies with not very great oblateness); (3) systems of globular clusters (for example, in our Galaxy); (4) compact clusters of galaxies.

1.1 Globular Star Clusters

The list of globular clusters belonging to our Galaxy is about 200, although in reality their number may be much greater (according to estimates of Saar up to 500, according to other estimates up to 2000 [122]). They form a system with a strong concentration toward the center of the Galaxy and with approximately spherical density distribution.

The counts of the number of stars in globular clusters show that the density n is a rather decreasing function with a radius of approximately $n \sim 1/r^3$. The typical dimensions of the globular clusters are $R \sim 10$ ps, the masses lie in the region of several hundred thousands of solar masses. For example, the mass of M3, estimated by Sandage, is $2.45 \cdot 10^5$ $M_\odot$ [122]. Johnson [243] has derived the lower limit for the mass of globular clusters from a rough model of the tidal equilibrium; it also proved to be of the order of 10^5 $M_\odot$.

1.2 Spherical Galaxies

Let us now describe (also very briefly) some observed data which refer to *spherical* and *elliptical galaxies*. The number of stars in giant galaxies of this type are, as in giant galaxies of other types, $N \sim 10^{11}$ (up to 10^{12}), the radius $R \sim 10$ kps. It is typical that they completely lack any structural details, except for small very condensed nuclei. It is believed that for all elliptical galaxies there is a common law of surface brightness distribution, and accordingly, stellar density distribution. In this connection the suggestion appears to be very natural that all elliptical galaxies are constructed according to the same general model, while individual objects differ only in their size, density, and degree of oblateness. The law of surface brightness distribution was found by Hubble; it has a very simple appearance:

$$B = B_0/(r + a)^2, \tag{1}$$

or

$$\log B/B_0 = -2 \log(r/a + 1). \tag{2}$$

These formulae describe very well the observations within the range $0.3 < r/a \lesssim 15$ and are satisfactory up to $r/a = 30$ (the brightness within this interval alters 1000 times).

1.3 Compact Galactic Clusters

According to Zwicky [146], all clusters of galaxies may be divided into three classes, from which the compact clusters reveal a nearly accurate spherical symmetry. Examples may be given for giant galactic clusters in Veronica's Hair and in the Northern Corona. Their observable dimensions are $R \sim 1$ mps, while the virial[1] masses are $M \gtrsim 10^{14}$ $M_\odot$. The observed radial velocity dispersions of galaxies reach $2 \cdot 10^3$ km/s in giant clusters. The most abundant clusters account for $N = 10^4$ members. The density distribution according to Zwicky is well represented by the isothermic model in (12), §1, Chapter III; other authors suggest some different models (see [32]).

§ 2 Classification of Unstable Modes in Scales

Rephrasing slightly the known expression by A. Eddington one can say that there is nothing simpler than spherical star *clusters*. Therefore it is natural that theoretical models of spherically-symmetrical systems are most numerous compared with flat or elliptical systems. With the accumulation of the

---

[1] That is, estimated by the formula $M \sim \overline{v^2} R/G$.

observational data these models get more complicated. As their components one uses astrophysical objects of a novel nature (for example: in the center of a spherical galaxy one places a black hole with a large mass [86[ad]]), introducing high anisotropy in the stellar velocity dispersion [61[ad]], and so on. With such complications of models, it is sometimes a feat to achieve satisfactory agreement with observational data. However, at times one forgets to satisfy the main condition—namely: the condition that the system may exist at all or, in other words, that the model be stable. The stability investigation of the models of spherical systems (Chapter III) shows that many of them are really unstable.

Possible instabilities may be conditionally (and roughly) divided in two classes—large-scale (with the characterizing scales λ having the order of the system size R) and small-scale ($\lambda \ll R$). Instabilities of these two classes act completely differently. The instabilities of most large scales may cause the visible alteration of the system's geometrical form. For example, it may turn an initially spherical system into an elliptical one. Small-scale instabilities cause such effects as, for example, smoothing of temperatures in different directions provided the high temperature anisotropy is originally present. At the same time these instabilities do not have any considerable influence on the form of the system.

The stability criteria for the small-scale perturbations obviously depend on many details of the equilibrium state. On the other hand, for large-scale perturbations, which include the system as a whole, stability or instability must depend only on some integral characteristics averaged over the system. Which are these characteristics? They are, for example, the total kinetic energy of the system T, potential (gravitational) energy W, the energy of rotation T_{rot}, and so on.

§ 3 Universal Criterion of the Instability

For global instabilities it is therefore possible to formulate universal stability criteria—that is so that the criteria remain valid for different models, including models vastly different each from other (for example: with completely different distribution functions over velocities, different densities $\rho_0(r)$, and so on). It is natural to look for these universal stability criteria in energy terms. At any rate, Peebles and Ostriker [301] formulated the universal stability criterion for highly flattened (disklike) systems with respect to elliptical deformations in such a way (see §5, Chapter IX).

The large-scale modes correspond to the widest instability region. Indeed, if the trajectories of all stars in the sphere are purely radial this sphere is unstable according to Jeans, with respect to the perturbations of any scale which are transverse to the radius (§5, Chapter III). If there is some thermal dispersion in the transversal velocities of particles $v_\perp$, this means the appearance of corresponding transversal Jeans size $R_D \sim v_\perp/\omega_0$ (ω_0 is some average Jeans frequency), and consequently all the disturbances with transversal scales $\lambda_\perp < R_D$ become stabilized. With an increase of $v_\perp$ the value R_D increases too so that more large scales become stable. It is clear that stabilization of the largest scales ($\lambda_\perp \sim R$) is the most difficult: for this the highest transversal velocities are necessary. If the velocity dispersion increases so that the system becomes isotropic over the velocities, one simultaneously gets complete stability (§2, Chapter III). All the systems with $v_\perp > v_r$ are also stable (besides some special cases—§3, Chapter III); at the limit $v_r \to 0$ we obtain here systems with purely circular orbits.

   The stability boundary lies between isotropic systems and systems with purely radial motions of stars. Quantitatively this boundary is determined in §7 of Chapter III, where we summed up stability investigations of the spherically-symmetrical collisionless systems (see Fig. 40). The main result consists in the formulation of the universal stability criterion which is probably valid for the very wide class of spherical systems:

$$\xi \equiv T_r/(T_\perp/2) < \xi_c = 1.70 \pm 0.25, \tag{1}$$

where T_r, $T_\perp$ are, respectively, total kinetic energies of the radial and transversal motions of particles, ξ is a global anisotropy.

§ 4 Specificity of the Effects of Small-Scale and Large-Scale Perturbations on the System's Evolution

Let us assume that we have a slightly unstable system, e.g., $\xi \gtrsim \xi_c$. We may then ask how such a system will evolve? Which perturbations begin to grow first of all? Possible forms of perturbations can be simply enumerated. First of all these perturbations must correspond to the most large-scale modes, e.g., those with the minimal number of radial nodes (or even those without nodes) and with maximally smooth angle dependence. This last is determined by an index l of spherical harmonics $Y_l^m(\theta, \varphi)$; $l = 0$ corresponds to radial perturbations; $l = 1$ and $l = 2$ correspond to perturbations with the symmetry of dipole and quadrupole types, respectively. But radial perturbations, for the distributions, decreasing with the energy (which are only interesting) are always stable. Therefore only two types of perturbations compete with each other: the elliptical deformation of the sphere (e.g., the mode of the quadrupole symmetry without nodes) and perturbations with $l = 1$ having a minimal number of radial nodes. But the nodeless perturbation with $l = 1$ is trivial—this is simply displacement of the sphere as a whole; all the remaining perturbations must have radial nodes. In all the cases considered in §6, Chapter III, the instability began from the elliptical deformation of the sphere ($l = 2$).

   Note that solutions of the problem of small perturbations of the sphere, which corresponds to the stability boundary ($\omega^2 = 0$), determine neighboring equilibrium states—collisionless ellipsoidal systems with small oblateness. It is possible to construct in this manner a large number of various stellar stationary ellipsoidal models from the results obtained in §6, Chapter III.

(One similar model was suggested by Hunter [69^{ad}].) In this connection we must note that even those galaxies which are usually considered as purely spherical (E0), in reality show a visible ellipticity of isophotes, moreover, the latter is increasing from the center to the periphery. If we assume that the average ratio of mass/luminosity is approximately constant then it will be necessary to allow such asymmetry in the mass distribution. All these peculiarities may probably be explained in a natural manner if we construct, by the method mentioned above, a model which is close to the spherically symmetrical model. As an initial spherical model one can take, for example, one of King's models [76^{ad}] which are isotropic at the center but have considerable anisotropy on the periphery.[2]

§ 5 Results of Numerical Experiments for Systems with Parameters Providing Strong Supercriticality

If an anisotropy ξ is essentially more than a critical one, then more small-scale perturbations also become unstable (apart from the ellipsoidal mode). The manner of evolution of the system in the case of initial high anisotropy may be complicated; this evolution cannot be predicted in full by the linear stability theory. However, the modeling of such systems by computer shows (see, for example, §5, Chapter III) that considerable elliptical distortion of the system's form often develops in this case also. It may be understood as follows. The spectra of initial perturbations include deformations corresponding to both the small-scale modes and to most large-scale modes. Moreover, though the small-scale instabilities develop more quickly, the heating due to these instabilities is far from sufficient to stabilize the large-scale modes, and principally to prevent elliptical deformation. Indeed, roughly speaking, one can assume that the development of the instability at a certain mode draw the system at the stability boundary of *this mode* (the stars get warm, so as to cause the suppression of instability). But we already noted that the widest region of instability just corresponds to the ellipsoidal mode ($l = 2$); hence this mode will increase, even when increasing of all the other modes has ceased.

In principle, there is the possibility of the formation of highly oblated configurations due to development of large-scale instabilities, of which the elliptical deformation of the system is the most apparent. In this connection we must note that some clusters of galaxies (for example, Coma) show very considerable oblateness [87^{ad}]. The same mechanism may be a cause of formation of highly oblate elliptical galaxies.

Analytical consideration of the stability problem of contracting collisionless cloud is very complicated. Here methods of statistical modelling

---

[2] It is assumed [77^{ad}, 86^{ad}] that King's models give a satisfactory description of the main observational properties for spherical galaxies and for ellipticals with small oblateness.

(different modifications of N-body methods) are more effective. In the work of Peebles [83[ad]] he examined the evolution of the initially cold spherical system with $N = 100$–300 pointlike masses, which must model the cluster of galaxies (Coma). In that work, however, the question of the form of the system obtaining under collapse is not especially considered. From the figure given in [83[ad]] one can see that the system forming during collapse has some oblateness. However, the oblateness can also be considerable: one projection (which is not in all probability most happy) is evidently not enough on which to judge the form of the system. Further examination of this interesting question is necessary. We can obtain the answer if we simultaneously examine several projections, follow the evolution of a quadrupole moment of the system, and so on.

In principle, the final picture can depend greatly on a spectrum of initial perturbation, in particular from the relative amplitude of the initial elliptical distortion of a system's form (or from the value of an original quadrupole moment). Generally, it is necessary to note that the problem of initial perturbations, and primarily the problem of its amplitudes, are decisive (and also the most indefinite) for all the considered tasks. In particular, the following question has principal significance. Is the cause of the initial deviations from exactly spherical form only purely thermal, statistical fluctuations, or are these deviations essentially stronger? (The last possibility, to our mind, is much more probable.)

If the level of fluctuations is purely thermal we must not expect formation of considerable ellipticity during the collapse of the future galaxy (when $N \sim 10^{10} \div 10^{12}$).[3] However, in the case of the formation of galaxy clusters (the process which was considered by Peebles [83[ad]]) the final oblateness can be considerable even for the purely statistical nature of the initial fluctuations.

§ 6 Example of Strongly Unstable Model

Thus, at the present time there is already a stability theory of spherically-symmetrical collisionless systems which is sufficient in order that one could, with a large degree of confidence, form a conclusion as to its stability or instability for practically any given model. For this purpose it is necessary to calculate the value of global anisotropy of the system and then compare it with critical 1.70. It is clear that strongly unstable models cannot be used as models of real systems.

Let us give an example of just such a situation. Recently, great interest was attached to the work of Sargent's group [86[ad], 90[ad]], in which an anomalously

---

[3] Note in this connection the paper by Miller [81[ad]] where it is shown in particular that during the first phase of the collapse of a quickly rotating cold spherical system of $N \simeq 10^5$ pointlike stars, the barlike instability has no time for considerable development. The reason is a rather small initial amplitude of elliptical deformation of the sphere under purely-statistical playing for such a number of points.

quick increase of the luminosity and stellar velocity dispersion at the central region of spherical galaxy M87 was discovered. An attempt at the explanation of these data within the frame of the standard spherical models (line to King's models), which are isotropic near the center, was not a success. After this the authors suggested that at the center of the above-mentioned galaxy, in the region the size of $r_0 \lesssim 100$ pc, there is a black hole or some other object with mass $M_h \sim 5 \cdot 10^9 \, M_\odot$. Below we will consider some problems appearing in connection with this possibility. It is, however, important that for the present we do not exhaust the more simple variants which the authors of other work [61^{ad}] brought to our attention. They suggest, for a description of the central region of galaxy M87, the highly anisotropic star velocity distribution instead of the standard isotropic one. The equilibrium state for the spherically-symmetrical, anisotropic collisionless system satisfies the equation (as we know from §1, Chapter III)

$$\frac{d}{dr}(\rho c_r^2) = -\rho \, \frac{GM(r)}{r^2} - \rho \left(\frac{2c_r^2 - c_\perp^2}{r} \right). \tag{1}$$

For the isotropic case the last term in (1) vanishes so that we obtain the equation which was employed by Sargent et al. [86^{ad}] under analyses of the mass distribution near the center of M87. If the tangential velocity dispersion at finite distances from the center r drops below the radial dispersion, the last term in Eq. (1) becomes negative. But this has evidently the same qualitative effect as adding a central mass to the first term on the right-hand side of Eq. (1). The effect is a steepening of the density gradient compared to the isothermal solution.

The specific computations were performed in [61^{ad}] for anisotropic Maxwellian distribution

$$f_0(E, L) = \text{const } e^{-(2E + k^2 L^2)/2c_r^2}. \tag{2}$$

The phase density (2) corresponds to the local anisotropy

$$c_r^2/(c_\perp^2/2) = 1 + k^2 r^2, \tag{3}$$

so that k^{-1} means the radius at which the anisotropy begins to be considerable.

Substituting the volume density $\rho_0 = \int f_0 \, d\mathbf{v}$ into the Poisson equation, we obtain

$$\frac{1}{x^2} \frac{d}{dx} \left(x^2 \frac{d\psi}{dx} \right) = \frac{e^{-\psi}}{1 + \tilde{k}^2 x^2}, \tag{4}$$

where we introduced the notation $\psi = \Phi_0/c_r^2$, $x = r/\alpha$ for the dimensionless potential radius, $\alpha = c_r (4\pi G \rho_c)^{-1/2}$ is the scale of length, ρ_c is the center density, $\tilde{k} = k\alpha$.

For the model with $\tilde{k} = 0.98$, authors of the work under consideration obtained very good agreement with observational profiles of $\rho(r)$ and $\sigma(r)$ up to $x \simeq 10$.

It is easily seen however that in spite of this such a model cannot correspond to any real system, because the model is very unstable. Indeed, the simple computation of the quantity $\xi(r) \equiv 2T_r(r)/T_\perp(r)$ where

$$T_{r,\perp}(r) = \frac{1}{2} \int_0^r d\mathbf{r} \int v_{r,\perp}^2 \, d\mathbf{v},$$

gives the following values

x	2	4	6	8	10
$\xi(x)$	2.39	4.23	5.87	7.33	8.67

Hence the instability of the model considered jointly with L. M. Ozernoy is obvious (recall that the critical value[4] for the global anisotropy $\xi_c < 2$).

§ 7 Can Lynden-Bell's Intermixing Mechanism Be Observed Against a Background of Strong Instability?

We talked above about some applications of the stability theory of stationary spherical collisionless systems. But it is no less interesting to ask which role possible instabilities may play in the evolution of the system initially far from the equilibrium. For example, one of the most important problems for the theory of galaxy formation is the problem of collapse of the spherical cloud of noncolliding particles. In this case practically all the energy released during the contraction converts into radial motion. Consequently, it is possible that strong instability may develop. Lynden-Bell, in the known work [286] devoted to the collisionless relaxation, probably bore in mind just such a process of collapse. However, his mechanism of mixing in the strongly nonstationary system is probably far less effective than the influence of instabilities which must be present here. The mixing mechanism of Lynden-Bell, in its pure form, acts effectively only when the instabilities are absent. (This aspect was emphasized by Kadomtzev [15^{ad}].)

§ 8 Is the "Unstable" Distribution of Stellar Density Really Unstable (in the Hydrodynamical Sense) in the Neighborhood of a "Black Hole"?

In conclusion we dwell upon effects which could be due to the presence of a "black hole" or another compact, massive body at the center of the stellar spherical system (we shall speak of "hole" in both cases for the sake of brevity).

---

[4] Note here that the scales $\lambda \lesssim 2$ may also be stable although $\xi(2) > 2$. Indeed, the stability criterion for perturbations with small scales must involve the parameter $\tilde{\xi}(x)$ determined by some different manner compared to $\xi(x)$: the contributions into T_r, $T_\perp$ from the stars with orbits, which move far away from the region of perturbation $(x_{max} \gg x)$, need to be omitted. However, for perturbations of the larger scales this effect is not essential.

Thus, let us suppose that at the center of the galaxy a "hole" really was formed. It is clear that a "hole" forms from those stars with small angular momenta and, hence, these leave the system of stars surrounding the central body. As a result, the immediate vicinity of a "hole" is filled by stars with nearly-circular orbits, which, in the region of sufficiently small angular momenta, leads to the condition $\partial f/\partial L > 0$. As noted in §4, Chapter VI, this circumstance may, at the appropriate sign of the wave energy, cause a kinetic loss-cone instability (for details see [39[ad]]). Here we will deal with possible hydrodynamical effects of the "hole" rather than with the kinetic effects. Since in the vicinity of the "hole" only stars with orbits close to circular are left (the rest of them are absorbed by the "hole"), there must appear in this vicinity a density distribution $\rho_0(r)$ with the increasing section near the "hole", as shown in Fig. 106(a).

In [106], it is shown that a spherically-symmetrical system with circular orbits of stars (Einstein's model, Fig. 2) can be unstable if its macroscopic density increases towards the edge (see §3, Chapter III), In [20], this instability was investigated for the first time in the system of two collisionless cylinders rotating in opposite directions. This instability may be caused by the centrifugal force, which, under the condition of density drop towards the system center, is not compensated by the gravitational force. In this case, it is likely that the presence of a massive compact body in the system center will stabilize the instability.

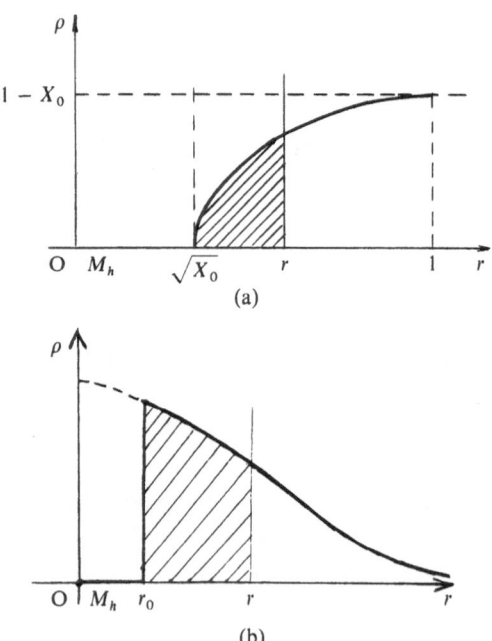

Figure 106. Stellar density near the "black hole": (a) the "initial" model is the homogeneous model of the second Camm series; (b) the "initial" model is the system with purely-circular orbits.

Thus, we should clear up the question: does a "hole" lead to such a distribution of density $\rho_0(r)$ in its vicinity which turns out to be unstable in the sense of [20, 106], or is the "hole" mass sufficient to stabilize this instability? We already investigated this problem in §3, Chapter III. Let us recall the main results. For the systems with purely-circular orbits the instability of the localized disturbances begin, provided that [20, 106]:

$$\frac{2\Omega}{\varkappa} \equiv \mu < 1, \qquad (1)$$

where $\varkappa$ is the epicyclic frequency, $\varkappa^2 = 4\Omega^2 + r(\Omega^2)'$; the criterion (1) corresponds to mode $l = 2$ which is the most unstable mode.

In Section 3.4, Chapter III the stability of the mode $l = 2$ was also investigated for the case when there is the finite radial velocity dispersion; then the picture represented in Fig. 20(a) occurs. The stability boundary $\mu = 1$ does not alter; all the scales become unstable for $\mu \to 1$. More obviously the instability condition (1) can be formulated as: for the instability of the perturbation localized near r it is necessary to fulfil the inequality

$$\rho_0(r) > \bar{\rho}, \qquad (2)$$

where the average density

$$\bar{\rho} = \frac{M(r)}{(4\pi/3)r^3} ; \qquad (3)$$

$M(r)$ is the total mass within the sphere of radius r. If at the center of the system the "hole" was absent, then for the density $\rho_0(r)$ increasing from the center we had the instability according to (2). However, it is essential that mass of the "hole" must be included in $M(r)$. Due to this, those mass distributions which form together with the "hole" are stable. It is especially evident in two limiting cases:

(1) purely-circular orbits (see Fig. 106(b)): a "hole" is, in this case, formed by the particles with $r < r_0$; from the figure it is evident that $\bar{\rho} > \rho$,
(2) nearly radial orbits; in this case a "hole" with a very large mass must form; the absence of an instability is rather obvious.

It may be shown that in the general case, for the mass distribution forming at the vicinity of the center, the instability criterion (2) does not satisfy.

We can present some results of the concrete calculations concerning a structure in the vicinity of the "hole". Let us take, for example, the simplest model—the "hole" in the initially homogeneous sphere with the distribution function (29) §1, Chapter III:

$$f_0 = \frac{\rho_0}{\sqrt{2}\pi^2} \left[\frac{L^2}{2} - E - 1 \right]^{-1/2}, \qquad (4)$$

where ρ_0 is the density, and the radius of the sphere we put to be equal to unity. The phase region of the system is represented in Fig. 107(a). For our

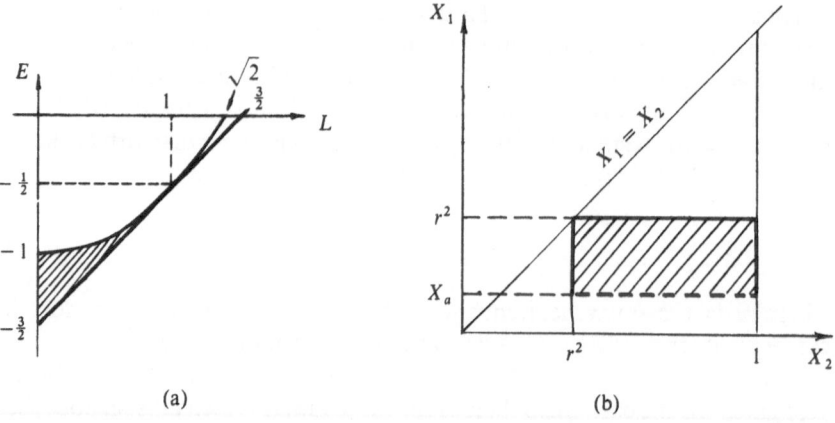

Figure 107. Phase region (shaded) for the homogeneous (in a density) model of the second Camm series: (a) in variables (E, L); (b) in variables (x_1, x_2).

aims it is natural to transform from E, L to other variables: $x_1 = r_{min}^2$, $x_2 = r_{max}^2$, where $r_{min}/r_{max}(E, L)$ are the minimum and maximum distances of the particle with given E, L from the center. The distribution function in these variables

$$F = 2\pi \left(\sqrt{\frac{1 - x_1}{1 - x_2}} - \sqrt{\frac{1 - x_2}{1 - x_1}} \right),$$ (5)

and the phase region of the system transforms into the triangle OAB in Fig. 107(b).

Suppose now that in this system a "hole" is formed due to falling into the center of those stars which had $r_{min} < r_0$, or $x_1 < r_0^2 \equiv x_0$. As a result, all the stars corresponding to the trapezium OCDB in Fig. 107(b), will form a "hole" with some mass M_h in the center O, and the remaining stars, occupying the triangle CDA on the phase plane (x_1, x_2), determine the modified distribution of the volume density $\rho(r)$.

Let us calculate M_h and $\rho_0(r)$. In this case the calculation is not complicated and may be carried out analytically up to the end. For the mass of the "hole" we obtain

$$M_h = \frac{8\pi}{3} \left(x_0 - \frac{x_0^2}{2} \right),$$ (6)

and for the density

$$\rho(r) = \frac{1}{r} \sqrt{(1 - x_0)(r^2 - x_0)}.$$ (7)

The latter is presented in Fig. 106(a). The mass $M_1(r)$ inside the sphere with radius r, corresponding to (7), is equal to

$$M_1(r) = \frac{4\pi}{3} \sqrt{1 - x_0} \, (r^2 - x_0)^{3/2}.$$ (8)

In this case we can convince ourselves directly that the instability condition
(2), which reduces to the inequality

$$\frac{1}{r}\sqrt{1 - x_0}\,(r^2 - x_0)^{1/2} > \frac{1}{r^3}\sqrt{1 - x_0}\,(r^2 - x_0)^{3/2} + \frac{2x_0 - x_0^2}{r^3}, \quad (9)$$

is not valid for any x_0 and r.

The second numerical example is a "hole" in the sphere which initially is
described by the Idlis model (32) §1, Chapter III. The results of calculations
which are completely analogous to those obtained above are given in the
form of the table in [39^{ad}].

CHAPTER X

Ellipsoidal Systems

§ 1 Objects Under Study

When speaking about astrophysical applications of the investigations of stationary states and stability of collisionless ellipsoids, we have borne in mind the different galactic systems. A roughly ellipsoidal shape is possessed, for example, by elliptical galaxies and the bars of the SB-galaxies. Elliptical galaxies obviously possess an axial symmetry while the bars of the SB-galaxies are, explicitly, not axially-symmetrical with respect to the axis of rotation.

All these galactic systems may with good accuracy be considered as collisionless: estimates given above for spherical galaxies are applicable to them, with minor modifications. We have in mind here only the stellar component of these systems (on the contrary, the interstellar gas is collisional).

§ 2 Elliptical Galaxies

2.1 Why Are Elliptical Galaxies More Oblate than E7 Absent?

Among the so far unresolved problems regarding the elliptical galaxies,[1] we would single out one already old but, in our opinion, very beautiful problem: that of explaining why there are no elliptical galaxies with an oblateness exceeding a definite critical value.

This fact was noted and repeatedly emphasized by Hubble, who believed that a limit type of elliptical galaxy is E6, when the small axis is 40% of the large axis (see [17]). Recall that the Hubble notation for the elliptical galaxies En means that the ratio of the semiaxes of its meridional cross-section $c/a = 1 - n/10$, $(n = 10[(a - c)/a])$. Now it is apparently assumed that the largest oblateness is possessed by the galaxies of type E7, for which $c/a = 0.3$ [35]. Irrespective of a specific number, the fact of the presence of some critical ratio c/a for elliptical galaxies is itself of some importance.

2.2 Comparison of the Observed Oblatenesses of S- and SO-Galaxies with the Oblateness of E-Galaxies

Below this limit we already have only spiral (S, SB) and lens-shaped (SO) galaxies, i.e., objects of a quite different kind which are readily distinguished from elliptical galaxies (see [35]).

Thus, the oblatenesses $q = c/a$ of elliptical galaxies, on the one hand, and of usual spirals or the SO-galaxies, on the other, lie in different regions. Comparatively recently, this has been confirmed in a detailed paper by Sandage *et al.* [318], using a very rich material (168E, 267SO + SBO, 254S—altogether 689 galaxies).

Due to this difference, the S- and SO-galaxies seemingly may not be evolutionarily associated with elliptical galaxies. Usually it is assumed [318] that the fundamental difference between the E and S (SO) systems lies in the differences of the original distribution in angular momenta, and primarily in the relative quantity of matter with low values of angular momentum, which may convert to stars already at the beginning of the compression of the protogalaxy (the rapidly rotating matter precipitates later in a gaseous form onto the equatorial plane).

2.3 Two Possible Solutions of the Problem

What benefit can be derived from investigations of equilibrium and stability of the collisionless systems, for the solution of the problem formulated in

---

[1] Such as, e.g., the problem of an adequate model, which would satisfactorily describe observations (in particular, the Hubble law of surface brightness variation).

Section 2.1? Firstly, it may turn out that the universal equilibrium states of elliptical galaxies which will be found, exist generally for only slightly oblate systems.

More probable, however, is another possibility connected with the stability loss of the system as the critical limit of oblateness is exceeded. We are aware in any event that such a situation takes place in the case of ellipsoids (spheroids) of Maclaurin: they become (dynamically) unstable when the eccentricity of the meridional cross-section exceeds the value $e_{cr} = 0.95289$, i.e. $(c/a)_{cr} \approx 0.31$. This result cannot of course be transferred directly to the real elliptical galaxies; we would like only to stress the possibility in principle of a similar situation.

At present, one may only speak seriously about the investigation of the stability of simple, in particular, *homogeneous*, ellipsoidal models (at least, by analytical methods). However, the Peebles–Ostriker criterion and some similar considerations described in Section 3.2, Chapter IV, are indicative of the fact that the results thus obtained (if they are formulated in relevant terms) may have a significantly broader region of applicability.

The rotation exerts a destabilizing influence on the system with respect to perturbations of a "global" character, at which the original axially-symmetrical shape of the system is violated, so that it takes on a "barlike" form. The Peebles–Ostriker criterion provides a quantitative formulation of this effect. The physical cause is due, as already noted (in Chapter II), to the decrease in effective gravity force on the system boundary, which facilitates the reconstruction of its shape. A convenient model for quantitative estimation of the critical value of oblateness and of the influence of the rotation velocity on stability of axially-symmetrical collisionless systems is the superposition of the Freeman spheroids rotating in the opposite direction (see end of Section 3.1, Chapter IV).

2.4 The Boundary of the Anisotropic (Fire-Hose) Instability Determines the Critical Value of Oblateness

If we propose that the equilibrium in highly oblate ellipticals must be produced by an anisotropy of star velocity dispersions, then a degree of anisotropy for the systems with maximum oblateness (types E6 and E7) must be also sufficiently high. It is not unlikely that these systems are just near the boundary of the anisotropic (fire-hose) instability which was in particular examined in §3, Chapter IV (for ellipsoidal systems). At any rate it is clear beforehand that the requirement of stability with respect to bending perturbations must lead to some restriction on maximum oblateness of stable elliptical galaxies (for example, because the infinitesimally-thin hot disks are obviously unstable—see Section 4.2, Chapter V). Then the following questions arise. Firstly, what, numerically, are the estimates for $(c/a)_{min}$ obtained from the stability theory; whether these values are sufficiently near the observational $(c/a)_{\min}$, so that we could say that in reality the fire-hose instability had some relation to the problem under consideration?

Secondly, what happens to the system if one has oblateness which is considerably higher than the critical value?

We have now only preliminary answers to the questions stated which are based on the analysis of strongly-simplified (homogeneous) models of elliptical galaxies.

The answer to the first question is positive. In §3, Chapter IV, we showed that, for example, the model of the ellipsoid of revolution hot at the plane of symmetry ((64), §1, Chapter IV) becomes unstable just for $(c/a) \lesssim 0.3$ (for the prolated spheroids at rest we had similar results—see Problem 9, Chapter IV). The preliminary results concerning the fire-hose instability of *nonhomogeneous* ellipsoidal systems are also in agreement with this estimate. As to the second question stated above, the certain answer is given in the results of the numerical experiments simulating an evolution of highly-oblate homogeneous ellipsoids at rest (these simulations are also described in §3, Chapter IV). These experiments show a rapid increasing of thickness in such systems, which is due to the excitement of unstable fire-hose modes. Thus, one can see that the ellipticals with an oblateness exceeding the critical value could indeed not exist: a growth of initial disturbances would ultimately draw the system onto the stability boundary.

2.5 Universal Criterion of Instability

It is clearly that extensive investigations of more realistic models are necessary for definite conclusions.[2] But the stability criteria relative to the large-scale disturbances may be universal provided that these criteria are formulated in some relevant terms. The model, which we used in §3, Chapter IV, for the investigation of the fire-hose instability, was composed of the homogeneous ellipsoid with mass M_d and with semiaxes a, c $(a > c)$, rotating with the angular velocity $\gamma\Omega_0$ in the immovable spherical halo with mass M_h. Let us introduce the parameters:

$$u_r = T_r/W, \qquad u_\varphi = T_\varphi/W, \qquad u_z = T_z/W, \qquad t = T/W, \qquad (1)$$

where $T_{r,\varphi,z}$ is the kinetic energy of the chaotic motion of stars along the axes r, φ, z, respectively, T is the energy of rotation of the galaxy, $W = |W_d| + 2U_0$ is the potential energy, where W_d is the energy of interaction of the stars of the galaxy with each other, U_0 is the energy of interaction of stars with the halo. For a uniform spheroidal halo $U_0 = \frac{1}{2}\int \rho_d\Omega_h^2(r^2 + z^2)\,dV$,

---

[2] Besides, the variations described above (which employ the requirements of the stability theory) do not, of course, exhaust all the possibilities for an explanation of the problem of maximum oblateness of ellipticals. And, finally, the explanations suggested say nothing about how in reality the elliptical galaxies are formed.

integration is over the volume of the galaxy, ρ_d is the stellar density. From the virial theorem, it follows

$$2(T_r + T_\varphi + T_z + T) = |W_d| + 2U_0; \tag{2}$$

therefore

$$u_r + u_\varphi + u_z + t = \tfrac{1}{2}. \tag{3}$$

The parameters $u_{r,\varphi,z}; t$ for our model are uniquely associated with the values μ, c/a, γ^2, and we may represent the instability regions on the plane $(u_\perp, u_z)$, where $u_\perp = u_r + u_\varphi = 2u_r$ (see Fig. 108).

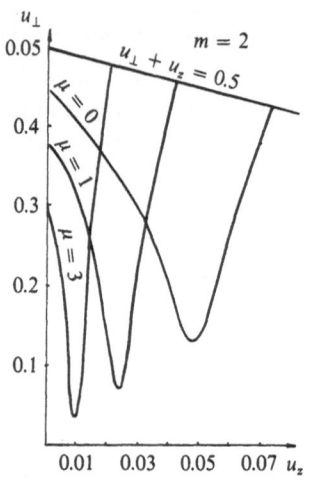

Figure 108. Marginal curves in variables $(u_z, u_\perp)$.

The basis of further applications of the criteria obtained to real systems serves the hypothesis that the galactic stability boundaries in the variables $u_\perp$ and u_z are dependent only on the parameter $\mu = M_h/M_d$ and are weakly dependent on the model. A similar suggestion was advanced by Ostriker and Peebles [301] as far as the barlike mode is concerned. In the case of bendings one can most simply formulate the stability criterion for the mode $m = 0$ ("bell");

$$\tfrac{1}{2}u_\perp + \beta(\mu)u_z > \alpha_{\mathrm{cr}}(\mu). \tag{4}$$

The value $\alpha_{\mathrm{cr}}(\mu)$ can be derived from the dispersion equation in the disk limit, the deviation of which is given in Section 2.1, Chapter V:

$$(\omega - m\gamma\Omega)^2 = \Omega_d^2(4\Gamma_n^m - 2) - \tfrac{1}{3}\Omega^2(1 - \gamma^2)$$
$$\times [(2n + m - 1)(2n + m) - m^2 - 2] + \Omega_h^2, \tag{5}$$

where $\Omega_h^2 = GM_h/a^3$, the expression for Γ_n^m is given in Section 2.1, Chapter V.

The "bell" mode has $m = 0$, $n = 2$. For this mode, from (5) we have $\omega^2 = \frac{5}{2}\Omega_d^2 - \frac{10}{3}(1 - \gamma^2)\Omega^2 + \Omega_d^2$, and for $\alpha_{cr} = \frac{1}{4}(1 - \gamma^2)_{cr}$ we get

$$\alpha_{cr} = \frac{3}{16}\frac{1 + 8\mu/15\pi}{1 + 4\mu/3\pi}. \tag{6}$$

For the $m = 1$ and $m = 2$ modes, the criteria are not written in the form of a simple linear dependence between $u_\perp$ and u_z. A similar dependence is roughly valid only for small u_z.

To confirm the above hypothesis about the universality of the stability criteria, the computations of the stability of the disklike models with different rotation curves for the bell-shaped mode were performed.

In the first model [227], the surface density of the disk had the form (see §1, Chapter V)

$$\sigma = \frac{M}{2\pi R^2} \sum_{k=1}^{n} b_k^{(n)} \xi^{2k-1}, \tag{7}$$

where $\xi = \sqrt{1 - r^2/R^2}$, M and R are the mass and the radius of the disk,

$$b_1^{(n)} = (2n + 1)/(2n - 1),$$

$$b_k^{(n)} = [4(k - 1)(n - k + 1)/(2k - 1)(2n - 2k - 1)]b_{k-1}^{(n)}.$$

The angular velocity of rotation of the disk is

$$\Omega^2(r) = \gamma^2 \frac{2n + 1}{4n} \frac{\pi GM}{R^2}(1 - \xi^{2n})/r^2, \tag{8}$$

where $\gamma^2 \leq 1$ is defined in the same way as above. In the case where $n = 1$, this model coincides with the model considered above, for which the instability criterion with respect to the bell-shaped mode yields $u_\perp = \frac{1}{2}(1 - \gamma^2) > \frac{3}{8}$, or $t < t_{cr} = 0.125$. It turns out that for $n \leq 7$,

$$0.101 \leq t_{cr} \leq 0.125. \tag{9}$$

For the second model [205] (cf. (39) in §1, Chapter V)

$$\sigma = \sigma_0 \exp(-\alpha r), \tag{10}$$

$$\Omega^2(r) = \gamma^2 \pi G\sigma_0 \alpha\left[I_0\left(\frac{\alpha r}{2}\right)K_0\left(\frac{\alpha r}{2}\right) - I_1\left(\frac{\alpha r}{2}\right)K_1\left(\frac{\alpha r}{2}\right)\right], \tag{11}$$

where I_n, K_n are the corresponding cylindrical functions. In this case the bell-shaped mode becomes unstable for

$$t < t_{cr} = 0.120. \tag{12}$$

Therefore, it is evident that t_{cr} (or u_{cr}) turns out to be approximately identical for very different models.[3]

---

[3] Note that the results described were applied [37[ad], 40[ad]] for estimating the upper boundary of the halo mass in a galaxy.

§ 3 SB-Galaxies

The SB-galaxies (intersected spirals, or the galaxies with a bar) have spiral arms going out from the ends of the bar, at whose center is the nucleus.

3.1 The Main Problem

Achievements of the SB-galaxy theory are so far very modest, especially when compared with the significant progress in understanding the structure of normal spiral galaxies (of the S-type) connected with the density wave theory (cf. next section).

One of the main difficulties consists in the construction of a sufficiently good collisionless model of the central, obviously not axially-symmetrical, region of SB-galaxies, first of all of the bar itself. In comparison with this, the second part of the problem (origin of spiral arms) now seems simpler.

3.2 Detection in NGC 4027 of Counterflows as Predicted by Freeman

If one deals with the quasistationary theory of spiral galaxies of the SB-type, one should note some important papers by Freeman and de Vaucouleurs [203, 206]. In [203], the self-consistent model of the homogeneous three-axis ellipsoid of stars (see §1, Chapter IV) is suggested as a model of a bar of the spiral galaxy. We have already noted one interesting characteristic of the macroscopic velocity field of such systems, namely the presence of "counterflows" for rather strongly oblate ellipsoids ($2b^2 < a^2$) in the rotation plane (see Fig. 46).

If the "inverse" motion is not only the property of the homogeneous models described, but indeed occurs in the bars of real SB-galaxies, it may be determined using its influence on the absorption lines in the spectrum of the stellar component of the bar. Freeman and de Vaucouleurs [206] made an attempt to perform special observations of the spiral galaxy with the bar, NGC 4027. The statement of the aims of the observations is as follows.

It is clear that the inverse average motion is possible only because of a high stellar velocity dispersion. At the same time, the velocity dispersion of the gaseous clouds, in which the emission lines arise, cannot be so large, since the time of collisions between the clouds is far less than the period of the bar's rotation. Therefore, the gaseous clouds must move only in one direction. Thus, the inclinations of the emission lines and absorption lines in the spectra, taken with the position of the spectrograph slit near the small axis of the bar orientated in an appropriate way, should be opposite: they should be directed according to rotation for the emission lines, and against the rotation for the absorption lines.

Such an effect was revealed in observations of the NGC 4027 galaxy. True, one seeming misunderstanding arose: in their value of inclination, the absorption lines proved to be greater (by a factor of approximately 4) than the inclination of the emission lines, although, one would think, it was to be expected that approximately equal (but opposite) inclinations would result. However, there is one simple explanation of this fact which is the following [344]. The mean velocity v satisfies the equation (§1, Chapter IV)

$$(\mathbf{v}\nabla)\mathbf{v} = -\operatorname{div} \boldsymbol{\sigma} - 2[\boldsymbol{\Omega}\mathbf{v}] - \frac{\partial \Phi}{\partial \mathbf{r}} + \Omega^2 \mathbf{R}. \tag{1}$$

For the inverse average motion, the centrifugal force and pressure (the last and the first terms of the right-hand side of (1)) are directed outwards, while the gravitational attraction and the Coriolis force are directed inwards. The resulting force must yield an acceleration directed inward to the system (the left-hand side of (1)). For a rapid mean motion, the most essential term in the right-hand side of (1) is the Coriolis force.

Let the curvature radius of the current line in a rotating system near the small axis of the bar be R_1, and the mean velocity v_1; then we have from (1)

$$v_1^2/R_1 \approx 2\Omega v_1, \qquad v_1 \approx 2\Omega R_1. \tag{2}$$

But $v_1 \approx 2\Omega R_1$ is far less than the rate of rotation Ωb, since $R_1 \gg b/2$ (in a similar way it is possible to show that the velocity at the ends of the bar is $v_2 \ll \Omega a$).

The finding of the counterflows is a strong argument in favor of the simplest bar model suggested by Freeman. As mentioned in a review by Freeman and de Vaucouleurs [344], if this fact can be confirmed by observations in other spirals with bars, then this will mean that the "present rudimentary ideas about the stellar dynamics of the SB-systems, are at least going in the true direction." In any event, the presence of the counterflows imposes strong restrictions on possible theoretical models of the SB-galaxies.

3.3 Stability of Freeman Models of SB-Galaxies with Observed Oblateness

Another important argument is obtained in [96] (see §2, Chapter IV), where the Freeman model is investigated for stability with respect to the largest-scale types of oscillations. Figure 48 shows that the oblateness of real bars (and, in particular, of NGC 4027) lies in a "stable" region.[4]

---

[4] More detailed stability investigation of prolate stellar systems was performed in [36[ad], 38[ad]].

CHAPTER XI

Disk-like Systems. Spiral Structure

Recently, the development of the equilibrium and stability theory of flat gravitating systems has occurred, mainly with the aim of understanding the galactic spiral structure. From the vast material accumulated here we have selected only some problems[1] which are, in our opinion, most closely related to the subject of the book.

Despite the fact that recently some essential progress has been outlined on the understanding of different mechanisms of the density wave generation and of the properties of their propagation, the problem of the origin of the spiral structure is still far even from qualitative solution.

§ 1 Different Points of View on the Nature of Spiral Structure

At present, quantitatively the most elaborate is the representation of spiral arms in the form of density waves rotating uniformly (as a solid body), independently of the galactic differential rotation.

This hypothesis was first advanced by Lindblad and then formulated in the form of linear theory of density waves by Lin and co-workers [267–272], Kalnajs [250], Contopoulos [187, 189], and others. The principles of the *nonlinear* theory (soliton theory) of density waves are stated in the previous chapter. A point of view on the nature of spiral arms which preceded that

---

[1] At the present time there are many reviews ([84, 188, 88[ad]], etc.) on this subject.

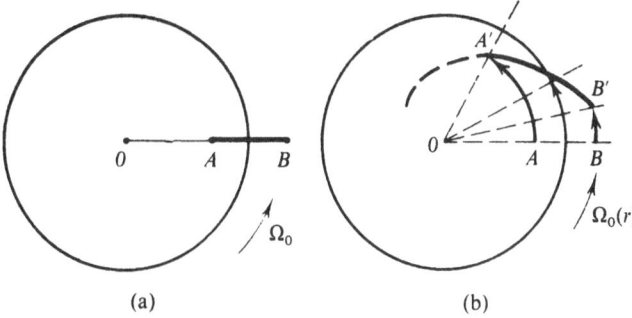

Figure 109. Stretching of the initial compression: (a) AB into the piece of the trailing spiral; (b) $A'B'$ in the differentially rotating disk with $\Omega'_0 < 0$.

mentioned above, was different: these were presented as compressions of stars and gas traveling at each point at the local rotation velocity of a galaxy. Such a point of view was called "material"—the name appears rather weak, because density waves consist of the same matter: compressions of stars and gas. It would be more correct to call such a theory "local".

It is easy to confirm that the "local" picture of the spiral pattern disappears for the time of the order of one revolution of the galaxy, $\sim 10^8$ years, see Fig. 109.[2] At the same time it is assumed that the age of the galaxies is $\sim 10^{10}$ years. It is hardly probable that the galaxies would all "agree" to be spiral simultaneously in the course of a very short period of time. But then one must conclude that the spiral structure must continuously or periodically be renewed in the galaxy.

We emphasize that the above statement is valid for purely gravitational theories; the presence of a sufficiently strong magnetic field, for example, may preserve the gaseous spirals from destruction by the differential rotation.

The most developed attempt to construct the regenerative theory of gaseous spirals is due to Goldreich and Lynden-Bell [210]. They suggest that the gaseous layer is rotating in a strong external gravitational field (which is assumed to be given and not subjected to perturbations). The theory is based on the local analysis of the hydrodynamical equations and the Poisson equation. There are then introduced the movable coordinate axes $\tilde{x}$, $\tilde{y}$, which are bent together with the differentially-rotating flow. An investigation is made of the temporal evolution of the perturbation, the spatial shape of which, $e^{i(k_x\tilde{x}+k_y\tilde{y})}$, is assumed to be unaltered in a movable system. This evolution resembles the origination, enhancement, and finally, the decay of spiral arms.[3]

---

[2] As seen from Fig. 109, any density compression moving together with a "cold" differentially rotating galaxy, would inevitably extend to form segments of trailing spirals and ultimately disappear by completely dissolving in the galaxy.

[3] The theory of Goldreich and Lynden-Bell was, however, subjected to criticism (cf. [233]).

We have not yet touched upon the question of the generation mechanism (or regeneration mechanism) of spiral arms. The purpose of Toomre's paper [333] was to verify the hypothesis about the formation of the spiral structure due to gravitational instability. Toomre calculated (see Section 4.1, Chapter V) the minimum radial velocity dispersion required for the full suppression of all axially-symmetrical instabilities, equal to $3.36G\sigma_0/\varkappa$. From this expression, by using the *observed* $\sigma_0 \sim 50 \div 65\ M_\odot/ps^2$ and $\varkappa \sim 27 \div 32$ km/s · kps one may calculate the minimum for the solar vicinity of our Galaxy; it proved to be about 35 km/s (a more accurate estimate was later obtained by Shu). But in the same region, there is also the mean velocity dispersion of stars, observed in the solar vicinity. In addition, as is seen from formula (50), Section 4.1 (Fig. 72), most difficult to stabilize are radial perturbations with wavelengths of the order of $\lambda \sim 0.55\lambda_T$, which in the solar vicinity is equal to $5 \div 8$ kps. At the same time, the radial perturbations of the stellar disk with $\lambda \sim 4$ kps (i.e., with the wavelength of the order of the observed distance between the arms) must obviously be *locally* stable.

§ 2 Resonant Interaction of the Spiral Wave with Stars of the Galaxy

The role of the resonant interaction of the spiral wave with the stars of the galaxy was investigated in detail by Lynden-Bell and Kalnajs [289]. Below we describe some results of their paper.

First (in Section 2.1), we give the derivation of exact formulae for the energy and angular momentum of the quasistationary wave, as well as expressions for the variation rate of these values. These formulae show the decisive role of the resonance stars.

Section 2.2 analyses in detail the physical mechanisms of enhancement (or damping) of the waves on all basic types of resonances in the galaxy.

2.1 Derivation of Expressions for the Angular Momentum and Energy of the Spiral Wave

We derive first of all an exact expression for *the angular momentum (and energy) of the spiral wave* [251, 289]. The calculations leading to this expression are similar to those which are performed in plasma physics in the clarification of the physical meaning of Landau damping (see, e.g., [86]). The details of the derivation of this formula, which we give in view of its importance, are as follows:

Consider the exchange of the angular momentum and the energy between the stars and the spiral wave. The equations of motion of a star in the disk plane in the cylindrical coordinates r, φ have the form

$$r - r\varphi^2 = -\frac{\partial}{\partial r}(\Phi_0 + \Phi_1), \tag{1}$$

$$\frac{d}{dt}(r^2\dot{\varphi}) = -\frac{\partial\Phi_1}{\partial\varphi}. \tag{2}$$

The system of (1) and (2) has for $\Phi_1 = 0$ the obvious laws of conservation:

of energy $\qquad\qquad E = \frac{1}{2}(\dot{r}^2 + L^2/r^2) + \Phi_0, \tag{3}$

of angular momentum $\qquad\qquad L = r^2\dot{\varphi}. \tag{4}$

In the consideration of the kinematics and dynamics of the flat gravitating systems, it is convenient to use, instead of the usual cylindrical coordinates $(r, \varphi, v_r, v_\varphi)$, the angle-action variables (J_1, J_2, w_1, w_2), which take into account the important property of the stellar orbits in the plane of an axially-symmetrical system—their double periodicity—and correspondingly, greatly simplify the description. Bearing in mind a further application of the perturbation theory, we shall denote the full quantities (unperturbed + perturbed) by primes, leaving the earlier notations (J_1, J_2, w_1, w_2) for the unperturbed quantities.

If the variables (J_1', J_2', w_1', w_2') are considered, then the system in (1) and (2) will be reduced to the following:

$$\dot{J}_j' = -\frac{\partial}{\partial w_j'}(H_0 + \Phi_1) = -\frac{\partial}{\partial w_j'}\Phi_1(J_i', w_i', t), \tag{5}$$

$$\dot{w}_j' = \frac{\partial}{\partial J_j'}(H_0 + \Phi_1) = \Omega_j(J_i') + \frac{\partial}{\partial J_j'}\Phi_1(J_i', w_i', t), \tag{6}$$

and the unperturbed motion in these variables is characterized by the equations

$$\dot{J}_i = -\frac{\partial H_0}{\partial w_i} = 0, \tag{7}$$

$$\dot{w}_i = \frac{\partial H_0}{\partial J_i} \equiv \Omega_i(J_1, J_2). \tag{8}$$

The meaning of w_i, J_i is most easily illustrated in the simple example of the nearly circular motion (epicyclic approximation). In this case, one easily obtains the following expression for the perturbed radius of the star (see §1, Chapter V):

$$r_1 = a\sin(\varkappa t + \alpha) + \cdots.$$

By using further the definitions of the angle-action variables, one may find in this approximation: $\Omega_1 = \varkappa$ is the epicyclic frequency; $\Omega_2 = \Omega$ is the angular velocity of the circular motion; $J_1 = \varkappa a^2/2, J_2 = L$. In the general case the interpretation of these variables is similar: w_1 is the phase of the radial oscillation of the star, J_1 is the function of the amplitude of this oscillation,

w_2 is the galactocentric angle of a uniformly moving epicenter (with angular velocity Ω_2).

The perturbed orbits of the stars are derived from the perturbation theory. The orbits of the first order are sought by solving Eqs. (5) and (6), into the right-hand side of which (i.e., in the calculation of the forces) we substitute the unperturbed orbits. For the correction of the first order $\Delta_1 J_j$ for J_j, then we get $\Delta_1 J_j = \partial \chi / \partial w_j$, where

$$\chi = \mathrm{Re}\left\{ (4\pi^2)^{-1} \sum_{l,m} \psi_{lm}(J_i) \frac{\exp[i(lw_1 + mw_2 - \omega t)]}{i(l\Omega_1 + m\Omega_2 - \omega)} \right\}, \qquad (9)$$

and $\psi_{lm}(J_i)$ are the coefficients of the Fourier expansion

$$\Phi_1(J_i, w_i, t) = \mathrm{Re}\left\{ (4\pi^2)^{-1} \sum_{l,m} \psi_{lm}(J_i) \exp[i(lw_1 + mw_2 - \omega t)] \right\}, \qquad (10)$$

The expansion in (10) is always possible due to the periodicity of $\Phi_1(J_i, w_i, t)$ with respect to angular variables w_i (with a period 2π).

Since $\Delta_1 J_i$ is periodical in initial phases w_1, w_2, the average change of the angular momentum of the system of stars which originally were uniformly distributed in phases, is zero: $\langle \Delta_1 J_j \rangle = 0$. Thus, the angular momentum (and energy) exchange between such a stellar group and the wave is of the second order with respect to the perturbed potential. The orbits of the second order are calculated, according to the forces in (5) and (6) as calculated for the orbits of the first order.

The result of the calculation is thus:

$$\langle \dot{L} \rangle = (2\pi)^{-2} \int_0^{2\pi} \int_0^{2\pi} \Delta_2 \dot{J}_2 \, dw_1 \, dw_2 = \tfrac{1}{2} \mathrm{Im}(\omega)$$

$$\times \exp[2 \, \mathrm{Im}(\omega)t] \, (2\pi)^{-4}$$

$$\times \sum_{l,m} m \left(l \frac{\partial}{\partial J_1} + m \frac{\partial}{\partial J_2} \right) \frac{|\psi_{lm}|^2}{|l\Omega_1 + m\Omega_2 - \omega|^2}. \qquad (11)$$

The total rate of variation of the angular momentum of stars, which initially had angular momenta $L_1 < J_2 \equiv L < L_2$ is

$$\dot{L} = 4\pi^2 \int_{L_1}^{L_2} \int_0^\infty \langle \dot{L} \rangle f_0(J_1, J_2) \, dJ_1 \, dJ_2. \qquad (12)$$

We integrate in parts. The total L may be split into the "volumetric" and the "surface" terms:

$$\dot{L} = -\frac{1}{8\pi^2} \, \mathrm{Im}(\omega) \exp[2 \, \mathrm{Im}\, \omega \cdot t]$$

$$\times \left\{ \int_{L_1}^{L_2} \int_0^\infty \sum_{l,m} m \left(l \frac{\partial f_0}{\partial J_1} + m \frac{\partial f_0}{\partial J_2} \right) \frac{|\psi_{lm}|^2 \, dJ_1 \, dJ_2}{|l\Omega_1 + m\Omega_2 - \omega|^2} \right.$$

$$\left. - \sum_{l,m} m^2 \int_0^\infty \frac{f_0 |\psi_{lm}|^2 \, dJ_1}{|l\Omega_1 + m\Omega_2 - \omega|^2} \Big|_{L_1}^{L_2} \right\} = \dot{L}_1 + \dot{L}_2. \qquad (13)$$

In this expression, the first integral ($\dot{L}_1$, "volumetric" term) corresponds to the variation in the angular momentum of the stars, which remain in the region (L_1, L_2), while the second integral ($\dot{L}_2$, "surface" or "convective" term) corresponds to the angular momentum which is convected through the boundaries of the region under consideration. If one takes the integral $\int_{-\infty}^{t} \dot{L}\, dt$ and takes into account only the "volumetric" term, we shall obtain the excess of the angular momentum for the stars in the chosen range of the values of L over that which they had in the absence of the wave:

$$\delta L = -\frac{1}{16\pi^2} \exp[2\,\text{Im}(\omega)t]$$

$$\times \int_{L_1}^{L_2} \int_0^\infty \sum_{l,m} m\left(l\frac{\partial f_0}{\partial J_1} + m\frac{\partial f_0}{\partial J_2}\right) \frac{|\psi_{lm}|^2\, dJ_1\, dJ_2}{|l\Omega_1 + m\Omega_2 - \omega|^2}. \tag{14}$$

The expression (14) thus found is the angular momentum, transferred by the stars to the wave, i.e., the angular momentum of the wave if integration over J_2 in (14) is extended to all the angular momenta of stars.

In the limit of a very slowly increasing wave, $\text{Im}(\omega) \to 0$, the variation of the angular momentum (and energy) of stars occurs, according to (13), only on resonances, where

$$l\Omega_1 + m\Omega_2 - \omega \to 0. \tag{15}$$

For $\text{Im}(\omega) \to 0$, we have, by using the identity

$$-\lim[\text{Im}(\omega)] \cdot |l\Omega_1 + m\Omega_2 - \omega|^{-2} \to \pi\delta(l\Omega_1 + m\Omega_2 - \omega),$$

and splitting the velocity $\dot{L}$ into the sum of the terms from different resonances:

$$\dot{L} = \sum_{l,m} \dot{L}_{lm},$$

$$\dot{L}_{lm} = -\frac{1}{8\pi} \iint m\left(l\frac{\partial f_0}{\partial J_1} + m\frac{\partial f_0}{\partial J_2}\right) |\psi_{lm}|^2 \delta(l\Omega_1 + m\Omega_2 - \omega)\, dJ_1\, dJ_2 \tag{16}$$

It is easy to see that contributions come only from the resonances, whose positions in the epicyclic approximation are defined by the formula $\Omega - \Omega_p = -l\varkappa/m$, where $\Omega_p = \omega/m$ is the velocity of the spiral wave. If one considers any single (m) component of the potential, then the resonances will be enumerated by one integer index l (positive, negative, or zero). The following are the three main resonances which have special names (let us recall them): $l = 0$ corresponds to the "*corotation*" resonance, or the "*particle*" resonance, on which $\Omega = \Omega_p$; the resonances corresponding to $|l| = 1$ are called *Lindblad* resonances. If one moves from the corotation resonance inward, toward the galactic center, the local angular velocity will increase, and one may finally (but not always) encounter a ring, on which Ω exceeds Ω_p by $\varkappa/|m|$. This resonance is called the *inner Lindblad resonance* (for it, $l = -1$ for $m > 0$ and $l = +1$ for $m < 0$). The other resonance $|l| = 1$ is located in the galaxy outwards from the corotation ring (the *outer Lindblad resonance*).

The resonances listed above take place where the frequency at which the star intersects the humps and hollows of the spiral wave potential, $|\omega - m\Omega|$, is either zero (i.e., the star is always in phase with the potential), or equals the oscillation frequency of the star $\varkappa$ near the circular orbit.

Resonances of a higher order are dynamically of less importance, and, in addition, it seems that all the outer resonances really lie outside the galaxies, while the inner ones are too close to the galactic nucleus.

For $|m| = 1$, the inner Lindblad resonance occurs only for the waves running in the direction opposite to the rotation of the system. For $|m| = 2$, the resonances are separated just roughly by the dimension of the galaxy, for $|m| \geq 3$, they are approaching the corotation radius. These facts all explain the preference of the two-armed pattern in our Galaxy (we have already discussed them in Section 3.2.).

Let us now clarify how, in a stationary spiral wave with a fixed number of arms, m, the variations of the angular momentum and energy of the star are interrelated with each other. In the frame of reference rotating together with the wave (at a velocity of $\Omega_p = \omega/m$), the total potential $\Phi_0 + \Phi_1$ is time-independent, so that each star conserves the energy in these axes (the so-called Jacobi integral)

$$\varepsilon_T' = \tfrac{1}{2}[v_r^2 + (v_\varphi - \Omega_p r)^2] + [\Phi_0 + \Phi_1 - \tfrac{1}{2}\Omega_p^2 r^2],$$

where the velocity of the star (v_r, v_φ) refers to the inertial system. In other words, $\varepsilon_T' = \varepsilon_T - \Omega_p L$, where ε_T is the energy in the inertial system, and the last formula itself is an ordinary relationship between the energies of the particle in the inertial (ε_T) and rotating (ε_T') coordinate systems (see, e.g., [69]). Since $d\varepsilon_T'/dt = 0$, then $d\varepsilon_T/dt = \Omega_p dL/dt$, $\delta\varepsilon_T = \Omega_p \delta L$, and if one sums up this equality over the stellar system with the total energy E and the angular momentum L, we shall obtain in a similar way

$$\frac{dE}{dt} = \Omega_p \frac{dL}{dt}, \qquad \delta E = \Omega_p \delta L.$$

The process of transfer of the angular momentum and energy by the spiral wave is illustrated in Fig. 110, taken from [289]. From this figure it is easy to see that the star increasing its angular momentum by δL at the corotation radius and, consequently, increasing its energy by $\Omega_p \delta L$, does not change the oscillation energy near the circular motion, since at this resonance $d\varepsilon/dL = \Omega_p$. The star losing δL at the inner Lindblad resonance, also loses the energy $\Omega_p \delta L$, but it comes to the state to which a still lesser value of circular motion energy would correspond. Consequently, in this case, the energy $\delta\varepsilon$ (see Fig. 110) is released into a noncircular motion.

Let us now find the resonances on which the stars increase their angular momentum and energy, and those on which they lose them. It turns out that the results may be formulated in a quite general form for the cases when the epicyclic consideration is applicable.

As is easy to see from (16), for $l = 0$ (corotation resonance), the resonant stars increase their angular momentum since $\partial f_0/\partial J_2 < 0$ for any reasonable

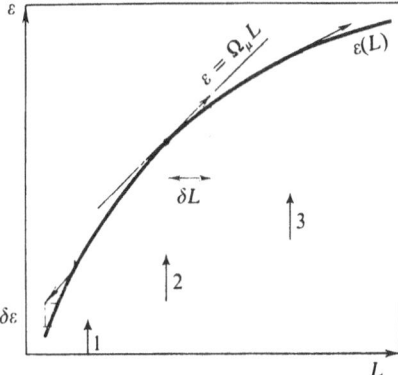

Figure 110. Transfer of the angular momentum δL and the energy $\delta E = \Omega_p \delta L$ by stars [289]; $\varepsilon = \varepsilon(L)$—the energy of the circular unperturbed motion with the angular momentum L; 1—the inner Lindblad resonance, 2—the corotation radius, 3—the outer Lindblad resonance.

distribution function, due to the general outward-fall-off of the surface density in galaxies (the orbits are assumed to be nearly circular, and larger r correspond to larger angular momenta L). For all the remaining resonances, the sign of the angular momentum and energy exchange between the stars and the wave is defined by the sign of lm, if it is assumed that $|\partial f_0/\partial J_1| \gg |\partial f_0/\partial J_2|$ (i.e., in other words, the validity of the epicyclic approximation; indeed, in this approximation $J_1 \sim \varkappa a^2$, $J_2 \sim \Omega R^2$, so that $J_2/J_1 \sim (R/a)^2 \gg 1$). Since at the inner Lindblad resonance $lm < 0$, so the stars at this resonance must give out their energy and the angular momentum.

In a similar way, with the same approximations, the stars at the outer Lindblad resonance absorb the angular momentum and energy.

Let us now determine *the sign of energy* of *the quasistationary spiral wave.* Transform expression (14) (with Im $\omega \to 0$) to the form

$$
\delta E = -\frac{\Omega_p}{16\pi^2} \iint \left\{ \sum_{l=1}^{\infty} \frac{4l^2 m^2 \Omega_1 (\Omega_2 - \Omega_p)(-\partial f_0/\partial J_1)}{|l^2 \Omega_1^2 - m^2 (\Omega_2 - \Omega_p)^2|^2} |\psi_{l,m}|^2 \right.
$$
$$
\left. - \sum_{l=-\infty}^{\infty} m^2 \left(-\frac{\partial f_0}{\partial J_2} \right) \frac{|\psi_{lm}|^2}{|l\Omega_1 + m(\Omega_2 - \Omega_p)^2|^2} \right\} dJ_1 \, dJ_2. \tag{17}
$$

In the epicyclic approximation $|\partial f_0/\partial J_1| \gg |\partial f_0/\partial J_2|$; therefore the first term in (17) is dominant, so that for $\Omega_2 > \Omega_p$ (i.e., inside the ring of corotation) $\delta E < 0$, and for $\Omega_2 < \Omega_p$ (outside the ring of corotation) $\delta E > 0$. A positive contribution to the energy δE is also made by the second term in (17) (which is nonessential in the epicyclic approximation), both for $\Omega > \Omega_p$ and for $\Omega < \Omega_p$. This may turn out to be important for the cases when the epicyclic approximation ceases to be valid, i.e., when the deviations from circular orbits are large. From the results of theoretical papers [93, 111, 252] as well as from numerical experiments [294] and especially those of Hohl [215, 220] it follows that these deviations should in fact be large for stable or nearly stable galaxies without a significant mass concentration toward the center (and in the absence of a halo).

We return now to the case of waves of negative energy as being more definite. It is obvious that, in the case $\delta E < 0$, the absolute value of $|\delta E|$ grows with increasing amplitude of perturbation (of a spiral wave), i.e., with the instability. Taking energy from perturbation, we will thereby excite it. And, on the other hand, by introducing energy we shall damp such perturbations. The waves of negative energy having such an "inverted" behavior, are well known, for example, in the theory of plasma instabilities (see [86]). The simplest example[4] of the medium in which oscillations of negative energy may exist is the cold moving plasma (with the velocity V). The scalar dielectric permittivity of such a system is

$$\text{Re } \varepsilon_0 = 1 - \frac{\omega_p^2}{(\omega - kV)^2},$$

so that the oscillation frequencies, which are defined from the equation $\text{Re } \varepsilon_0 = 0$, are $\omega_k = kV \pm \omega_p$. Hence it follows that

$$\omega_k \frac{\partial \text{ Re } \varepsilon_0}{\partial \omega_k} = 2\left(1 \pm \frac{kV}{\omega_p}\right),$$

and therefore for $Vk/\omega_p > 1$ and for the solutions with the minus sign (see Fig. 111) the oscillation energy is negative:

$$W_k = \omega_k \frac{\partial \text{ Re } \varepsilon_0}{\partial \omega_k} \frac{|E|^2}{8\pi} < 0.$$

The sign of oscillation energy is of course not invariant with respect to the change of the frame of reference (the growth rate of oscillations is, for example, independent of the reference system). However, in some cases, some one definite frame of reference is distinguished among the others. In the case of

---

[4] This, as well as other simple examples, may be found in the monograph [86], where the energy classification of slowly increasing perturbations in plasma is given.

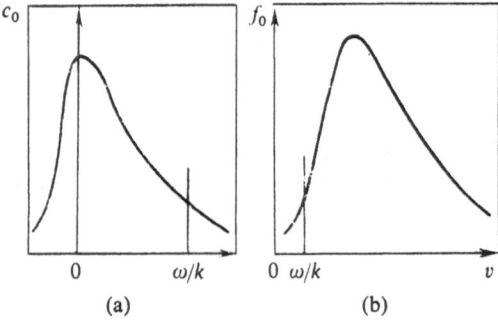

Figure 111. Waves of (a) positive and (b) negative energies for the simplest velocity distribution of particles [86].

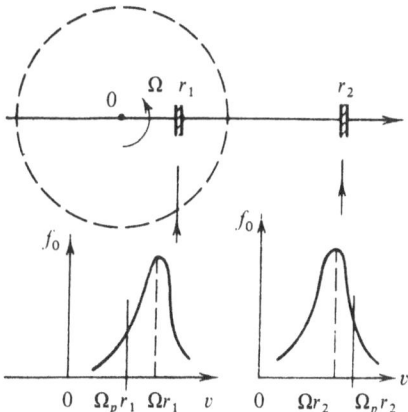

Figure 112. Illustration for the determination of the sign of energy density of the spiral wave.

interest, of differentially rotating gravitating systems, the inertial frame of reference is evidently such a distinctive one. Consider an arbitrary area of the disk (Fig. 112) at a distance r_1 from the center inside the radius of corotation $(r_1 < r_c)$. One easily notices the similarity between the situations presented in this figure and in Fig. 111.

For our Galaxy, according to Lin *et al.* [271], $\Omega = \Omega_p$ at $r = r_p \approx 14$ kps. At the same time, the observed spiral pattern in the Galaxy extends up to $r = 14 \div 15$ kps (from observations at $\lambda = 21$ cm, i.e., for neutral hydrogen HI). Therefore, in this case, only the internal region is essential, inside which the spiral wave has negative energy. Such a situation is normal for most of the theoretically *investigated* [204, 240] galaxies: our Galaxy, M31, M51, M81; however, for the galaxy M33, according to data of [240], the corotation ring lies rather close to the galactic center (i.e. Ω_p is rather large), so that in this case[5] the internal and external regions may contain a comparable amount of the matter, and be dynamically equally important.

As a more general example, it is easiest to take the barlike modes of the uniformly rotating disks, treated in detail in [254]. A direct calculation by formula (14) leads to the following conclusions. The mode rotating in the negative direction has positive energy and negative angular momentum. For "direct" modes, these quantities have the same sign, and for the more rapid mode they are positive, while for the slower one they are negative, until $\gamma < 0.5072\ldots$. At $\gamma = 0.5072\ldots$ these two modes are merging and then transform into the increasing and decreasing pair of modes, and each of them has a zero angular momentum and a zero energy.

Consider further the problem of the redistribution of the angular momentum, which must occur during the period of the growth of the barlike

---

[5] And, apparently, also in all cases when the density is not very much concentrated toward the galactic center, so that the rotation rather resembles a uniform rotation rather than the strongly differential rotation characteristic of our Galaxy.

mode, by using the calculations (up to the values of the second order of smallness), presented above. The marginally-stable $\gamma = 0.5072$ case is typical. The equilibrium distribution is the function of the angular momentum $L_z = J_2$ and radial action J_1, which in this case is $(E - |J_2|)/2$, if one assumes the origin of reference to be the energy of the star, being at rest at the center. The boundary between the "suppliers" and "consumers" of energy is the straight $J_1/J_2 = 0.4859 \ldots$. To determine this boundary, we present the potential $\Phi_1 = (x + iy)^2 = r^2 e^{2i\varphi}$ in the action-angle variables, i.e.,

$$\Phi_1 = \frac{1}{4\pi^2} \sum_{\substack{l \\ (m=2)}} \psi_l(J_1, J_2) \exp[i(lw_1 + mw_2) - i\omega t]. \tag{18}$$

The action variables

$$J_1 = \frac{1}{2\pi} \oint p_r \, dr = \oint \sqrt{2E - 2\Phi_0 - L_z^2/r^2} \, \frac{dr}{2\pi} = \tfrac{1}{2}(E - |L_z|)$$

and $J_2 = (1/2\pi) \int p_\varphi \, d\varphi = L_z$ have already been given. It is also easy to calculate the angular variables w_1 and w_2, i.e., the phase angles in the rth and φth movements of the particles, by the formulae

$$w_1 = \int^r \frac{dx}{\sqrt{4J_1 x - (x - J_2)^2}} \, ;$$

$$w_2 = \varphi + \frac{1}{2} \int^r \frac{(x - J_2) \, dx}{x\sqrt{4J_1 x - (x - J_2)^2}}. \tag{19}$$

Finally, one can obtain

$$\psi_0 = J_1 + J_2; \qquad \psi_{-1} = -2\sqrt{J_1^2 + J_1 J_2}; \tag{20}$$

$$\psi_{-2} = J_1; \qquad \psi_{+2} = \psi_{+1} = 0.$$

Thereafter, one should make use of formula (11) for the variation of the angular momentum of the particle; in the given case, this formula is reduced to the form

$$\langle \dot{L} \rangle = - \operatorname{Im} \omega \exp[-2 \operatorname{Im}(\omega) \cdot t] \cdot (2\pi)^{-4} \sum_l , \tag{21}$$

where

$$\sum_l = 2 \frac{\partial}{\partial J_2} \frac{|\psi_0|^2}{(\omega - 2)^2} + \left(-\frac{\partial}{\partial J_1} + 2 \frac{\partial}{\partial J_2} \right) \frac{|\psi_{-1}|^2}{\omega^2}$$

$$+ \left(-2 \frac{\partial}{\partial J_1} + 2 \frac{\partial}{\partial J_2} \right) \frac{|\psi_{-2}|^2}{(\omega + 2)^2}$$

$$= 4 \left[\frac{J_1 + J_2}{(\omega - 2)^2} - \frac{J_2}{\omega^2} - \frac{J_1}{(\omega + 2)^2} \right]. \tag{22}$$

The boundary between the "suppliers" and "consumers" of the angular momentum is determined, apparently, from the equality $\sum_l = 0$. If one takes $\gamma = 0.5072$ ($\omega = \sqrt{5/6} \approx 0.91$) it is then possible to obtain $J_1/J_2 \approx 0.4859$. This relation J_1/J_2 refers to the relation $E/J_2 \approx 1.972$, which implies the eccentricity of the orbit $e \approx 0.95$ or the ratio of the epicycle size $\rho = (a - b)/2$ to the mean radius $(a + b)/2$ (a and b are the sizes of the major and minor semiaxes of the elliptical orbit of the particle) equal to 0.5718. The directly rotating stars with an eccentricity less than that determined give their angular momentum to other stars, irrespective of their position on the disk.

There is also another way of distinguishing between the "suppliers" and "consumers." Since the rotating torque acting on the star from the side of the perturbed field is proportional to r^2, then it is necessary to calculate the time-average of its angular velocity $\dot\varphi$ with a weight r^2, i.e., $\Omega_\tau \equiv \overline{r^2 \dot\varphi}/\overline{r^2}$. For the disk $\Omega_\tau = L_z/E$ (since $\overline{r^2 \dot\varphi} = L_z$, $\overline{r^2} = E$), and the "consumers" are the stars with $\Omega_\tau > \gamma$, and "suppliers" are the stars with $\Omega_\tau < \gamma$. This criterion is exactly equivalent to the former one ($L_z/E = \gamma = 0.5072$ corresponds to the value $E/L_z = 1.972$, which we have encountered above).

2.2 Physical Mechanisms of Energy and Angular Momentum Exchange Between the Spiral Waves and the Resonant Stars [289]

2.2.1. Lindblad Resonances. Note first of all that, of course, exactly circular orbits receive none of the gravitational torques. The effect arises under the action of the perturbed gravitational field on the perturbed orbit (the field must "catch on" the irregularities of the orbit). Since the torque is small, of second order, determined by the product of perturbed forces in the radial g_{1r} and azimuthal $g_{1\varphi}$ directions $g_{1\varphi} \cdot g_{1r}$, it is sufficient to calculate the displacement from the unperturbed (circular) orbit to first order.

If the perturbed potential is represented in the form

$$\Phi_1 = -S \sin(kr_1 + m\varphi + \omega t), \tag{1}$$

then the star moving exactly along the resonance orbit will suffer the force at the epicyclic frequency $\varkappa$. Indeed, the frequency is evidently equal to $|\omega - m\Omega|$, and at the Lindblad resonance $|\omega - m\Omega| = \varkappa$. If one assumes that $S \approx$ const, then the radial (f_r) and the transversal (f_φ) forces will act on the star approximately in phase, and one may write

$$f_r = F_r \cos(\varkappa t + \gamma), \qquad f_\varphi = F_\varphi \cos(\varkappa t + \gamma), \tag{2}$$

where F_r, F_φ are the amplitudes of these forces, and $F_r \gg F_\varphi$. By linearizing the equations of motion and taking into account the definition of the epicyclic frequency, we shall obtain the following equation of perturbed motion:

$$\ddot r_1 + \varkappa^2 r_1 = \frac{2\Omega_0}{r_0} L_1 + F_r \cos(\varkappa t + \gamma),$$

$$\dot L_1 = r_0 F_\varphi \cos(\varkappa t + \gamma), \tag{3}$$

$$\dot\varphi_1 = \frac{L_1}{r_0^2} - 2\Omega_0 \frac{r_1}{r_0}.$$

By integrating (3) with the zero initial conditions, we shall obtain:

$$L_1 = r_0 F_\varphi \varkappa^{-1}[\sin(\varkappa t + \gamma) + \sin \gamma],$$
$$r_1 = -\Omega_0 F_\varphi \varkappa^{-2} t \cos(\varkappa t + \gamma) \qquad (4)$$
$$+ \tfrac{1}{2} F_r \varkappa^{-1} t \sin(\varkappa t + \gamma),$$

and, in the expression for the radial displacement r_1, only the time-increasing (secular) terms are written, which arise due to the resonance between the free radial oscillations of the star and the perturbing force produced by the spiral arms. The solution for r_1 consists of two parts, which are produced by the radial and tangential forces F_r and F_φ.

Consider first the influence of the radial forces, which give in (4) the dominating term ($F_r \gg F_\varphi$). The time-increasing displacements corresponding to them lag the forces by one-quarter of the period $2\pi/\varkappa$, i.e., the major axis of the perturbed orbit coincides with the azimuth, on which the spiral structure (the maximum of the density σ_1, or ($-\Phi_1$)) reaches the resonant circle. We shall further consider the case of the two-armed spiral, $m = 2$. For the *inner Lindblad resonance* (Fig. 113), the major axis is displaced (from the position on the circle) toward the position outside this circle, where it slightly leads the arm (if one takes into account the trailing character of the arms, as well as the additional effect from the tangential forces, see below). As a result, there the torque arises, which pulls the arm forward, and the orbit backward. The minor axis (see Fig. 113) is located slightly behind the region where the arm structure has a "negative density." Accordingly, this region repels orbits. In both cases, the angular momentum and the energy are taken from the orbit and fed into the spiral wave.

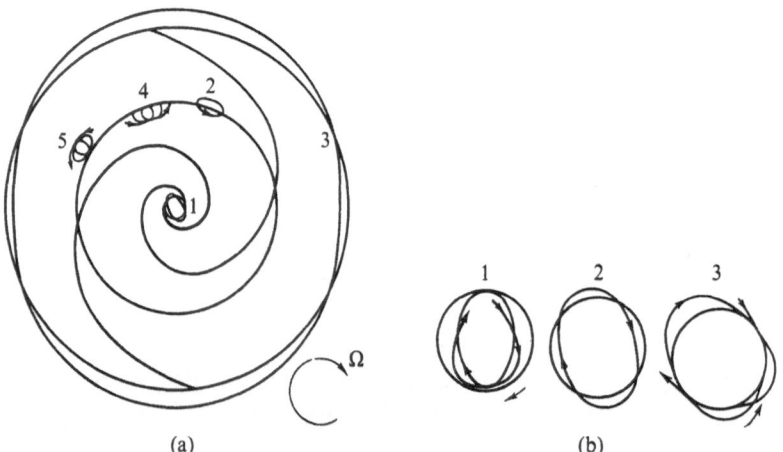

(a)   (b)

Figure 113. Unperturbed resonant orbits at the three main resonances [289]; (a) 1—the inner Lindblad resonance; 2—the corotation resonance; 3—the outer Lindblad resonance; 4, 5—nearly resonant orbits close to the corotation; (b) the region of the inner Lindblad resonance $r = r_L$; arrows show the rotation of the major axis for the nearly resonant orbits: 1—$r < r_L$; 2—$r = r_L$; 3—$r > r_L$; the system of reference rotates together with the spiral wave.

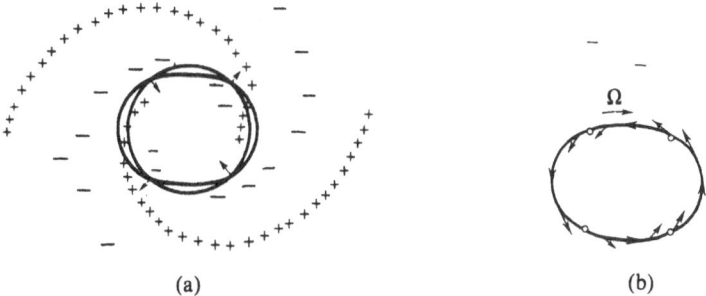

(a)   (b)

Figure 114. Perturbations of the circular orbit under the action of radial forces (which are shown by arrows) from (a) the spiral at the inner Lindblad resonance and (b) the excess transversal forces at the perturbed orbit [289]; plus (+) and minus (−) signs show the location of the spiral wave (maxima and minima of density).

At *the outer Lindblad resonance*, on the contrary, the minor axis lags the positive arms, while the major axis leads the negative arms. Therefore, here the angular momentum and the energy are transferred from the spiral wave to the stellar orbits.

Consider now the additional small effect of the tangential forces. In order to isolate the action of these forces, assume that the radial forces are not present. Since the tangential forces, according to (3), cause a radial displacement which is in counter-phase with the force, the major axis due to these forces alone will tend to lead the spiral structure by $\pi/4$ (Fig. 114).

Thus, under the combined effect of radial and tangential forces, the major axis will slightly lead the spiral structure. But the main effect of the tangential forces consists not in the eccentricity of the orbits, which is produced by them, but primarily in the slowing down or acceleration of the particles at different azimuths. In the wave system, the azimuthal angular velocity of the star is

$$|\dot{\phi}_A| = |\dot{\phi} - \Omega_p| = |Lr^{-2} - \Omega_p|. \tag{5}$$

The secular effects in $|\dot{\phi}_A|$ occur only due to the secular growth of r. Therefore, at the inner Lindblad resonance, where $Lr^{-2} > \Omega_p$, these effects give a slowing down, when r is large (apocenter), and a speeding up when r is small (pericenter). At the outer Lindblad resonance $Lr^{-2} < \Omega_p$, and, correspondingly, the situation is reversed. At the inner resonance, the major axis (corresponding, we recall, to the tangential forces only) lies on $\pi/4$ in front of the spiral structure. The density excess at these azimuths is attracted backwards, towards the arm. In the same way, the lack of density near the minor axis is attracted backwards, to the lack of density in the spiral structure. Therefore, the resulting torque connected with the tangential force reinforces the torque caused by the increasing (mainly, due to radial forces) eccentricity of the orbit. It may easily be shown that a similar amplification takes place also for the outer resonance.

In conclusion, we pay attention on one apparent contradiction. The effect following from the calculations made increases with time, while the effect described by formulae (16) and (17) (Section 2.1) is independent of time. The paradox is explained, of course, by the fact that we have considered (for the sake of simplicity) only the exactly resonant stars. In reality, of course, there is no exact resonance. The stars in near-resonant (but not quite resonant) orbits, make a contribution of a definite sign during the course of a long time interval, but, however, the sign of the effect is ultimately reversed. As a result, if we integrate over all nearly-resonant orbits, we shall obtain exactly the total effect as in (16) and (17) which is independent of time. The physical mechanism leading to small (additional) terms in formula (16), dependent on $\partial f_0/\partial L$, is discussed below, after the explanation of the mechanism of exchange of the momentum at the corotation resonance. We shall see that these mechanisms are in many respects similar (in particular, this is due to the fact that in both cases they are defined by the derivative $\partial f_0/\partial L$).

2.2.2. Corotation Resonance. The situation here very much resembles the situation well studied in plasma physics (Landau damping). Therefore we shall mention first of all the ordinary qualitative explanation of the plasma wave damping. Consider the monochromatic wave of small amplitude in a homogeneous plasma. Assume that the wave propagates in the direction of the positive axis x with a phase velocity $v_{ph} = \omega/k$. It is clear that the particles, having the velocities close to the wave velocity v_{ph}, interact strongly with the wave. This interaction leads to contrary effects for the particles a little faster than the wave, and the particles somewhat slower than the wave. Take first of all a uniform (at the initial time $t = 0$) distribution of the particles which have a velocity slightly exceeding the velocity of the wave. The particles trying to climb out of a potential well, lose their energy and are decelerated, while the particles on the downhill slope are accelerated by the wave. Therefore in the case of interest, the following statement is true:

(1) There is an excess of particles on the uphill climb (they are accumulated here in accordance with the continuity equation) and a deficit on the downhill slope of the potential well.

The particles on the uphill climb of the potential well push the wave in the direction of its propagation, and, consequently, feed their energy into the wave, while the particles on the downhill slope, on the contrary, are pushed by the wave and therefore take energy from the wave. Therefore, the following statements are also true:

(2) The particles going just faster than the wave, feed their energy into the wave. In a similar way:
(3) The particles going just slower than the wave, take energy from the wave. And finally:
(4) If (as, e.g., in the state of thermodynamical equilibrium) the particles moving more slowly are larger in number than those moving more

rapidly, the effect of wave damping (3) exceeds the amplification effect (2) so that the wave energy decreases with the decrement de-wave, give energy and momentum to the wave. And finally, the concluding result:

Let us return now to the corotation resonance in the galaxy. We shall consider the interaction of the wave with the particles, whose orbits are close to the corotation circle. The unperturbed trajectories of these stars are schematically depicted in Fig. 113 (in the frame of reference connected with the rotating spiral wave; this system is coincident with the system of the particle at the corotation circle). In the system of the wave, the stars with epicenters lying inside this circle are moving, on the average, with an angular velocity slightly greater than the angular velocity of the wave, while the stars whose epicenters lie outside the corotation circle have angular velocities slightly lesser in comparison with the wave.

At first sight, it appears that it would be quite logical to assume that here also the statements similar to those in (2) and (3) are valid. Then we shall immediately obtained that, at the corotation resonance, the energy must be transferred from the particles to the wave (indeed, in galaxies the density increases toward the center and, consequently, there is a somewhat larger number of particles with an angular velocity slightly greater than the velocity of the wave in comparison with particles with a lower velocity). But this conclusion would be incorrect: for example, it contradicts the formal con-clusion following from (16) (Section 2.1) which we already established. The cause of the error lies in the fact that in this case the statement in (1) is wrong and, consequently, also the statements in (2) and (3), connected with it. In order that this might be understood, it is necessary to consider the situation more carefully. In reality, the following takes place. The unperturbed orbits close to the corotation circle (cf. Fig. 113) have very small average motions in the system of the wave: on the internal side, this slow drift is directed forwards and on the external side, backwards. Under the action of the force acting from the direction of the spiral arm, the forward moving star will convert to the epicycle with a slightly larger angular momentum so that its mean drift motion, on the contrary, will slow down rather than speed up. According to the expression by Lynden-Bell and Kalnajs [289], the stars in their motion in the azimuth are acting like donkeys: they slow down when they are pushed forward, and speed up, when they are pulled backward. Therefore instead of the statement in (1), in this case another statement (contrary to (1) in its meaning) takes place:

(1) There is an excess of stars on the downhill slope and a deficit on the uphill climb; correspondingly, also the statements in (2) and (3) are reversed;

(2) There will be slightly more stars on the downhill slope with an angular velocity somewhat larger than the velocity of the wave so that they will take energy and angular momentum from the wave;

(3) The stars with angular velocities slightly less than the velocity of the wave give their energy and momentum to the wave. And, finally, the concluding result;

(4) Since normally (see above), the number of particles with larger angular velocities is somewhat larger, then at the corotation resonance both energy and momentum of the wave are absorbed and transferred to the stars.

We consider finally the mechanism leading in formula (16) (Section 2.1) to the terms $\sim \partial f_0 / \partial L$ at the Lindblad resonances. Since this effect disappears for the circular unperturbed orbits, it is necessary to consider the noncircular unperturbed orbits, close to the Lindblad resonances (see Fig. 114). We shall not take into account the effects connected with eccentricity induced by the spiral wave (these effects are considered above). Only the orbits which lie exactly at these resonances are strictly closed in the system of the wave (they intersect the resonant circle). Each orbit near the inner Lindblad resonance is connected to the density excess near the ends of the major axis (due to the decrease of the azimuthal velocity of the stars in this region).

The orbits located completely inside (or outside) the resonant circle are no longer closed. If they are close to the resonant circle, however, then they may be considered as closed orbits slowly rotating forwards (respectively, backwards). Such a consideration is natural just because of a great difference in angular velocities: of a rapid rotation of the particles in their orbit and a slow drift of the orbit itself (see Fig. 114). If the forward rotating major axis is subjected to the action of the rotating torque also pushing forward, then the particles in their orbits will slightly increase their angular momenta, and, as a result, the precession of the orbit itself will slow down. Thus, the major axes of the orbits will again behave like donkeys (compare with the discussion above). Therefore, one may literally repeat the statements under the preceding item, which pertain to the "donkeys-stars" at the co-rotation resonance, for this case of density excesses connected with the major axes.

In the region, where the rotating torque tries to speed up the motion of the major axis, it does indeed slow down, and vice versa. As a result, we get that outside the resonant circle there is a slight excess of density of the major axes on the azimuths slightly lagging the spiral structure, while inside the resonant circle, the larger axes have a slight excess somewhat ahead of the spiral structure (see Fig. 114). Therefore, for the density falling outside one may expect that the mechanism under consideration ($\sim \partial f_0 / \partial L$ in formulae (16) and (17), Section 2.1) should lead to absorption of energy and angular momentum at the inner Lindblad resonance.

In papers dealing with the Landau damping of plasma waves, the effect of the trapping of *particles* in the potential wells of the wave is investigated in detail. In a similar way, one may expect also trapping ("alignment") of orientations of the *major axes* near the resonant orbits (Lindblad). In the *nonlinear* consideration, not only the resonant axes should be trapped, but also the closely lying axes, if they cannot overcome the hump of the perturbed potential. This may be true for a wide region of the disk, when the value $\Omega - \varkappa/2$ does not vary very rapidly with radius.

The effect of alignment is considered by some workers to be important in the problem of the origin of bars in galaxies. For example, according to the opinions of Lynden-Bell and Kalnajs [289], the bars are a quasistationary standing wave. For that reason, the problem of origin of the bars is associated with the problems of wave theory. In the internal parts of the galaxy (where the presence of a bar is possible) the eccentricities of the stellar orbits should be large to ensure the stability of the axially-symmetrical modes. But the influence of the resonances becomes small when the stellar orbits are eccentric, while the modes of the system without resonances should satisfy the antispiral theorem, so that the main "two-arm" perturbation, according to [289], is a bar.

The trend toward the formation of a bar, which is revealed by a number of authors in linear theory, may be traced further and developed (already in nonlinear theory) just by means of considering the trapping of major axes of the orbits. Following Lindblad, consider a galaxy in which $\Omega - \varkappa/2$ changes insignificantly with radius. Then, the nonlinear potential perturbation may trap the major axes, making them oscillate near the azimuth of the potential well (a similar influence was found earlier by Contopoulos near the Lindblad resonance [189]). The density associated with these trapped orbits will increase the potential and further increase the trapping. The eccentricities of such trapped orbits are large at the inner Lindblad resonance, so that nearly circular orbits are rare. Thus, similarly to Lindblad, Lynden-Bell and Kalnajs [289] believe that the bars "are made" of stars in eccentric orbits with aligned major semiaxes. The angular velocity of such a bar will increase due to the action of its gravitation on the stellar orbits, but will remain significantly less than the angular velocity of the stars composing it.

Note another conclusion [254] following from the considered picture of the interaction of resonance particles with density waves, and concerning the boundaries of applicability of such general stability conditions as the Peebles–Ostriker criterion (see Section 3.2, Chapter IV). It is clear, for example, that the stable disks A and B_0 from the number of the composite models considered earlier (Section 4.4, Chapter V) having $t = 0.125$ and $t = 0.086$ may be slightly modified in such a way that they will become slightly unstable. For example, the stable barlike mode of negative energy rotating in the direct direction, will obviously become slightly unstable, if we place a small number of stars on nearly circular orbits around the region of the outer Lindblad resonance, where these stars act as the absorbers of (positive) energy and angular momentum [254]. Therefore, one may arrive at the conclusion that although some simple criterion (of the type of $t \leq 0.14$) may be a sufficient condition of the lack of instabilities with growth rates comparable with the orbital frequencies of stellar revolutions, it is questionable that a strict stability criterion would exist for nonaxially-symmetrical perturbations.

§ 3 The Linear Theory of Stationary Density Waves

3.1 The Primary Idea of Lin and Shu of the Stationary Density Waves

Lin and Shu [267] called attention to the following circumstance. If the Toomre criterion is applied to the region comparatively close to the center of the Galaxy, $r \sim 3 \div 5$ kps, then the minimum velocity dispersion of stars, which makes this region *locally* stable, there will result an *unnaturally* large value $c_r \sim 90$ km/s. Indeed, such "hot" stars must, during the time period $\sim 10^9$ years, have reached the solar vicinity of the Galaxy, but in reality, as we are aware, there are no such stars present here (in any appreciable number). Hence, the conclusion should be drawn that the velocity dispersion near the center is less than $c_{r\,min}$ and this region is therefore locally unstable. At the same time, as shown by Toomre, the solar vicinity of the galactic disk is locally stable.

The primary idea of Lin and Shu [267] was that in a system with stable and unstable spatial regions there may exist *stationary* density waves. Without discussing the correctness of this statement,[6] note that the idea itself proved to be very fruitful. Its elaboration has finally led to some theory of the *stationary* density waves in galaxies (based on the analysis of short-wave perturbations of the disk).

The spiral density waves are collective oscillations of the disk of the form

$$\sigma_1 \sim \tilde{\sigma}_1 \exp\{-i(\omega t - m\varphi + \psi(r))\}, \tag{1}$$

imposed on the stationary background $\sigma_0(r)$. Here $\tilde{\sigma}_1$ is the amplitude, ψ is the phase, ω is the frequency, m is the azimuthal number equal to the number of arms. From (1) it is easy to see that the spiral wave rotates with an angular velocity ω/m, without changing its shape with time. Since in this theory, the wave frequency ω is considered as a constant, the local dispersion equation defines the wave number as a function of radius. For example, for a cold disk, according to Section 2.2, Chapter V,

$$k(r) = \frac{\varkappa^2(r) - [\omega - m\Omega(r)]^2}{2\pi G \sigma_0(r)}. \tag{2}$$

As we already know from Section 4.1, dispersion Eq. (2) on which the primary analysis of Lin and Shu [267] was based, was essentially improved in subsequent papers by introducing various types of "reduction" factors for stars and by taking into account the gaseous constituent of the system.

---

[6] The paper [114^{ad}] contains a statement that in such a system intermediate wavelengths may be excited. However, the proof of this statement is absent (as is also the statement's proof in [267]).

Comparison of the theory with observed data on our Galaxy is performed by Lin *et al.* [271] and with data on some other galaxies by Shu *et al.* [326]. We shall give a brief summary of this comparison somewhat later.

3.2 The Spiral Galaxy as an Infinite System of Harmonic Oscillators

One may distinguish two aspects in the consideration of the evolution of initial density perturbations [251]:

(1) kinematic evolution, in whose consideration we neglect those forces associated with the perturbation itself (in the purely gravitational theory, this is the neglection of self-gravity);

(2) dynamical evolution, which takes into account the influence of the perturbed gravitational field.

Let us start with the former, kinematic, aspect of the evolution. As we shall now see, this evolution in stellar systems may in principle occur in a quite different way to that in gaseous systems (see Fig. 109). Our intuition, based on hydrodynamical examples with which we are well accustomed, implies that any initial perturbation must be twisted and disappear due to the differentiality of the galactic rotation, for the time of the order of one rotation. We shall, however, see that the real fate of arbitrary perturbations in a galaxy is not so evident, and the customary intuition here sometimes fails.

It ignores the following essential difference between the gaseous and collisionless systems [251]: the viscous forces, for example in a tea cup, make liquid elements move in circular orbits, while the stars in the galaxy, apart from rotation about the center, may still oscillate in the radial direction. The cause of twisting and "dissolution" of the original picture of motion in the tea cup is of course the difference in angular rates of rotation of liquid elements at different radii. The same mechanism must have acted also in the galaxy if there were no radial oscillations. But the presence of radial oscillations makes the situation in the galaxy more complicated and also less obvious, despite the fact that the radial velocities in galaxies are normally an order of magnitude less than the circular velocities. The point is that for the phenomenon of "twisting," the velocities are not essential but, rather, for the oscillation frequencies and, primarily, the full width of the frequency "spectrum" (different for stars at different distances from the galactic center, etc.), these values are comparable for radial and circular unperturbed motions in galaxies. At any fixed point of the galaxy, there exist all the possible linear combinations of these frequencies, which just complicates the resulting motions in the galaxy. It turns out that there may be very diverse and unexpected possibilities. One may, for example, build up in a galaxy perturbations of a certain type, which have motions opposite to the galactic rotation, so that the differential rotation makes them leading structures. This example is in complete contradiction to our initial intuition, which "predicts" trailing spirals.

We turn now to the quantitative description of the kinematics, by using the action-angle variables (see [289]). The action variables are introduced in the following way:

$$J_1 = \frac{1}{2\pi} \oint p_r \, dr = \frac{1}{2\pi} \oint \sqrt{2E - 2U - \frac{L^2}{r^2}} \, dr, \tag{1}$$

$$J_2 = \frac{1}{2\pi} \oint p_\varphi \, d\varphi = L_z \equiv L. \tag{2}$$

They are the functions of integrals of motion E, L, and therefore are themselves integrals of motion. The Hamiltonian is dependent only on J_i, so that the equations for the angular variables w_i are thus:

$$\dot{w}_i = \frac{\partial H_0(J_1, J_2)}{\partial J_i} \equiv \Omega_i(J_1, J_2) = \text{const.} \tag{3}$$

The angles w_i change by 2π for one period of oscillation. If the orbits are nearly circular ($J_1/J_2 \ll 1$), they may be represented by the epicycles [289], and then $\Omega_1 \approx \varkappa$, to the epicyclic frequency, $\Omega_2 = \Omega$, the angular rate of rotation.

Any single-valued g function of a point in the phase space must be periodical in angular variables w_1, w_2 (with periods 2π). Consequently, the Fourier expansion

$$g(J_i, w_i) = \frac{1}{4\pi^2} \sum_{l, m = -\infty}^{\infty} g_{lm}(J_1) \exp[i(lw_1 + mw_2)], \tag{4}$$

is valid, where

$$g_{lm}(J_i) = \int_0^{2\pi} \int_0^{2\pi} g(J_i, w_i) \exp[-i(lw_1 + mw_2)] \, dw_1 \, dw_2.$$

If $g(J_i, w_i)$ is the perturbation of galaxy at the time $t = 0$, then the subsequent (kinematic!) evolution is a simple transfer of this perturbation by the stars along their unperturbed orbits. This inference can be obtained also as a formal consequence of the kinetic equation, which in this case has the form $df/dt = 0$, where d/dt is the Lagrange derivative along the unperturbed trajectories of stars. Take the Euler description, i.e., observe the change of the $g(J_i, w_i, t)$ function, being at a fixed point of the phase space and not following the phase trajectories of stars (as with the Lagrange approach). Then the value g at a selected point at a later time $t > 0$ will be equal to the value of g in that element of the phase space at the time $t = 0$, which at the time t has arrived at the point considered. From (4) it follows that at the time $t = 0$ this element was at the point $(J_i, w_i - \Omega_i t)$. Therefore

$$g(J_i, w_i, t) = g(J_i, w_i - \Omega_i t)$$

$$= \frac{1}{4\pi^2} \sum_{lm} g_{lm}(J) \exp[il(w_1 - \Omega_1 t) + im(w_2 - \Omega_2 t)]. \quad (5)$$

Hence it is obvious that a galaxy may be considered as an infinite system of harmonic oscillators; moreover, with the kinematic description of the evolution of perturbations, i.e., neglecting self-gravity, these oscillators are free (the coupling is realized just by means of the self-consistent perturbed gravitational potential). Oscillators may be "numbered" by the four numbers $(J_1, J_2, l. m)$, from which the two former (J_1, J_2) may change continuously, while the latter two are the pairs of integers (positive or negative). The amplitude and the phase of such an oscillator $((J_1, J_2, l, m)$-oscillator) is defined by the $g_{lm}(J_1, J_2)$ function, while the frequency is $(l\Omega_1 + m\Omega_2)$. The density wave corresponding to the oscillator rotates with an angular rate $(l/m)\Omega_1 + \Omega_2$; m is the angular periodicity of the wave, while l defines the radial structure.

3.3 On "Two-Armness" of the Spiral Structure

As already noted, the idea of spiral *waves* of density was advanced by Lindblad. Recently, however, for a number of reasons, his role in the creation of the galactic spiral structure theory is frequently understated. Kalnajs [251] recalls the decisive contribution of Lindblad to the theory and explains his ideas by using a more sophisticated terminology. In particular, he notes that the theory (including that of Lin and others) is actually due to Lindblad for the explanation of the preference of the two-armed structure.

Assume that in the galaxy a certain *smooth* perturbation is imposed. It is obvious that in most cases (excepting only very special ones) the presence of such perturbation is equivalent to excitation of continuum of the oscillators of types described above. The evolution of density will be characterized by the cuttings of the original smooth picture into still finer scales. "The dissolution" of the original distribution may be described as the interference or phase intermixing of different harmonic components (a similar phenomenon is studied, for example, in plasma physics). The characteristic time of this process is, roughly, the mean inverse angular velocity of the perturbation in the region of the phase space in question. The only effect remaining after the full intermixing from the perturbation is associated with the g_{00} term in (5), Section 3.2.

Consideration of the frequency range associated with a typical flat galaxy, demonstrates that the process of intermixing *for most* of the perturbations has, as a rule, the time scale of the order of one period of rotation, which apparently corresponds exactly with our intuition.

It is however important that this rule has an exception which corresponds to the $(l = -1, m = 2)$-terms in (5), Section 3.2. Lindblad demonstrated [273] that the linear combination $\Delta\Omega \equiv \varkappa - 2\Omega$ remains roughly constant for a significant part of the Galaxy (moreover, $\Delta\Omega \ll \Omega$). Therefore, the two-armed perturbations may exist in the Galaxy in the course of many rotations (unlike all other perturbations). Thus it is obvious that our intuition in this case has not taken into account the possibility of a close coincidence of different harmonics of radial oscillations and the circular rotation of the stars of the Galaxy. Of course, this remarkable fact would have appeared to us *a priori* somewhat occasional, and we should have omitted it. Nevertheless, just such an "occasional" situation is realized in reality, and the explanation of this fact is one of the most important (and not yet resolved) problems of the future theory of the origin and evolution of spiral galaxies. The relative constancy of $\Delta\Omega \equiv (\varkappa - 2\Omega)$ is due to the corresponding distribution of density in flat galaxies.

In the now existing theory of the established spiral waves, this fact is in no way explained and is taken as given. It is likely to provide an explanation of why the two-armed shapes prevail in these systems. Requirement of the stability of the galaxy with respect to axially-symmetrical perturbations [333] limits the effect, which may be exerted by a given force field, especially on the shortest spatial scales. The decrease in the value of the effect may be compensated for by a longer time, in the course of which it may result in a given response. Therefore, the stable nature of two-armed perturbations, and their long-term resistance to differential rotation, makes them to be the most probable self-consistent perturbations. The self-gravitation of waves in this case may be comparatively modest in order to overcome the influence of the remaining weak "shear" and to provide the uniform rotation of the spiral pattern.

An example which appears to contradict the intuition, is any one-armed structure formed by the oscillators ($l = 1, m = 1$). Such a structure rotates at a velocity of $\Omega_p = \Omega - \varkappa \approx -\Omega$, i.e., in the "inverse" direction, and is twisted in the *leading* direction.

In the Lin and Shu theory, the distinct nature of the two-armed perturbations formally follows from the condition $\nu^2 < 1$, i.e.,

$$\Omega - \frac{\varkappa}{m} < \Omega_p < \Omega + \frac{\varkappa}{m} \qquad (1)$$

($\Omega_p = \omega/m$ is the angular velocity of the wave), which singles out the "main" part of the spiral picture. If $\Omega(r)$, $\varkappa(r)$, and $\Omega \pm \varkappa/2$ from the Schmidt model [319] (Fig. 115) are used, then for $m = 2$ the spiral pattern in (1) will occupy the range from $r = 4$ kps to $r \geq 20$ kps for $\Omega_p = 11$ km/s · kps. At the same time,[7] for all $m > 2$, the main part of the spiral pattern will have a quite small spatial extension, moreover, independent of the selection of Ω_p. It is evident that this cause of the distinct nature of the $m = 2$ mode is in essence the same as in the previous discussion.

---

[7] The one-armed perturbations need special consideration.

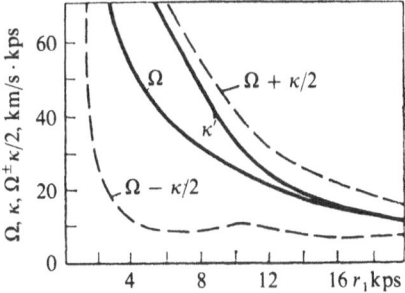

Figure 115. Rotation curve of the Galaxy according to Schmidt's model [271]; epicyclic frequencies $\varkappa$ and Lindblad's combinations $\Omega \pm \varkappa/2$ are also presented.

3.4 The Main Difficulties of the Stationary Wave Theory Of Lin and Shu

3.4.1. Antispiral Theorem. The validity of the gravitational theories of spirals was questioned by Lynden-Bell and Ostriker [287] who proved the so-called antispiral theorem, which states that the spiral shape cannot exist as a neutral mode of oscillations of differentially-rotating and nondissipative gaseous systems.

The interpretation of this difficulty of the Lin and Shu wave theory is contained in a paper by Toomre [334] (as well as by Shu [325], see below, subsection 3.4.2), who has demonstrated that the waves of a spiral form (both in the gaseous and in the stellar disk) propagate in the radial direction with a rather high group velocity, so that even the existing spiral waves must finally disappear.

To begin with, turn to the "direct" derivation of the antispiral theorem [324] in the case of a stellar disk, similar to the derivation for the gaseous case [287]. The basic integral equation for normal modes of the stellar disk (derived by Shu [324] and Kalnajs [249]) is Eq. (38), Section 4.3, Chapter V. We are further interested in the case of neutral oscillations in the absence of resonances. In this case, the kernel $K_{m,\omega}(r, a)$, according to (38) and (39), Section 4.3, will be real. Then the antispiral theorem of the type proved by Lynden-Bell and Ostriker [287] will take place. Taking the equation complexly-conjugate to Eq. (38), Section 4.3, Chapter V, it is easy to show that if $\tilde{\sigma}_1(r)$ is the solution, then also the complexly-conjugate $\tilde{\sigma}_1^*(r)$ will also be a solution. If ω is not degenerate, then $\tilde{\sigma}_1^*(r)$ may differ from $\tilde{\sigma}_1(r)$ only in the complex constant (with the unity modulus): $\tilde{\sigma}_1^*(r) = e^{-2i\chi}\tilde{\sigma}_1(r)$, where χ is some real constant. Equating the arguments in this equality, we shall obtain $\arg\{\tilde{\sigma}_1(r)\} = \chi$. Since the phase $\tilde{\sigma}_1(r)$ is strictly constant, this normal mode has no spiral shape, it has the appearance of a "cart wheel." If, however, ω is degenerate and $\tilde{\sigma}_1$, $\tilde{\sigma}_1^*$ are linearly independent, they must correspond to spirals of opposite twisted shape. Therefore, one may assume as linearly independent the purely real solutions (and, consequently, nonspiral)

$$\tfrac{1}{2}(\tilde{\sigma}_1 + \tilde{\sigma}_1^*), \qquad \frac{1}{2i}(\tilde{\sigma}_1 - \tilde{\sigma}_1^*). \tag{1}$$

If, however, there are resonant stars, such an analysis is not valid, and the arguments leading to the antispiral theorem are inapplicable. This is not surprising. The antispiral theorem is mainly a reflection of the temporal reversibility of the equations of motion. If we reverse the direction of time and simultaneously "turn" the galaxy over (reflect all the motions in the meridional plane), $\varphi \to -\varphi$, then we shall arrive at a state where we impose the perturbation with the opposite direction of spiral twisting on the same stationary background. The oscillation frequencies in the original and transformed states should be coincident and there are no reasons to give preference to any one of them. The respective solutions for the equations for normal modes must be, generally speaking, antispiral, It may be said that for neutral oscillations there is no "time arrow" and there is no prevailing direction of spiral twisting with respect to the motion of matter.

With the resonances existing, it should be taken into account that the effects of the interaction of the wave with the resonant stars began in the past: this defines the "time arrow." In a similar way, the instability may also introduce a difference between the leading and the trailing spirals.

Thus, the "antispiral theorem" similar to the "gaseous" theorem of Lynden-Bell–Ostriker [287] is applicable in the linear theory to all neutral modes, for which there are no resonances.

3.4.2. Wave Packet Drift.

Here we shall now consider perhaps the most essential difficulty of the original theory of the spiral density waves of Lin and Shu, which was indicated by Toomre [334]. He paid attention to the fact that the Lin and Shu wave packets (with frequencies lying within the interval ω, $\omega + \Delta\omega$) must be drifting in the radial direction with a group velocity $c_g = d\omega/dk$. More specifically, Toomre's argument is as follows. Assume that the dispersion equation $\omega = f(k, r, m)$ is known. Then, substituting (as in the geometrical optics) in this equation $\omega \to \partial\Phi/\partial t$, $k \to -\partial\Phi/\partial r$, where $\Phi(r, t)$ is the WKB phase of the wave packet, we shall obtain $\partial\Phi/\partial t = f(-\partial\Phi/\partial r, r, m)$. Therefore, by differentiating with respect to t and r we find the equations

$$\frac{\partial\omega}{\partial t} + \left(\frac{\partial f}{\partial k}\right)_r \frac{\partial\omega}{\partial r} = 0, \tag{2}$$

$$\frac{\partial k}{\partial t} + \left(\frac{\partial f}{\partial k}\right)_r \frac{\partial k}{\partial t} = -\left(\frac{\partial f}{\partial k}\right)_k. \tag{3}$$

Equations (2) and (3) show that the information, regarding the frequency and the wave number in a slowly evolving wave packet, propagates along the radius at a velocity $dr/dt = (\partial f/\partial k)_r = c_g(r, k, m)$. The characteristics of these equations have at a given point (r, t) an inclination equal to $dr/dt = c_g$. Along each characteristic curve, according to (2), the frequency ω remains constant.

We introduce, as in [334], the notation $|v| = N(|\zeta|, Q)$ for the functions whose plots are given in Fig. 75. Then the Lin and Shu dispersion equation will be written thus

$$\omega = m\Omega(r) + \operatorname{sgn}(v)\varkappa(r)N(|\zeta|, Q). \tag{4}$$

Hence

$$c_g = \frac{dr}{dt} = \frac{\partial\omega}{\partial k} = \operatorname{sgn}(\zeta v)\left[\frac{\varkappa(r)}{k_T(r)}\right]\frac{\partial N}{\partial|\zeta|}, \tag{5}$$

and the dimensionless wave number $\zeta = k/k_T(r)$ changes along the characteristic curve according to

$$\frac{d\zeta}{dt} = -\frac{\partial\omega}{\partial r} = \frac{\varkappa(r)}{k_T(r)}\left(\frac{dv}{dr} - \operatorname{sgn}(v)\frac{\partial N}{\partial Q}\frac{dQ}{dr}\right). \tag{6}$$

In order to obtain representation on the typical characteristic curves, consider [334] a simple model of the Galaxy, in which $\Omega(r) = V/r$, $\sigma_0(r) = V^2/2\pi Gr$, $V = $ const. Assume also that the Toomre stability parameter Q is independent of the radius: $Q \neq Q(r)$. For such a model $\varkappa(r)/k_T(r) = V/\sqrt{2}$ so that the dependence $v(r)$ is linear: $v = m[(r/r_c) - 1]/\sqrt{2}$ where r_c is the radius of corotation, on which $\omega/m = \Omega(r)$. In addition, from Eq. (6) it follows that in this case the dependence of the dimensionless wave number ζ on time t is also linear: $\zeta = (\omega/2)t + $ const. Therefore, the problem of obtaining the characteristic curves is here reduced simply to the substitution of the notations of axes v, ζ in Fig. 75, which shows the dispersion curves for different Q, by r, t.

Figure 116 is a system of characteristic curves, corresponding to $Q = 1.2$. The inner Lindblad resonance $v = -1$ corresponds to $r/r_c = 1 - 1/\sqrt{2} \approx 0.293$, while the outer one $v = +1$ corresponds to $r/r_c \approx 1.707$. Between

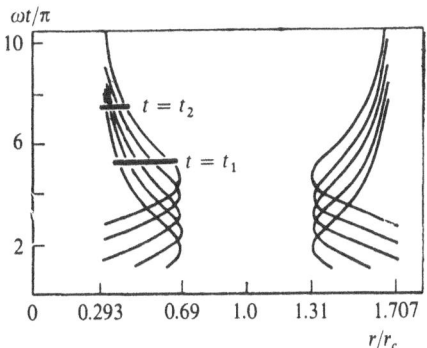

Figure 116. Some characteristic curves for the disk with $Q = 1.2$ according to Toomre, $m = 2$ [334].

these points, there is the main part of the spiral structure (according to the Lin and Shu terminology). Outside this region ($|v| > 1$) there are no solutions of the WKB type oscillating along the radius. From Figure 116 it is seen that at $Q > 1$, there is still one prohibited region, which in this case lies within the interval $0.69 < r/r_c < 1.31$. It is, as we have seen, the consequence of the fact that at $Q > 1$ the local self-gravitation is not able to reduce $|v|$ below a certain minimum level (cf. Fig. 75).

We now make a quantitative estimate of c_g for the solar vicinity of our Galaxy. Assume, according to Lin and Shu [271], that $\Omega_p = \omega/2 \approx 12.5$ km/s · kps, so that the corotation radius ($v = 0$) is equal to $r_c \approx 17$ kps, i.e., the region corresponding to $v > 0$ lies in fact outside the Galaxy, and it might remain so without being considered. Since, according to [271], the dispersion $c_r \approx 35$ km/s roughly corresponds to the stability boundary ($Q \approx 1.0$), therefore $\varkappa/k_T \approx (0.2857)^{-1/2}(c_r/Q) \approx 65$ km/s and $\lambda_T \approx 13$ kps. Since $\Omega_p \approx 12.5$ km/s · kps corresponds, for the solar vicinity of the Galaxy, to $\lambda = 3 \div 4$ kps, then $\zeta = k/k_T \approx 4$. The inclination of the curve $N(|\zeta|)$ (Fig. 75) at the point $\zeta = 4$: $\partial N/\partial|\zeta| \sim 0.15$. The data thus obtained allow one to obtain the required estimate $c_g \approx -10$ km/s, where the minus sign means that the wave packet should propagate toward the center of the Galaxy.

Thus, for the time $\sim 10^9$ years (i.e., only for four revolutions of the Galaxy about its center—galactic years) the initial spiral perturbation should be transferred from the periphery to the center, covering a distance of ~ 10 kps.

For a more detailed treatment of the points discussed above, it is necessary to solve the kinetic equation and the Poisson equation within an accuracy of up to two orders of magnitude with respect to small parameters $1/kr$ and $c_r/r\Omega$, i.e., the correct account for the amplitude of perturbation (pre-exponent in the WKB-method). The necessary work was performed by Shu [334] and Mark [290].

Let us first of all consider the Poisson equation. Let the perturbed potential in the disk plane be $\Phi_1(r, \varphi, z = 0, t) = \Phi(r) \exp[-i(\omega t - m\varphi)]$. This potential has a form of a short-wave (tightly wound) spiral, if in the representation $\Phi(r) = A(r) \exp[i\psi(r)]$ ($A(r)$ and $\psi(r)$ are the real functions), the rate of change of the phase $\psi(r)$ is high in comparison with the rate of change of the amplitude $A(r)$. Thus, we require that there be $|r\psi'(r)| \gg 1$. Then we may show (detailed calculations, see Section 7.1, Appendix) that within an accuracy of up to two orders of magnitude with respect to $1/kr$, there is the following relation between the perturbed potential and the surface density $\sigma_1(r, \varphi, t) = \sigma(r) \exp[-i(\omega t - m\varphi)]$:

$$\sigma(r) = -\frac{|k|\Phi(r)}{2\pi G}\left\{1 - \frac{i}{kr}\frac{d}{d \ln r}\ln[r^{1/2}A(r)]\right\}. \tag{7}$$

In the lowest approximation, from Eq. (7) it follows the old result: maxima of the surface density correspond to minima of the potential.

Let us now consider the response of the stellar disk to a given perturbation of the potential. Calculations [325, 290] are rather cumbersome, and therefore they are shown in Section 7.2 of the Appendix. Below, we re-stricted ourselves to only a description of the assumed approximations and the results obtained.

As the distribution function of the unperturbed potential Shu [325] takes the Schwarzschield modified distribution

$$f_0(E, L) = \begin{cases} P_0(r_0) \exp[-\varepsilon/c_0^2(r_0)], & \varepsilon < -E(r_0), \\ 0, & \varepsilon > -E(r_0). \end{cases} \tag{8}$$

The "epicyclic" integrals r_0 and ε are defined as a function of E and L from the equations

$$r_0^2 \Omega(r_0) = L, \qquad \varepsilon = E - E_c(r_0), \qquad E_c(r_0) = \tfrac{1}{2} r_0^2 \Omega^2(r_0) + \Phi_0(r_0),$$

where $r_0 \Omega^2(r_0) = \partial \Phi_0/\partial r_0$. The meaning of $E_c(r_0)$ is obvious: this is the energy of the particle on an exactly circular orbit with a radius r_0; cor-respondingly, ε is the deviation of the exact energy of the particle E from E_c. The form of the $P_0(r_0)$- and $c_0(r_0)$-functions may always be selected so that any reasonable surface density and radial velocity dispersion are satisfied.

In the immediate vicinity of the resonances, where $v(r_0)$ is an integer or zero, there might be essential absorption (or, on the contrary, enhancement) of the density waves. These effects were considered in detail in §2. Below we shall restrict ourselves to the "main region" of the spiral structure $r_{-1} < r < r_{+1}$ ($v(r_{-1}) = -1$, $v(r_{+1}) = 1$, assuming the influence of the resonances inside this region to be negligible [325]. Besides, it is assumed that the second (after $1/kr$) dimensionless parameter of the problem $\varepsilon = c_0(r_0)/r_0 \varkappa(r_0)$ is also small in comparison with unity (which is equivalent to the assumption about the smallness of the peculiar velocities as compared to the circular velocity). Assuming then the hypothesis that $\varepsilon \sim |kr|^{-1}$ (thus, for the solar vicinity of the Galaxy $\varepsilon \sim 0.1$, while $|kr|^{-1} \sim 0.06$) one can obtain (see Section 7.2, Appendix) within an accuracy of up to two orders of magnitude with respect to the parameters ε, $|kr|^{-1}$:

$$\sigma(r) = -\frac{k^2 \sigma_0 \Phi}{\varkappa^2 (1 - v^2)} \mathscr{F}_v(x) \left\{ 1 - \frac{i}{kr} D_v(x) \frac{d}{d \ln r} \ln \left(\frac{\sigma_0 k}{\varkappa^2} \frac{\mathscr{F}_v}{1 - v^2} D_v r A^2 \right) \right\},$$

$$\tag{9}$$

where $\mathscr{F}_v(x)$ is the reduction factor of (22), Section 4.1, Chapter V, $x = \varepsilon^2 k^2 r^2 = k^2 c_r^2/\varkappa^2$, and

$$D_v(x) = -(1 - v^2) \frac{v\pi}{\sin v\pi} G_v'(x)/\mathscr{F}_v(x) = \frac{\partial}{\partial \ln x} \ln[x \mathscr{F}_v(x)],$$

$$G_v(x) = \frac{1}{2\pi} \int_{-\pi}^{\pi} \cos vs \, \exp[-x(1 + \cos s)] \, ds. \tag{10}$$

We finally take into account the requirement of the self-consistency of the density waves. Assuming v^2 to be a real number, we equate the real and imaginary parts in Eqs. (7) and (9):

$$\frac{k_T}{k}(1 - v^2) = \mathscr{F}_v(x), \tag{11}$$

$$\frac{1}{2}\frac{d \ln(rA^2)}{d \ln r} = D_v \frac{d}{d \ln r} \ln\left[\frac{|k|}{k_T}\frac{\mathscr{F}_v}{1 - v^2} D_v rA^2\right]. \tag{11'}$$

The former equation is the Lin and Shu dispersion equation, and we are already aware of it. For a marginally-stable disk, this equation gives the relation between v and λ/λ_T, shown in Fig. 73.

Equation (11') defines the radial variation of the wave amplitude. Using (11') it is easy to demonstrate that the density amplitude $S(r)$ of the stationary wave with the frequency v must satisfy the equation

$$\frac{d}{dr}\left[\frac{rS^2(r)}{k^2(r)}R_v(x)\right] = 0, \quad \text{i.e., } rA^2 R_v(x) = \text{const}, \tag{12}$$

where A is the amplitude of the potential perturbation ($A \sim S/k$), $R_v(x) = -\{1 + 2\partial \ln \mathscr{F}_v(x)/\partial \ln x\}$ (Fig. 117). The variation of the amplitude of the perturbed surface density, following from (12), for the marginally-stable disk is

$$\frac{|S(r)|}{\sigma_0} = \text{const} \cdot r^{-1/2}(\lambda_T \varkappa)^{-2}S_v(x),$$

$$S_v(x) = \frac{\lambda_T}{\lambda}[|R_v(x)|]^{-1/2} = \frac{(1 - v^2)}{\mathscr{F}_v(x)\sqrt{|R_v(x)|}}. \tag{13}$$

The value $|S|$ becomes large (formally, according to (13), infinite) near the corotation ($v = 0$) and Lindblad ($v = \pm 1, \ldots$) resonances (Fig. 118). Of course, in the immediate vicinity of the resonances, the derived relations are not valid (for more detail, see §2). In particular, the above singularity is also fictitious.

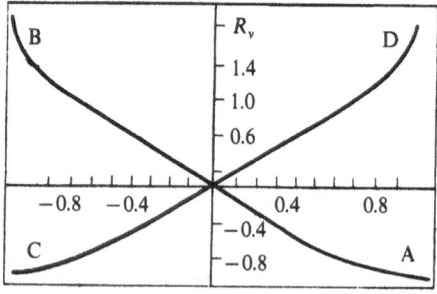

Figure 117. The R_v function [325]; A, B, C, D correspond to the notations in Fig. 73.

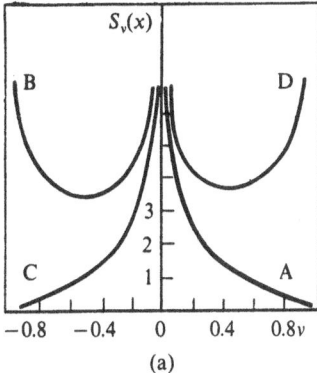
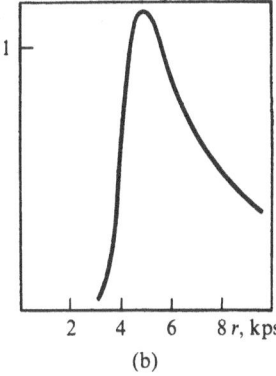

(a)   (b)

Figure 118. Amplitude of the perturbation of the surface density; (a) the S_v function [325]; A, B, C, D correspond to the notations in Fig. 73; (b) the dependence of the quantity $\sigma_1 \cdot \mathrm{Re}\, k$ on the radius with accounting for the resonant absorption for Schmidt's model ($\Omega_p = 13.5\ \mathrm{km/s \cdot kps}$) [290a].

As noted by Shu [325], Eq. (12) is so expressive that it would be surprising if it did not admit a simple interpretation. Such an interpretation was obtained by Toomre [334]. In combination with the dispersion equation and the expression for the group velocity $c_g = d\omega/dk$, (12) means that

$$\frac{d}{dr}\left(\frac{rc_g \tilde{E}}{\omega - m\Omega}\right) = 0, \tag{14}$$

where the positive value $\tilde{E}$ is the density of the wave energy for the observer in a locally rotating system:

$$\tilde{E} = \frac{\pi G}{2k_T(r)}\, S^2 v\, \frac{\partial}{\partial v}\left(\frac{\mathscr{F}_v(x)}{1 - v^2}\right). \tag{15}$$

This energy $\tilde{E}$ may be calculated [334] as the work per area unit performed between $t = -\infty$ and $t = 0$ by some outward, axially-symmetrical and very slowly increasing ($s \to 0$) mass distribution

$$\sigma(x, t) \sim S e^{st} \cos(kx \pm vkt),$$

in the plane of the initially unperturbed infinite stellar disk.

We shall not give here this elementary derivation, but refer to the available direct analogy between the case in question and similar formulae in plasma physics (see, e.g., [86]). Expression (15) can be obtained directly by analogy with the well-known expression for the energy of potential perturbations in the plasma

$$W_k = \omega_k \frac{\partial}{\partial \omega_k}\, \varepsilon_0(\omega_k)\, \frac{|E|^2}{8\pi}, \tag{16}$$

which equals the sum of the energy of nonresonant particles (for which, according to the definition, $\omega' \equiv \omega - kv \gg \gamma$, γ is the growth rate) and the electrostatic energy. Here the value ε_0 (the scalar dielectric permittivity) determines in the following way the response of the system to the potential, if one considers the electrostatic (electron Langmuir) waves:

$$\rho = en = -\Phi \frac{k^2}{4\pi} (\varepsilon_0 - 1). \tag{17}$$

Equation (16) may be compared to expression (15) for the energy of the spiral wave used by Toomre and constructed, as is seen, in a quite similar way. The "reduction" factor $\mathcal{F}_\nu(x)$ (more accurately, $\mathcal{F}_\nu(x)/(1 - \nu^2)$) enters the expression for the gravitational response of the disk to the potential Φ similarly to $(\varepsilon_0 - 1)$ in (17):

$$S(r) = -\Phi \frac{k^2}{\varkappa^2} \sigma_0 \frac{\mathcal{F}_\nu(x)}{(1 - \nu^2)}. \tag{18}$$

It is clear that, by knowing the "plasma" expression in (16), we could have at once written by analogy the corresponding "gravitational" expression in (15). The use of the "internal" frequency of the wave $\nu = (\omega - m\Omega)/\varkappa$ instead of ω in a "plane" plasma case is also, of course, quite natural.

Formula (14) refers to the case of a single temporal harmonic (the monochromatic wave, $\omega = $ const). For the superposition of such harmonics, from (14) one can derive[8] also the corresponding statement for the slowly evolving perturbation

$$\frac{\partial A}{\partial t} + \frac{1}{r} \frac{\partial}{\partial r} (c_g A) \equiv \frac{\partial A}{\partial t} + \text{div}(c_g A) = 0, \tag{19}$$

where $A = \tilde{E}/(\omega - m\Omega)$. Hence, it is seen that the density of the "wave action" A (rather than the wave energy density $\tilde{E}$ itself) propagates with a group velocity c_g. Kalnajs [251] has ultimately clarified the interpretation by demonstrating that the energy density E and the angular momentum density L, referring to the inertial system, can be expressed through the "action density" in the following way (see above, §2):

$$E = \omega A, \qquad L = mA. \tag{20}$$

The relationship between $\tilde{E}$ (energy in the system, which rotates at a local angular velocity Ω) and E, L is

$$\tilde{E} = E - \Omega L. \tag{21}$$

This relation is a usual law of transformation of energy for the transition from the inertial to the rotating system. The law of conservation of the wave action in (19) together with Eq. (21) describe therefore the obvious fact of the conservation of the wave energy and the angular momentum referred to the inertial system:

$$rc_g E = \text{const}, \qquad rc_g L = \text{const}. \tag{22}$$

---

[8] The integrals of the type $\int S(r, \omega) \cos[\Psi(r, \omega) - \omega t] \, d\omega$, arising as a result, are calculated by the stationary phase method.

§ 4 Linear Theory of Growing Density Waves

4.1 Spiral Structure as the Most Unstable Mode

This section deals with a different approach to the problem of galactic spiral structure. It suggests that a more natural and satisfactory explanation of the origin and existence of the spiral density waves can be obtained if one assumes them to be unstable modes of flat galaxies (rather than stationary wave packets). Such an approach was investigated by many authors, in particular, by Kalnajs [250] (see also, e.g., [96]).

In general, the following picture of excitation and maintenance of the spiral pattern in galaxies emerges. The central region, with reasonable assumptions about the amounts of the stellar velocity dispersions, remains (in the linear approximation) unstable with respect to the "global" (first of all, the "barlike") mode (see Section 4.4, Chapter V). An oval distortion of the shape, a barlike standing wave, is there produced, the frequency of which Ω_p is defined by the equilibrium parameters of this region. In turn, the bar excites, mainly in the flattest and the coldest subsystems of the galaxy, a trailing spiral density wave, having strong twisting. The established stationary amplitude of the spirals is defined, apparently, by the nonlinear or dissipative (for example, the production of shock waves in a gaseous subsystem, etc.) effects.

It should further be noted that recently there have appeared some approaches giving a certain synthesis (or versions) of the original picture of Lin and of the picture described here, see, e.g., Sections 4.2 and 4.3.

Kalnajs [250] studied numerically the stability of some simple model of galaxy M31. For distances $r > 4$ kps, the stellar velocity dispersion adopted by him was sufficient to make the model stable with respect to axially-symmetrical perturbations. However, for stabilization of the internal part, large eccentricities of the orbits are required. Instead, to extrapolate the epicyclic orbits to the eccentricities larger than 0.2, Kalnajs correspondingly decreases the response by assuming that only a part of the stars takes part in collective modes. Assuming further that the perturbations are small in amplitude, he finds the eigenmodes by solving the integral equation in (38), Section 4.3, Chapter V.

The two-armed perturbations ($m = 2$) are preferred for reasons which we have already explained (see §2). Kalnajs investigates numerically [250] the so-called "largest" mode; by the "largest" mode it is understood that the gravitational interactions associated with it are "strongest." In turn, the "strength" of the interaction is measured by the shift of the angular velocity of the spiral pattern from its kinematic value $\Omega - \varkappa/2 \approx 10$ km/s · kps (the latter is obtained if one neglects the gravitational effect of the perturbation).

The amplitude of such a spiral increases by e^2 times for 10^9 years. The energy and the angular momentum of the entire disk are, of course, preserved, but they redistribute and are carried away outwards: the stars inside the corotation radius, moving more rapidly than the waves, transfer them to the stars at the outside.

Since the spiral structure is seen more distinctly in objects with the lowest velocity dispersions, Kalnajs calculated the response of the density of the objects with a zero dispersion (see Section 4.5). In the calculations, it was assumed that the surface density of the gas is constant on the disk; however, the smooth variation of σ_0 with radius will probably not change the picture to any extent. An interesting result of these calculations is the strong dependence of the perturbed density of the subsystem on its velocity dispersion.

The density wave in a stellar disk is in essence a barlike distortion of the central galactic region, which acts on the gas. The tightly wound picture of the spirals and the large contrast of the density in the gaseous constituent are due, according to Kalnajs, to the presence of resonances. The location of the resonances is defined by the internal part of the model, while the growth rate depends mainly on density of the outer resonance. The decrease in the latter produces a slower growth rate of perturbations.

Since the model is defined by the curve of rotation, the results attained by Kalnajs [250] for M31 may, in principle, be applicable to our Galaxy also (the rotational curves of these two galaxies are alike, at least from $r \approx 4$ kps outward).

To conclude this section, we sum up the main merits and demerits of the two approaches to the theory of spiral structure considered so far. The main advantage of the theory of Lin and co-authors is a good agreement of the predictions made with its help with observations of galaxies (above all, of course, our Galaxy). As far as unstable modes (of the type of those calculated by Kalnajs) are concerned, they now appear to be inconsistent with observations in the Galaxy.[9] It should however be noted that this refers only to the traditional interpretation of observations. In the papers [250, 93, 96] a different interpretation, which will possibly lead to a picture consistent with the theoretical one, was suggested. In any event, it obviously appears correct that the usual tacit suggestion that the spiral gravitational potential is coincident with the observed spirals (which are indicated by young stars, neutral gas, regions of ionized hydrogen HII, etc.) is not a necessarily needed consequence of the gravitational spiral theory. The observed tightly wound spirals may also be a response to the more open spiral gravitational field. The theory of unstable modes is not yet sufficiently developed, above all in the part concerning the comparison with observed data. At the same time, a large number of papers is devoted to an "intense" fitting to observations of the Lin theory.

---

[9] Two "open" spiral arms result. It should be noted, however, that all the versions of the galactic spiral arms theories at present available are not complete. For instance, the role of massive halos (hidden mass) is quite insufficiently clarified so far (see subsequent sections, as well as the epigraph for this Chapter).

On the other hand, at least, the original version of the Lin theory encounters some fundamental difficulties: it is, for instance, the drift of the wave packets connected with the suggested continuous spectrum of the real frequencies ω.

Unstable modes seem not to suffer such a difficulty; the eigenvalues of such modes are fixed according to definition, and are sufficiently separated from each other (see Section 4.3, Chapter V).

To "save" the Lin program, a mechanism is needed that generates waves. A fairly large number of different versions is suggested [272, 289]. We shall deal with some of them in Sections 4.2–4.5.

4.2 Gravitational Instability at the Periphery of Galaxies

To begin with, we turn to the interpretation of the large-scale spiral structure suggested by Lin himself [272]. The picture drawn by Lin and described below seems to be realized in galaxies not having any strong inner Lindblad resonance. We have already mentioned (e.g., at the end of Section 4.1, Chapter V) that in the range of the inner Lindblad resonance there is absorption of the spiral wave, which under definite conditions (as, for example, in our Galaxy [290]) may become strong decreasing the wave amplitude to a negligible level. For galaxies not having such a resonance, i.e., possessing a comparatively smooth dependence of the surface density $\sigma_0(r)$ on the radius, without any strong concentration toward the center, the short spiral density wave propagating from the corotation circle inward to the center, will have near the center a sufficiently increased amplitude (thanks to conservation of the action, see subsection 3.4.2) to produce an oval (barlike) distortion of the mass distribution in the central region. In turn, this oval configuration (rotating at an angular velocity of the spiral pattern Ω_p), by acting on the galactic disk by its own gravitational field, will produce a response reaction, and its influence will be especially strongly felt in the external regions of the galaxy, where the circular velocity is close to Ω_p. Lin [272] believed that just here, on the periphery of galaxies, was the sources of spiral waves. Thus, the circuit of feedback is closed, which may lead up to maintenance of a stationary spiral pattern.

Thus is briefly the ideology of the approach to the interpretation of the spiral structure as suggested by Lin [272]. It is easy to see in what way this approach is different (with all its obvious similarity) from that considered in the previous section. Maybe, above all, this is a different location of the wave source, on the galactic periphery. Another essential difference is the suggestion (in the Lin picture) about the propagation of a quasi-stationary wave group inward to the center, unlike the self-excitation of one sole "globally"-unstable standing wave in the alternative picture.

We shall consider below one of the possible specific mechanisms of the generation of spirals (also suggested by Lin [272]), the gravitational instability of the external regions of galaxies. An essential addition to the general outline of the picture of the theory contained in [272] is the paper by

Feldman and Lin [199], in which it is shown that the barlike structure rotating at the galactic center produces in the vicinity of the corotation circle a trailing spiral wave (for more detail see Section 4.5).

The gravitational instability of the gas on the galactic periphery. As the mechanism of the initial initiation of the spiral structure, Lin considered first [272][10] the Jeans instability in external parts of the galactic disk.

The external, peripheral parts of the disk are probably indeed unstable since the percentage of stars there is lower and the available stellar velocity dispersion (as well as the degree of turbulence of the gas) is seemingly insufficient to stabilize the Jeans instability. The latter produces structural irregularities, which, owing to a strong differential rotation of the galaxy, are extended to form segments of the trailing spiral arms rotating mainly with an angular velocity of the galaxy (see §1). It is possible that such irregularities do indeed exist. As the observational evidence for this statement, Lin gives the connecting links between the main spiral arms frequently observed in the external parts of many galaxies. In particular, the Orion arm in our Galaxy is, according to Lin, just one of such interarm branches.

The perturbation in the form of a segment of a "corotating" spiral arm should of course exert an influence on the other parts of the galaxy and can initiate the density waves. However, in the general case, its influence should be limited, as is to be expected, if there is no resonance of any form.

We expand the perturbation in a series over angular harmonics $\sim e^{im\varphi}$ ($m = 0, 1, 2, \ldots$) and concentrate attention on any one of them (m). The stars, being at a distance r from the galactic center, will feel the perturbed gravitational field (of this harmonic) at an angular frequency

$$f = m[\Omega(r) - \Omega_0],$$

where Ω_0 is the angular velocity in the place of location of the perturbing segment of the spiral. There is resonance in the case if this frequency is equal to the epicyclic frequency $\varkappa(r)$; in this event one may expect a strong influence of the perturbation on a given radius r.

Write the resonance condition as follows:

$$\Omega_0 = \Omega(r) - \frac{\varkappa(r)}{m}. \tag{1}$$

Strictly speaking, condition (1) may be satisfied for only one particular value of r. However, the quasiresonance may take place for a broader range of the values of r, if the right-hand side of (1) is nearly constant. We already know that the value $\Omega(r) - \varkappa(r)/2$, i.e., (1) for $m = 2$, is really roughly constant throughout the galactic disk (Lindblad). It should therefore be concluded that the perturbation with $m = 2$ may exert an essential influence on the whole galaxy provided that it presents at such a distance from the center

---

[10] Later [199] he noted that this is only one of the many possible mechanisms. Moreover, the origin of the spiral structure might have been, for example, simply the initial irregularity of the galaxy.

where Ω_0 is equal to the nearly constant value of the quantity $\Omega(r) - \varkappa(r)/2$. For our Galaxy this implies that $\Omega_0 = 11 \div 13 \text{ km/s} \cdot \text{kps}$ and that the perturbations must be generated near $r_0 \simeq 15 \text{ kps}$. In this region, there is indeed a lesser percentage of stars, and the system therefore can be gravitationally unstable.

Let us have on the periphery a group of trailing spiral waves with an angular velocity $\Omega_p \approx 11 \div 13 \text{ km/s} \cdot \text{kps}$. What does occur as the group propagates in the radial direction? Of course, it would be too much to expect that these short trailing waves would naturally lead to the quasi-stationary spiral picture. For example, the energy supplied by perturbations, occasionally produced in the external regions, should be limited. In addition, it scatters during the time of its propagation.

These difficulties, according to Lin, are, however, resolved if there is a feedback mechanism, mentioned at the beginning of this section, acting in the system. In the stationary case, according to the principle of the conservation of the wave action, $rc_g A = \text{const}$, where the density of the action is $A = \tilde{E}/f$ (see subsection 3.4.2). Near the point of the spiral wave origin, the angular frequency f is very low ($f \approx 0$; roughly "corotating" waves). Consequently, a small amount of energy $\tilde{E}$ is required in order to produce a substantial amount of action A. As the wave group is propagating inward to the galactic center, the energy density must be increasing both as a consequence of the increase of f (since $d\Omega/dr < 0$) and due to a decrease in $|c_g|$ and r (an increase in energy density with decreasing r is of course an obvious consequence of the cylindrical geometry of the system). When the wave group reaches the center, its very much increased amplitude must be enough to subject the galactic nucleus to a slight distortion to form a short bar (rotating with an angular velocity Ω_p).

There is therefore a gravitational field rotating with the same angular velocity Ω_p and propagating outwards (the long-wave mode). Its influence will be especially strongly felt in the outer regions of the galaxy, where the circular velocity is Ω_p, i.e., just where the waves become initiated. Thus, the cycle is closed, and the stationary state may be established even in the presence of losses, since there is an essential enhancement of the energy when the short waves are moving inwards but there is no respective energy loss when the long waves, with a scale of the order of the galactic radius, are propagating outwards.

In the case of a sharp Lindblad resonance (for example in NGC 5364 or in our Galaxy), the waves cannot penetrate into the center. In [272] Lin suggests for this case a possibility of the reflection of the spiral waves already from the resonant circle where the stars may be located in orbits collectively forming an oval structure [189], which substitutes the bar in the preceding discussion of the reflection mechanism. We have seen, however, that in reality the short waves are absorbed on the inner resonance. For this case in the next section we shall consider one of the possible mechanisms of maintenance of the spiral pattern (suggested by Lynden-Bell and Kalnajs [289].

4.3 Waves of Negative Energy Generated Near the Corotation Circle and Absorbed at the Inner Lindblad Resonance — Lynden-Bell–Kalnajs' Picture of Spiral Pattern Maintenance

In [289] another possible picture of maintenance of the spiral pattern of the Galaxy was suggested: the waves of negative energy are emitted near the corotation circle and are absorbed at the inner Lindblad resonance.

Such an arrangement of the emitters and absorbers of energy at the resonances in galaxies correlates with the picture of Lin and co-workers, if one takes into account the negativity of the wave energy and assumes the direction of the group velocity to be toward the center, in accordance with Toomre's [334] and Shu's [325] conclusions.

Lynden-Bell and Kalnajs show that the energy of the system may be decreased by means of the transfer of the angular momentum from the central region to the periphery. This may be understood in the following simple example. Consider the movement of two particles in a fixed potential. Denote the masses, angular momenta and energies (per 1 g) of particles as m_1, m_2; L_1, L_2; $\varepsilon_1, \varepsilon_2$. The problem is, what is, the minimal value of the energy

$$E = \sum m_i \varepsilon(L_i), \tag{1}$$

for the fixed value of the angular momentum $\sum m_i L_i$? To answer this question, one should minimize

$$E = m_1 \varepsilon(L_1) + m_2 \varepsilon(L_2) \tag{2}$$

with the limitation

$$m_1 L_1 + m_2 L_2 = L. \tag{3}$$

It is evident that

$$dE = m_1 \, dL_1 \varepsilon'(L_1) + m_2 \, dL_2 \, \varepsilon'(L_2), \tag{4}$$

where $m_1 \, dL_1 + m_2 \, dL_2 = 0$, i.e.,

$$dE = m_1 \, dL_1(\varepsilon'(L_1) - \varepsilon'(L_2)) = m_1 \, dL_1(\Omega_1 - \Omega_2). \tag{5}$$

In transforming the expressions in (5), the equilibrium condition was used. From (5), it is seen that energy can be reduced by exchanging angular momentum between the particles, such that the orbit with a lower angular velocity acquires an additional momentum. This means that $dE < 0$, if $dL_1 < 0$ (for $\Omega_1 > \Omega_2$). Since for galaxies the angular velocity Ω decreases toward the periphery, the energy decreases if the angular momentum is transferred from the center outwards.

Although this result has been obtained so far for the system consisting of only two particles, in reality it has a general meaning since, for example, introducing friction into any system leads, apparently, to the transfer of angular momentum outwards, and to the disappearance of energy due to dissipation. Thus, the galaxy, in order to transit to lower energy states, should find a mechanism of transfer of the angular momentum outwards. This cannot be made by axially-symmetrical motions of a stellar system since they do not give any gravitational couple between the internal and external parts. In order to see what form of gravitational disturbance is necessary, one may introduce the tensor of gravitational tensions by expressing the force density as minus divergence of the stress tensor. The force density in case of gravitation is ($\psi \equiv -\Phi$)

$$\rho \nabla \psi = -(4\pi G)^{-1} \Delta \psi \nabla \psi$$
$$= -(4\pi G)^{-1} [\text{div}(\nabla \psi \nabla \psi) - (\nabla \psi \nabla)(\nabla \psi)]$$
$$= -(4\pi G)^{-1} \text{div}[\nabla \psi \nabla \psi - \tfrac{1}{2} I(\nabla \psi \nabla \psi)]$$
$$= -\text{div}[gg/(4\pi G) - (g^2/8\pi G)\hat{I}]; \tag{6}$$

i.e., $\rho \nabla \psi = -\text{div } T$ where

$$T = gg/4\pi G - (g^2/8\pi G)\hat{I}; \qquad g = \nabla \psi.$$

The gravitational torque, acting on the external part of the system from the internal part, is calculated in the following way. Divide all the space by a right circular cylinder of a certain radius (with its center on the axis of the system). Then the torque is calculated by the formula

$$\mathbf{M} = \int \mathbf{R} \times T \cdot d\mathbf{S}, \tag{7}$$

where integration is performed over the surface of the cylinder, $d\mathbf{S}$ is directed along the outward normal of the cylinder, $\mathbf{R} = \mathbf{R}(x, y, 0)$. The component M_z of interest to us is

$$M_z = (4\pi G)^{-1} \int R g_\varphi g_r \, dS. \tag{8}$$

From (8) it is easy to see that $M_z > 0$, i.e., the momentum is transferred outwards, provided that $g_{1\varphi} g_{1r} > 0$. Consequently, there must be $g_{1\varphi} > 0$ (Fig. 119), and the equipotential surfaces corresponding to such a picture are due to be "trailing." This, according to the expression of the authors, determines the "cause" and "purpose" of the existence of the trailing spiral waves in the galaxy: they promote its evolution.

It would be logical to assume that the wave ceases to increase when the velocities of the perturbed motions caused by the wave exceed its phase velocity. The radial displacement of the star, due to the force kS (per unit mass) is $kS[\varkappa^2 - (\omega + m\Omega)^2]^{-1}$, and so the condition of saturation may be written as

$$\varkappa k S[\varkappa^2 - (\omega + m\Omega)^2]^{-1} < (\omega + m\Omega)/k. \tag{9}$$

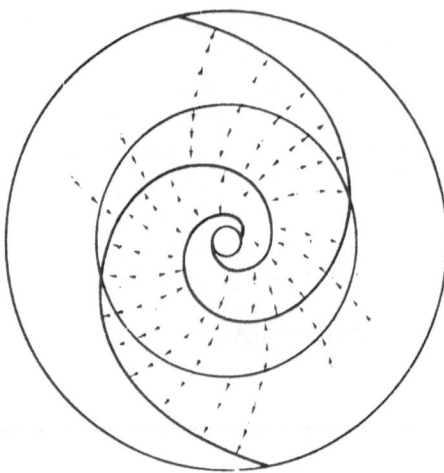

Figure 119. Spiral density wave (solid lines) and corresponding gravitational forces (arrows) [289].

Writing $v = (\omega + m\Omega)/\varkappa$ we obtain from (9) the following estimate for the time which is necessary for transfer of a substantial fraction of the total angular momentum from the central region to the periphery ($MR^2\bar{\Omega} = \bar{L}$):

$$\frac{L}{M_z} = MR^2\Omega \bigg/ \left(\frac{1}{4}\frac{mRS^2}{G}\right) = \frac{2\Omega^3 R^4}{S^2} = \frac{2\Omega^3(kR)^4}{\varkappa^4 v^2(1-v^2)} \sim \frac{(kR)^4}{\Omega}. \quad (10)$$

Hence it follows that the distribution of the angular momentum will substantially change after $(kR)^4/2\pi$ rotations. For 100 rotations there must be $kR < 5$, i.e., the inclination of the wave $i = \arctan(m/kR) > 23°$. The "so open" waves may, consequently, greatly change the distribution of the angular momentum of galaxies. At the same time, the shape of the galaxy will also change: its external parts will expand, while the internal parts will contract. The standing waves of large amplitude may arise at the galactic center (they, according to [289], correspond to bars). The galactic evolution should follow the scheme SA → SAB → SB (Sc → Sb).

4.4 Kelvin–Helmholtz Instability and Flute-like Instability in the Near-Nucleus Region of the Galaxy as Possible Generators of Spiral Structure

Spiral waves in galaxies could in principle be maintained by some local instability of a nongravitational nature, for example, the beam or gradient-temperature instabilities (Chapter VI). However, for these instabilities, the "longitudinal" wavelength, as a rule, is improbably large. Therefore, in real galactic systems, such instabilities are unlikely to develop (except for, say, needle-like galaxies).

Since, in spirals, a large percentage of the mass falls on the gas, one may suggest that the spiral structure appears due to excitation of some hydro-dynamical instability (see §3, Chapter VI) in a gaseous subsystem of the galactic disk. This point of view was investigated in [98]–[100]. Unlike other nongravitational instabilities, the increase in thermal dispersion does not stabilize the hydrodynamical instabilities but, on the contrary, leads to an increase of the growth rate. However, these instabilities, unlike the above-mentioned nongravitational instabilities, do not require any special con-ditions for their development.

In [93], we paid attention to the region of a sharp drop in the velocity of rotation of galaxies $v_0(r)$ in the near-nucleus region, as found by the recent astronomical observations [315]. They can be connected with the presence in spiral galaxies of very flattened nuclear formations (nucleus, bulge, bar). If for the sake of simplicity one represents the thin nuclear lens of mass M in the form of a homogeneous spheroid with a large semiaxis a, and eccentricity e, then the equilibrium gravitational potential Φ_0 in the region external with respect to this lens, in its equatorial plane $(z = 0)$, may be written in the form [64, 147]

$$\Phi_0 = \frac{3GM}{4ae}\left[\left(\frac{r^2}{a^2e^2} - 2\right)\arcsin\frac{ae}{r} - \frac{r}{ae}\sqrt{1 - a^2e^2/r^2}\right], \qquad (1)$$

where r is the distance from the center, $r > a$. The epicyclic frequency cor-responding to (1) is $\varkappa^2 = \partial^2\Phi_0/\partial r^2 + (3/r)(\partial\Phi_0/\partial r)$. For example, in the disk limit $(e = 1)$

$$\varkappa^2 = \frac{3GM}{2a^2}\left\{\frac{4}{a}\arcsin\frac{a}{r} - \frac{1}{\sqrt{r^2 - a^2}} - \frac{a^2}{r^2\sqrt{r^2 - a^2}} - \frac{3}{r^2}\sqrt{r^2 - a^2}\right\}. \qquad (2)$$

Hence, it is easy to see that near the edge of the disk $(r \approx a)$ $\varkappa^2$ is a large negative value (as $r \to a$, $\varkappa^2 \to -\infty$). In such a situation it is obvious that for a sufficiently strongly flattened (having a "sharp edge") nucleus, $\varkappa^2$ in its immediate vicinity is defined only by the parameters of this nucleus and are independent of the mass distribution in other parts of the system (though most of the total mass is contained in it).

So, the circular orbits near such a nucleus should be unstable $(\varkappa^2 < 0)$.[11] According to [147], for an isolated flattened spheroid, instability takes place for $e > e_{cr}$, where $e_{cr} \approx 0.834$, in the range $a \leq r \leq ae/e_{cr}$, so that the maximally broad region of instability (for the disk, $e = 1$)

$$a \leq r \lesssim 1.2a. \qquad (3)$$

In reality, of course, this near-nucleus region of instability of circular orbits may be still narrower if one takes into account that real systems are not isolated. We shall not consider further the estimation of the size of the instability region, which will inevitably not be very reliable. We restrict

---

[11] The question of the possible linkage of this instability with radial flows of gas in central regions of galaxies is of interest.

ourselves only to the indication of a possibility of principle of such a "general" explanation of sharp drops in the curves of the rotational velocity of spiral galaxies, which implies that the property indicated may be inherent to many systems, and not only to Andromeda nebula (M31), for which it is likely to be reliably established.

A rapid variation of the velocity of rotation is favorable for excitation of instability of Kelvin–Helmholtz. A detailed analysis of such a possibility is performed in [99] and [100] (see §3, Chapter VI).

The gas density in the plane of the Galaxy has a form resembling the shape of a nonsymmetrical bell. If one assumes that the temperature near the gas density maximum is a monotonous function of the coordinate, one can show [98–100] that the necessary condition of excitation of the flute-like instability, also leading to the formation of spirals, is satisfied.

The hypothetical mechanism of the spiral arm formation, investigated in the above papers, allows one to explain the nature of multitier spirals as galaxies with several regions (with respect to the number of spiral tiers) of a rapid decrease of the rotational velocity or several extrema of gas density (or both at the same time).

4.5 The "Trailing" Character of Spiral Arms

We have already mentioned (see subsection 3.4.1) the rather general character of the antispiral theorem. It is valid, in particular, both for collisionless stellar systems and for the gaseous medium. The cause is not "saved" by, for example, the radial electric currents in the absence of azimuthal currents (the latter would lead to the *initial* asymmetry of the system). In this case, in proving the antispiral theorem by the method used at the end of subsection 3.4.1, one also needs to add the charge inversion to the ordinary reflection. In general, the presence of the magnetic field (toroidal or poloidal without the primordial asymmetry) does not violate the theorem.

The radial gas flows could in principle participate in the formation of trailing spiral arms, and such a possibility was investigated in the literature.

First of all, it should be said that the "problem" of the antispiral theorem has a different urgency in the two main versions of the wave theory. In Lin's picture, as we are aware, the wave-generating mechanism is required, which *per se* may possess a needed asymmetry. In a new interpretation of Lin considered in Section 4.2 such a mechanism is the local gravitational instability on the periphery of galaxies, which leads to excitation of "segments" of trailing spirals. On the other hand, as shown by Feldman and Lin [199] (see subsection 4.5.1) the response of the system to the bar-like distortion of the shape of density distribution in the central region has in the region of the corotation radius an appearance of trailing arms. Thereby is shown the distinctive nature of the trailing waves at the "extreme" points of the spiral pattern. At the same time, for the "propagation region", i.e. for the main part of the disk galaxy, a simple explanation of the preference of trailing spirals is suggested by Mark [290]. Both the leading and the trailing waves are damped in the direction of their propagation. Since the trailing wave propagates inward to the center, it must be initiated in the outer areas, so that it has a sufficiently extended propagation region up to the inner Lindblad resonance. On the contrary, the leading wave coming out of the central region will be practically completely absorbed on the inner resonance, not having reached the region of free propagation.

The antispiral theorem presents serious difficulty in the case when the spirals are considered as unstable modes of a disk system. Neutral oscillations, according to this theorem, are "antispiral"; therefore very "open" spirals result in not very strong unstable systems. The problem, thus, is to find the mechanism that builds up the tightly twisted spirals in systems with a comparatively weak instability.

We have already mentioned this question in Section 4.1, where it was noted that the way out may be an analysis of the response to a relatively open spiral potential of the flattest and coldest subsystems; this response has the appearance of a much tighter wound spiral. Below, in subsection 4.5.2., this question is discussed using the example of several simple models, which allow an exact solution.

4.5.1. Excitation of Trailing Spiral Density Waves by the Rotating Barlike Structure at the Galactic Center.

Feldman and Lin [199] study the influence of the barlike center in the framework of the model consisting of three components: spherically-symmetrical "stellar" nucleus, uniformly rotating "stellar" bar and "gaseous" disk (or cylinder). The rotation axis of the bar passes through the center of the nucleus. For the sake of simplicity, it is assumed that the nucleus and the bar are not perturbed, while the stars and gas interact only via their gravitational fields.

The consideration of a purely gaseous disk as a model of the real system consisting of stars and gas, may be justified by the fact that in the vicinity of the corotation radius (which is of interest to us, above all) the gas and the stars behave in a similar way.

Assuming that the gaseous flow is stationary in the frame of reference rotating with the bar, we find that all the physical values (velocities, density, pressure, gravitational potential) are independent of time if they are expressed through the coordinates r, $\bar{\varphi}$, z, where $\bar{\varphi} = \varphi - \Omega_p t$, and r, φ, z are the ordinary cylindrical coordinates with the z-axis directed along the rotation axis of the galaxy, Ω_p is the angular velocity of the bar.

Without taking into account the influence of the bar the gas flow is assumed to be axially-symmetrical and circular, i.e.,

$$\mathbf{v} = (v_r, v_\varphi, v_z) = (0, r(\Omega - \Omega_p), 0), \tag{1}$$

v_r, v_φ, v_z are the cylindrical components of the velocity in the rotating frame of reference, while $r\Omega(r)$ is the unperturbed velocity in the inertial system.

Further considered is the two-dimensional gravitating gaseous system, the behavior of which is defined by the hydrodynamical equations and the

Poisson equation. The gas is assumed ideal and to satisfy the polytropic law ("barotropic") $P = K\rho^\gamma$, $\gamma > 1$. Introduce the enthalpy

$$\eta = \frac{\gamma}{\gamma - 1} K\rho^{\gamma-1} \qquad \left(d\eta = \frac{dP}{\rho}\right)$$

as well as the dimensionless variables R, ω, ω_p, u, v, h, ψ (Φ is the potential):

$$r = r_0 R, \qquad \Omega = \Omega_0 \omega, \qquad \Omega_p = \Omega_0 \omega_p, \qquad v_r = (r_0 \Omega_0) u,$$
$$v_\varphi = (r\Omega_0) v, \qquad \eta = (r_0 \Omega_0)^2 h, \qquad \Phi = (r_0 \Omega_0)^2 \psi, \tag{2}$$

where r_0 is the corotation radius, and Ω_0 is equal to the wave frequency Ω_p. By assuming that the bar only slightly changes the initial axially-symmetrical circular gas flow, the Euler equation may be linearized, which leads them to the form

$$(\gamma - 1)h_0(u_1 + Ru_1' - imv_1) + [rh_0'u_1 - im(\omega - \omega_p)Rh_1] = 0, \tag{3}$$

$$im(\omega - \omega_p)u_1 + 2\omega v_1 = h_1' + \psi_1', \tag{4}$$

$$(2\omega + R\omega')u_1 - im(\omega - \omega_p)v_1 = im/R(h_1 + \psi_1), \tag{5}$$

where the prime denotes differentiation with respect to R, $\omega^2 = (h_0' + \psi_0')/R$. Introduce the dimensionless parameters

$$E^2 = \frac{1}{(\gamma - 1)h_0}; \qquad H = \frac{1}{\gamma - 1}\frac{d \ln h_0}{d \ln R} = \frac{d \ln \rho_0}{d \ln R},$$

$$v = -\frac{m\Omega_0}{\varkappa}(\omega - \omega_p), \tag{6}$$

where $\varkappa^2/\Omega_0^2 = 4\omega^2(1 + r\omega'/2\omega)$ (note that E is an inverse dimensionless sound velocity). Eliminating the components of the velocity from the hydrodynamical equations, we shall find

$$h_1'' + Ah_1' + (B + C)h_1 + \psi_1'' + A\psi_1' + B\psi_1 = 0, \tag{7}$$

where

$$A(r) = \frac{1 + H}{R} - \frac{d}{dR}\ln[(1 - v^2)\varkappa^2], \tag{8}$$

$$B(r) = -\frac{m^2}{R^2} + \frac{2\omega}{R^2(\omega - \omega_p)}\left\{H - R\frac{d}{dR}\ln\left[\frac{\varkappa^2(1 - v^2)}{\omega}\right]\right\}, \tag{9}$$

$$C(r) = -E^2 m^2 \varkappa^2 (1 - v^2)/\Omega_0^2. \tag{10}$$

Let us further consider the potential as a sum of two terms, one of them being due to stars, while the other is due to the gas: $\psi = \psi_s + \psi_g$, where

$$\psi_0 = \psi_{0,\text{NUCL}} + \psi_{0,g}, \qquad \psi_1 = \psi_b + \psi_g, \tag{11}$$

ψ_b is due to the bar, while ψ_g is due to the perturbation of the gas density. Determine now the $D(r)$ function:

$$D(r) = -(\psi_b'' + A\psi_b' + B\psi_b), \tag{12}$$

then Eq. (7) will be written in the following manner:

$$h_1'' + A h_1' + (B + C) h_1 + \psi_g'' + A \psi_g' + B \psi_g = D. \tag{13}$$

The A, B, C functions depend only on the stationary parameters of the system and the frequency ω_p, but are independent of perturbations. The value of D is defined by the bar.

Make an estimate of the order of magnitude of different coefficients in Eq. (13): $r_0 \approx 10$ kps, $\Omega_0 \approx 10$ km/s · kps, $c_s \approx 10$ km/s, so that $E \approx 10$. Instead of the value $E \gg 1$, it is easier to deal with the inverse value:

$$\delta \equiv \frac{1}{E(1)} = \frac{c_s(1)}{r\Omega_0} \ll 1,$$

then one may write

$$A = a_1(R), \qquad B = \frac{a_2(R)}{R - 1}, \qquad C = -\delta^{-2} a_3(R), \qquad D = a_4(R)/(R - 1),$$
$$\tag{14}$$

where all $a_i(R)$ are of the order of 1, regular and nonzero in the vicinity of $R = 1$.

Determine now a new coordinate $\xi = (R - 1)/\delta$. The coefficients in Eq. (13) written in terms of ξ, are regular power series with respect to δ. By expanding the perturbations in powers of δ

$$h_1 = h_1^{(0)} + \delta h_1^{(1)} + \cdots, \qquad \psi_g = \psi_g^{(0)} + \delta \psi_g^{(1)} + \cdots, \tag{15}$$

and taking into account that $a_3(1) = 1$, we find

$$\frac{d^2\psi_g^{(0)}}{d\xi^2} + \frac{d^2 h_1^{(0)}}{d\xi^2} - h_1^{(0)} = 0, \tag{16}$$

$$\frac{d^2\psi_g^{(1)}}{d\xi^2} + \frac{d^2 h_1^{(1)}}{d\xi^2} - h_1^{(1)} = \frac{a_4(1)}{\xi} - a_1(1)$$

$$\times \frac{d}{d\xi}(h_1^{(0)} + \psi_g^{(0)}) - \frac{a_2(1)}{\xi}(h_1^{(0)} + \psi_g^{(0)}) + \xi a_3'(1) h_1^{(0)}. \tag{17}$$

Above, we have assumed that the gas flow is two-dimensional. Accordingly, one may consider two models: the cylinder and the disk. We shall restrict ourselves below only to a simple analytical model of the cylinder (in [199] it is shown that qualitatively the results for the two models are coincident). In this case $\partial^2\psi_g/\partial z^2 = 0$, and the Poisson equation yields

$$\psi_{0, \text{gas}}'' + \frac{1}{R} \psi_{0, \text{gas}}' = \frac{4\pi G \rho_0}{\Omega_0^2}, \tag{18}$$

$$\psi_g'' + \frac{1}{R} \psi_g' - \frac{m^2}{R^2} \psi_g = \Lambda h_1, \tag{19}$$

where

$$\rho_0 = \left[r_0^2 \Omega_0^2 \frac{\gamma - 1}{\gamma K} h_0 \right]^{1/(\gamma - 1)}, \tag{20}$$

$$\Lambda = \frac{4\pi G \rho_0}{(\gamma - 1) h_0 \Omega_0^2} = \frac{4\pi G \rho_0 r_0^2}{(c_s)_0^2} = (k_{cr} r_0)^2 = \left(\frac{2\pi r_0}{\lambda_{cr}} \right)^2, \tag{21}$$

λ_{cr} is the Jeans wavelength, and $(c_s)_0$ is the sound velocity in an unperturbed matter. The estimate of the value Λ (for the solar vicinity of the Galaxy) leads [199] to the relation $\Lambda \sim \delta^{-2}$, therefore it is convenient to define the function $L(R)$: $\Lambda = \delta^{-2} L(R)$; L is of the order of unity, $L(1) > 1$. With these assumptions we find

$$\frac{d^2 \psi_g^{(0)}}{d\xi^2} = L(1) h_1^{(0)}, \tag{22}$$

$$\frac{d^2 \psi_g^{(1)}}{d\xi^2} = L(1) h_1^{(1)} + \xi L'(1) h_1^{(0)} - \frac{d\psi_g^{(0)}}{d\xi}. \tag{23}$$

By combining these equations with (16), (17), we shall obtain

$$\frac{d^2 h_1^{(0)}}{d\xi^2} + \alpha^2 h_1^{(0)} = 0, \tag{24}$$

$$\frac{d^2 h_1^{(1)}}{d\xi^2} + \alpha^2 h_1^{(1)} = \frac{a_4(1)}{\xi} + F_{cyl}^{(0)}, \tag{25}$$

where

$$F_{cyl}^{(0)} = \xi[a_3'(1) - L'(1)] h_1^{(0)} + [1 - a_1(1)] \frac{d\psi_g^{(0)}}{d\xi}$$

$$- a_1(1) \frac{dh_1^{(0)}}{d\xi} - \frac{a_2(1)}{\xi} [h_1^{(0)} + \psi_g^{(0)}], \tag{26}$$

$$\alpha = [L(1) - 1]^{1/2}. \tag{27}$$

Equation (25) contains the singular term $a_4(1)/\xi$. As always in such cases, the required neutral mode should be considered as the limit of increasing modes. Let us assume that ξ, Ω_p, ω_p are "slightly complex,"

$$\Omega_p = \Omega_0(1 + is), \qquad \omega_p = 1 + is \qquad (s = 0^+ \text{ or } s = 0^-)$$

and take the perturbation of the form $h_1(\xi) e^{-im(\varphi - \Omega_p t)} e^{-ms\Omega_p t}$. Then $ms < 0$ will correspond to the increasing modes. For complex ω_p, the pole $[\omega(\xi) - \omega_p]^{-1}$ lies at the point $\xi = i\varepsilon$, such that $\varepsilon = s/\omega'(0)$. Since in any reasonable model $\omega'(0) < 0$, then ε has the sign opposite to s, therefore m and ε should have the same sign. Below, we assume that $m > 0$, so that also $\varepsilon > 0$, i.e., the pole lies above the real ξ-axis. Accordingly, in the limit $s \to 0^-$, $\varepsilon \to 0^+$, in integrating over ξ, one should go round the point $\xi = 0$ from below. Consider now, as in [199], the equation

$$\frac{d^2 y}{d\xi^2} + g_1 \frac{dy}{d\xi} + g_2 y = \frac{1}{\xi}. \tag{28}$$

Equation (25) is a special case of (28) for $g_1 = 0, g_2 = \alpha^2$. Assume that the roots of the corresponding characteristic equation are purely imaginary (as in the case of (25)), so that the two solutions of the homogeneous equation are oscillatory:

$$y_1 = e^{i\alpha_1 \xi}, \qquad y_2 = e^{i\alpha_2 \xi} \tag{29}$$

$(\alpha_1, \alpha_2$ are real). If we integrate in the complex plane over the contour (C) going along the real axis and bypassing the singularity $\xi = 0$ from below, we find the partial solution of the nonhomogeneous equation

$$y = \frac{2\pi}{\alpha_1 - \alpha_2} [e^{i\alpha_2 \xi} I(\xi, \alpha_2) - e^{i\alpha_1 \xi} I(\xi, \alpha_2)], \tag{30}$$

$$I(\alpha, \xi) = \frac{1}{2\pi i} \int_\xi^\infty \frac{e^{-i\alpha\xi}}{\xi} d\xi, \tag{30'}$$

and the integral is taken along C. From (30') it is clear that $I(\xi, \alpha) = O(1/\xi)$ as $\xi \to +\infty$. It is also easily proved that, as $\xi \to -\infty$,

$$I(\xi, \alpha) = \begin{cases} O(1/\xi), & \alpha > 0, \\ \frac{1}{2} + O(1/\xi), & \alpha = 0, \\ 1 + O(1/\xi), & \alpha < 0. \end{cases} \tag{31}$$

For $m > 0$ and for the solution of the form $h_1(r) \sim e^{ikr}$ the spiral is trailing, if $k < 0$ and leading if $k > 0$. The solution of equations in (24), (25) may be presented in the form:

$$h_1 = k_L e^{i\alpha\xi} + k_T e^{-i\alpha\xi} + \delta k_D f_D(\xi) + O(\delta^2), \tag{32}$$

where the first two terms are the "free" solutions (the indices L and T denote "leading" and "trailing"), while the last term is due to the action of the bar,

$$k_D = \frac{\pi}{\alpha} a_4(1), \tag{33}$$

$$f_D(\xi) = e^{-i\alpha\xi} I(\xi, -\alpha) - e^{i\alpha\xi} I(\xi, \alpha). \tag{34}$$

Since $I \to 0$ as $\xi \to +\infty$, and $I \to e^{-i\alpha\xi}$ as $\xi \to -\infty$, so it may be concluded that $k_D f_D$ vanishes at $+\infty$ and behaves as the trailing spiral at $-\infty$ (the "leading" part f_D vanishes as $\xi \to -\infty$).

4.5.2. Simple Models Which Allow Exact Solutions.

Take a model, exactly calculated in linear theory, of the collisionless ellipsoid (11), §1, Chapter IV. As is well known, the spirals consist of young stars and gas of the flat subsystem. In accordance with this, we single out from the whole set of stars, those being close to the equatorial plane. The partial density of these stars is $\tilde{\rho}_0 \sim (1 - r^2)^{-1/2}$. From the equations of multiflows hydrodynamics (see, e.g., [86]), for these stars (i.e., for a "flow" with $v_z \simeq 0$) one can obtain the following partial perturbed density (see (11), Section 2.2, Chapter V):

$$\rho_1 \sim \frac{1}{r} (r\varepsilon\Phi'_1)' - \frac{m^2}{r^2} \varepsilon\Phi_1 - \frac{2m}{r\omega_*} (\varepsilon\Omega_0)'\Phi_1, \tag{35}$$

where $\varepsilon = \tilde{\rho}_0/(\omega_*^2 - 4\Omega_0^2)$; Φ_1 is the total perturbed potential, $\omega_* = \omega - m\Omega_0$. Substituting $\tilde{\rho}_0 \sim (1 - r^2)^{-1/2}$ into (35), we obtain (assuming $m = 1$)

$$\rho_1 \sim \Phi_1'' + \frac{\Phi_1'}{r(1 - r^2)} - \left[\frac{1}{r^2} - \frac{2}{\omega_*(1 - r^2)}\right]\Phi_1. \qquad (36)$$

Represent the partial perturbed density in the form

$$\rho_1 \sim \exp[i(\psi(r) + \varphi)],$$

so that the equation of spiral

$$\psi(r) + \varphi = \text{const.}$$

Denoting $\omega_1 = \text{Re } \omega$, $\omega_2 = \text{Im } \omega$, $\delta_1 = \text{Re } \delta$, $\delta_2 = \text{Im } \delta$, for the mode in (19), Section 3.1, Chapter IV, we obtain

$$\tan \psi(r) = \frac{\omega_2(8 + \delta_1 - 5r^2) + \delta_2(\omega_2 - 2)}{\omega_1[6(1 - r^2) + 3] - [(2 - \omega_1)(r + \delta_2) + \omega_1] - \delta_2\omega_2}. \qquad (37)$$

By using the data of calculations, one can verify that the solution with the amplitude increasing in time always gives the monotonically decreasing function $\varphi(r) = -\psi(r) + \text{const}$, which corresponds to the *trailing* spiral. For the example given above, the twisting of the spiral at a distance of the disk radius makes up $\sim 25°$. At the same time, the total density (and the potential) in this case have the form of "weakly *leading*" spirals. The subsystem of the stars with $v_z \approx 0$ considered above evidently constitutes only a small part of all the stars, even in the plane $z = 0$. It is clear that this consideration also remains valid for all the subsystems having velocities v_z in the plane $z = 0$, lower than a certain boundary velocity v_0, provided that $v_0 \ll \omega_0 c$. This boundary velocity corresponds to a certain thickness of the flat subsystem h (which must be far less than c). The ratio of the mass of the flat subsystem, defined in this way, to the total mass will be $\sim h/c$. Though this ratio is small, the "trailing" spirals of the flat subsystem can be observed: they can be defined physically; for example, due to the luminosity of the youngest stars being in the flattest subsystems of the Galaxy or from the 21 cm emission of hydrogen forming the gaseous trailing spirals. Note that the description of these latter is automatically included in this scheme since the cold gaseous component at $z \approx 0$ may be described with equal validity both in the framework of hydrodynamics and in kinetic theory.

At each given moment, in the $z \approx 0$ plane, there are also stars with high velocities v_z. They pertain to other subsystems, consist of "old" stars of moderate luminosity and form a background having no spiral shape. As is shown, only stars and gaseous clouds of the flattest subsystems which are constantly near the $z = 0$ plane, get wound to form trailing spirals.

For the disks considered in Section 4.4, Chapter V, the situation is similar: the partial density of stars of the "coldest" subsystems provides a clear picture of trailing spirals, while the total density and potential have an "antispiral" shape.

§ 5 Comparison of the Lin–Shu Theory with Observations

Irrespective of any future theory of the origin and evolution of the galactic spiral structure (it is still to be created) the now available "semi-empirical" density wave theory of Lin and Shu may prove to be useful in the interpretation of observations.

As we have seen, the theory contains one free parameter, the velocity of the spiral wave Ω_p. The observational test of the theory, therefore, consists in comparison of a large amount of data. Some of them may be used to determine the angular velocity of the wave, the others provide a test of the predictions of the theory. Lin *et al.* [271] make an estimate of the agreement of the theory and observations within $\sim 20\%$.

5.1 The Galaxy

The first rather detailed comparison of the inferences of the Lin and Shu theory with observed data on our Galaxy was performed in [271]. The comparison is performed for the following items: (a) the distribution of atomic hydrogen; (b) the systematic movement of the gas; (c) the distribution of young stars; and (d) the migration of moderately young stars. It is noted that there is good agreement in all the cases if the angular velocity of the spiral pattern is assumed to be of the order of 11–13 km/s · kps, while the spiral gravitational field is assumed to be equal to approximately 5% from the axially-symmetrical field.

5.1.1. Main Parameters of the Galaxy. Choice of the Value of the Spiral Wave Angular Velocity and Estimation of the Velocity Dispersion of Stars. The application of the density wave linear theory of Lin and Shu to the Galaxy substantially depends on the adopted (theoretical) equilibrium model. The point is that direct astronomical observations yield so far rather meager information, and we have fairly reliable measurements of parameters only for the *solar vicinity* of the Galaxy nearest to us. The values commonly used for estimation of the parameters essential in the Lin and Shu theory are as follows: the surface density $\sigma_0 \sim 50 \div 65 \ M_\odot/ps^2$, the mean velocity dispersion of stars in the rotation plane $c_r \sim 30 \div 40$ km/s, the epicyclic frequency $\varkappa \sim 27 \div 32$ km/s · kps.

At the same time, application of the theory requires full knowledge of the equilibrium state of the Galaxy: the surface density of stars and gas *at each point* of the disk of the Galaxy, the angular velocity $\Omega(r)$, the epicyclic frequency $\varkappa(r)$, the velocity dispersion of stars and the effective sound velocity of the gas (also at each point). For this purpose, at first a model is built in which distributions of mass $\sigma_0(r)$ and the rotation velocity $\Omega(r)$, related to each other, are calculated. The most elaborate of such models of the Galaxy is believed to be the familiar model of Schmidt [319], in which the Galaxy is presented in the form of superposition of several subsystems inserted in each other, with different degrees of oblateness (they simulate different real, physical subsystems of the Galaxy). Lin and Shu [271], as well as many of their followers, used this Schmidt model.

It should be noted, however, that the use of this or a similar model does not yet completely determine the equilibrium state of the Galaxy or, in particular, the velocity dispersion $c_r(r)$ at each point of the disk, knowledge of which is also necessary for the application of the Lin and Shu theory. In [271] it is simply assumed that the velocity dispersion is such as to ensure the marginal stability of the disk of the Galaxy with respect to radial perturbations (according to Toomre).

The spiral pattern may be calculated by using the dispersion equation in (21), Section 4.1, Chapter V, if the equilibrium model of the Galaxy is known. If one assumes the Schmidt model [319], as in [271], then for $\Omega_p = 11 \div 13$ km/s $\cdot$ kps we have the distances between the arms consistent with observations. Here one should bear in mind that generally not a single wave is excited but a whole group of waves with frequencies close to Ω_p, while the comparison of observations with theory in [271] is performed, for the sake of simplicity, for the sole wave. Therefore, it cannot be expected that the agreement will be too accurate, however the ensuing agreement may be considered as satisfying.[12]

An important test of the general concepts of the theory is the investigation of the velocity dispersions of stars, predicted by the criterion (21), Section 4.1, Chapter V. In this paper, the disk is assumed to be a purely stellar one and having no thickness. In such assumptions, the velocity dispersion for the solar vicinity of the Galaxy $c_r \approx 52$ km/s turns out to be too large (at least by 25%). The contradiction still increases if the presence of the gaseous constituent is taken into account. However, Shu [323, 325] showed that this discrepancy disappears if the finite thickness of the disks of stars and gas is taken into account. The velocity dispersion in the solar vicinity should then be ≈ 37 km/s (or somewhat less) for the stability from the local collapse. The estimate of the velocity dispersion obtained by Shu is in reasonable agreement with observations. Shu revealed also that the relative contributions of the gas and stars are roughly identical despite the fact that the mass of stars greatly exceeds the mass of gas. This is explained by the fact that the gaseous disk of the Galaxy is far thinner than the stellar disk.

5.1.2. Relationship of Large-Scale Systematic Noncircular Motions of Stars in the Galaxy with the Gravitational Field of Arms. Observers long ago noticed the wave-shaped variations on the rotation curves of galaxies, but at the beginning these variations were thought of as a possible consequence of the gas loss by the interarm regions. This effect, however, turns out to be small [271], and the correct interpretation of the variation of the rotation curve is provided by the density wave theory. It is evident that the component

---

[12] The details of selection of the velocity of wave Ω_p are discussed in detail by Yuan [359, 360]. This selection proves to be limited by rather narrow limits.

of the perturbed velocity of the particles in the azimuthal direction (due to the presence of spiral arms) should cause changes in the observed curve of rotation of the galaxy. The quantitative study of the systematic movement of the gas performed by Yuan [359, 360] had led to the required estimates for the amplitude of variation of the velocity, of the order of 8–10 km/s. This may be caused by the spiral gravitational field $\sim 5\%$ from the axially-symmetrical field acting on the gas with a turbulent velocity of ≈ 7 km/s (the mean square value of one component of velocity) and with the magnetic field $H \sim 5 \cdot 10^{-6}$ gs. Note also here the natural explanation [271] of the difference between the northern and southern curves of rotation of the Galaxy following from the stability theory of flat rotating systems. In all likelihood, the cause of this difference is the oval (barlike) distortion of the shape of the Galaxy which, as we have seen (in §4, Chapter V), is especially difficult to stabilize by the velocity dispersion of stars.

5.1.3. Birth and Migration of Moderately Young Stars. Stars are born in places of the highest gas density, i.e., inside the gaseous arms. Finally, these stars must migrate from the arms since the stars rotate at the angular velocity of the matter, which is different from Ω_p (the angular velocity of arms). For ten million years (the age of young O- and B-stars), in the solar vicinity of the Galaxy, such stars must have been separated from the gaseous arms by approximately 1.2 kps. However, since the inclination of the spiral branches is small (they are tightly wound in the Galaxy), the *radial* distances will constitute only one-tenth of the indicated distance, so that the young stars must actually lie within the gaseous arms, which is just supported by observations: the blue bright stars of the O- and B-types are, as is well known, the optical indicators of the spiral structure clearly outlining the gaseous branches.

At the present time, the methods for the determination of the ages of the "moderately young" stars [329] are rather sophisticated, in order that the problem might be settled of finding out the places of birth of these stars and the history of their migration might be reconstructed. The present locations and velocities of stars necessary for calculations are also known with a good degree of accuracy.

Already a preliminary treatment performed by Lin *et al.* [271] has shown that even a small spiral field ($\sim 5\%$ of the axially-symmetrical field) can provide an essential effect. Yuan [359, 360], by experimenting with different choices of the frequencies of the spiral pattern and field strengths, found that a good choice corresponds to $\Omega_p \approx 13.5$ km/s · kps and to the field strength $\sim 5\%$ of the axial-symmetrical one. He has investigated the paths of twenty-five stars by using the data of Strömgren [329]. If the spiral gravitational field is not taken into account, then the positions of these stars at the time of their birth does not fit into any structure known from radio observations. At the same time, as the spiral field is switched on, these stars fall on the locations of the spiral arms. Just such a prediction should of course be given by a true theory, since the stars are formed inside the gaseous spirals.

The authors in [271] note that good agreement (probably the most impressive of those considered) in the problem of the migration of moderately young stars will result, in spite of the presence of a large number of factors which could have confused the results.

5.2 M33, M51, M81

5.2.1. Short Characteristics. In a paper by Shu *et al.* [326], on the basis of the density wave theory, spiral patterns of three galaxies (M33, M51, and M81) are investigated. In each of these galaxies there is a clear-cut two-armed spiral. The rotation of M33 is nearly uniform for a significant part of the disk, while the galaxies M51 and M81 rotate with a nearly constant linear velocity. Accordingly, the mass distributions also differ. They are strongly concentrated toward the center in the case of M51 and M81, while for M33 the distribution is "smooth." The main difference between M51 and M81, apart from their sizes, consists of different relative gas contents. The M51 galaxy ("Whirlpool") is also of additional interest since it is associated by one of its arms with a close satellite which may in principle play an important role in excitation (or, vice versa, in destruction) of the spiral structure.

For the sake of simplicity, Shu *et al.* consider only the stars. The inclusion of a small amount of gas ($\sim 10\%$) should not strongly alter the determined characteristics of the spiral patterns.

5.2.2. Models of Equilibrium States of Galaxies. First of all, on the basis of the given curves of rotation, the mass models of galaxies with a finite, though small, thickness are constructed. At each point of the system, a modified Schwarzschield peculiar velocity distribution is assumed. In the vertical direction, the density varies as $\text{sech}^2(z/z_0)$. Here the parameter $z_0(r)$ is the local scale of the galaxies in the z-axis and is expressed through the mean-square velocity v_z^2 by the usual relation: $z_0 = v_z^2/\pi G \sigma_0$, where $\sigma_0(r)$ is the local surface density. The velocity dispersions in the modified Schwarzschield distribution should satisfy certain constraints following from the equilibrium and stability conditions. From the equilibrium condition, as we are aware, in the epicyclic approximation we have the Lindblad connection between the velocity dispersions of stars in the radial and tangential directions: $c_\varphi/c_r = \varkappa/2\Omega$. The ratio $z_\parallel = c_z/c_r$ is probably equal to 1 in central areas of spiral galaxies, where a "well-mixed" equilibrium state [281] should prevail. In the outer regions of the galaxy, $z_\parallel$ can be essentially less than unity (for example, $z_\parallel = 0.5 \div 0.6$ in the solar vicinity of the Galaxy). In the models, adopted in [326], $z_\parallel$ decreases monotonically from unity at the galactic center to 0.5 at the most external regions (Fig. 120). The choice of the rate of decrease is of course rather arbitrary, but the calculated spiral pattern proved to be insensitive to specific choice.

Thereafter, it remains to determine c_r as a function of radius. The minimum values of c_r ensues from the requirement that the stellar disk be stable with

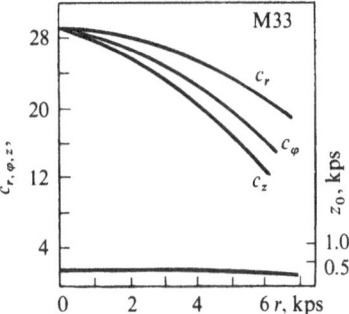

Figure 120. The velocity dispersions and the vertical scale for M33 [326]; the velocity dispersions are calculated by assuming that they are equal to the minimum level necessary for the suppression of the Jeans instability. The vertical scale is connected with the balance between gravitational forces and peculiar motions in the z-axis.

respect to Jeans instability (of radial perturbations). Toomre [333] considered the case of infinitely thin disks. Figure 121 shows the results of [326] generalizing the Toomre criterion in (28') §4.1, Chapter V, on the stellar disks with a finite thickness $z_0 = z_0(r)$; $k_T = c_{r\,min}/\varkappa$ and c_z/c_r are given as the $k_T z_0$ functions ($k_T = \varkappa^2/2\pi G\sigma_0$). To illustrate the use of data in Fig. 121, let us take the solar vicinity of the Galaxy. Assume that the level c_r is the minimum necessary for stability. Following Schmidt [319], take that $\varkappa = 32$ km/s · kps. Assume also the following reasonable estimates: $\sigma_0 = 90\ M_\odot/\text{ps}^2$ and $z_0 = 300$ ps. These estimates correspond to the volumetric density in the central plane equal to $\sigma_0/2z_0 = 0.15\ M_\odot/\text{ps}^3$. Hence we obtain: $k_T = 0.42\ \text{kps}^{-1}$ and $k_T z_0 = 0.126$. From Fig. 121, then we shall obtain: $c_z/c_r = 0.60$, $c_r = 0.42\ \varkappa/k_T = 32$ km/s. These two values are in good agreement with observations [326]. A somewhat more realistic estimate which takes into account that about 10% of the mass falls on the interstellar gas with an effective speed of sound $D \approx 8$ km/s, raises the estimate just obtained by about 3 km/s.

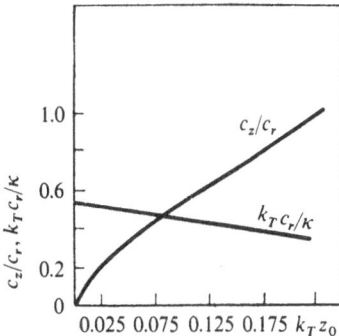

Figure 121. Criterion of the marginal stability [326]; $k_T z_0$ is the dimensionless thickness of the disk.

The coincidence between the velocity dispersion actually existing in the solar vicinity, and the minimum one necessary for stability, is advanced by the authors of [271] and [326] as a decisive argument in favor of the suggestion that the Toomre stability index $Q = c_r/c_{r\,min}$ is unity throughout the disk of the Galaxy, except for probably the most central areas. For the same reason (and for the sake of simplicity) they assume that $Q = 1$ also for the disk models of all galaxies studied in [326].

The velocity dispersions calculated in agreement with the assumptions considered above, are decreasing functions of r, such that at the center $c_r = c_\varphi = c_z$, but everywhere for $r \neq 0$, $c_r > c_\varphi > c_z$.

5.2.3. Spiral Patterns. After completion of the construction of the equilibrium state model, one may start studying the local properties of the density waves in these galaxies. For the stellar disk which possesses a modified Schwarzshield distribution, we have already given the local dispersion equation in Section 4.1, Chapter V. The dependence of the wave amplitude on the radius for an infinitely thin disk was derived in [325, 334] in the second order of the WKB approximation (*vide supra*, subsection 3.4.2), but these calculations have not yet been generalized for the case of a disk of finite thickness. Qualitatively true results are obtained in the following way [326]. Denote by $g = (g_r^2 + g_\varphi^2)^{1/2}$ the amplitude of the spiral gravitational field averaged with the mass weight and express it as a fraction F of the mean gravitational field $r\Omega^2$; then we obtain (see Section 4.1, Chapter V):

$$F = g/r\Omega^2 = const(k^2 r^2 + m^2)^{1/2}(rR_v)^{-1/2}/r^2\Omega^2,$$

$$R_v = \left[-1 + \frac{\partial}{\partial \ln k} \ln(\mathscr{F}_v I) \right].$$

One of the problems of the paper under consideration was the test of Lin's hypothesis about the formation of spirals under the action of the gravitational collapse in the outer regions of spiral galaxies (cf. Section 4.2). The two-armed spiral waves may propagate only in the region where the horizontal line $\Omega = \Omega_p$ lies above $\Omega - \varkappa/2$ and below $\Omega + \varkappa/2$ (Fig. 122). However, if Lin [272] is right then only in the region where $(\Omega - \varkappa/2) < \Omega_p < \Omega$ may there exist an *organized* spiral pattern. Then one may assume that the velocity of the spiral pattern should be approximately equal to the velocity of rotation of outermost HII regions (i.e., regions of ionized hydrogen formed round young stars under the action of their powerful shortwave emission). The yields for M33, $\Omega_p \approx 16$ km/s · kps, while for M81, $\Omega_p \approx 21$ km/s · kps, so that the corresponding radii are 6.8 and 11.2 kps. In case of probable destruction of the outer regions of M51 by a nearby satellite, this method cannot give a correct estimate of the corotation radius for M51. In such circumstances it is assumed arbitrarily that: $\Omega_p = 33$ km/s · kps (the corresponding corotation radius is 4.5 kps) as a value not being in disagreement with Lin's proposal.

The corotation radius lies slightly outside the most external HII-regions, if the formation of stars is initiated only by the mechanism of spiral galactic

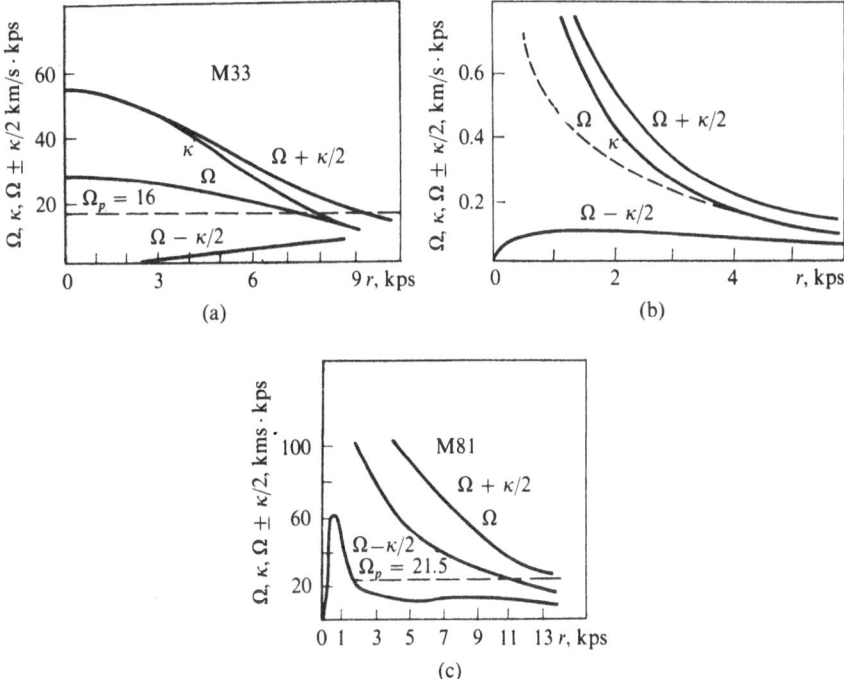

Figure 122. Dependences of the angular velocity Ω, the epicyclic frequencies $\varkappa$, and the Lindblad combinations $\Omega \pm \varkappa/2$ on the radius r according to the mass models of (a) M33, (b) M51, (c) M81 [326].

shock waves [311], and is coincident with these HII-regions, if the collapse of the interstellar gas itself causes star formation. The spiral pattern associated with our Galaxy appears to be satisfying Lin's criterion [326]. It is interesting to verify whether it is also satisfied for other galaxies. The test consists in whether the theoretical pattern, with a wave velocity estimated by the technique thus described, provides a good fitting of the observed spiral structure inside the corotation radius.

The spiral patterns calculated for such Ω_p coincide well with the observed ones. Since Ω is nearly constant throughout the disk of M33, the curve $\Omega - \varkappa/2$ is very flat. Therefore the spiral waves may pass into the very center of this galaxy (and disturb it by forming thereby a barlike structure). One may expect that this is a general feature of galaxies which do not possess a strong concentration of the mass toward the center. At the same time, for the chosen velocities, the waves in M51 and M81 should encounter "barriers" at the inner Lindblad resonances.

Connected to these remarks is the trend for galaxies of the type of M33 to have more open spiral arms. The waves in such galaxies should probably be everywhere far from the Lindblad resonance, the consequence of which is the fact that the dispersion equation will never yield very short waves for the self-consistent spirals.

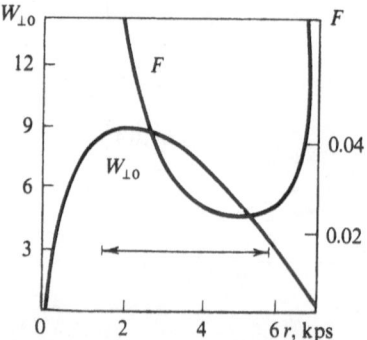

Figure 123. Dependence of the unperturbed velocity $W_{\perp 0}$ and the relative amplitude of the spiral gravitational field F on the radius r for short waves in M33 ($\Omega_p = 16$ km/s · kps) [326].

The authors of [326] also show the boundaries (Fig. 123) in which the calculations are adequate, i.e., the conditions of applicability of the WKB approximation are satisfied. In each case, they cover nearly the whole galaxy, but of course, they do not include the regions close to the resonances.

Observations do not allow one to construct a single unambiguous galactic model. In this connection, in [326] two somewhat different models of M51 (and for two different Ω_p: $\Omega_p = 33$ and $\Omega_p = 43$ km/s · kps) are considered. From this treatment the following conclusions are obtained.

5.2.4. Conclusions. 1. Except for the localization (but not the existence) of the inner Lindblad resonance, the main features of the spiral pattern, for a given Ω_p, are *insensitive to the details* of the adopted equilibrium model. In particular, the pitch angle $i = \arctan(m/|k|r)$ is finally defined by the *main* features of the model.

2. The calculated picture is *sensible* to the variations in Ω_p (i.e., to the *localization of the corotation radius*); the 30% increase in Ω_p (from 33 to 43 km/s · kps) for M51 leads up to the pattern with the pitch angle increased roughly speaking also by 30%, while the radii of corotation and the outer Lindblad resonance are decreased by about 20% (the curve of rotation at these distances is nearly Keplerian, $\Omega \sim r^{-3/2}$).

The calculated pictures have a tendency toward a somewhat stronger twisting as compared with the observed one, so that the velocities of waves in the future might be slightly raised (but probably by not more than 30%).

The general conclusion arrived at in [326] implies that the conception of Lin of initiation of density waves by the Jeans instability in the outer regions of normal spiral galaxies agrees with the study of the spiral structures of M33, M51, and M81. It is obvious, however, that the described results are of a more general nature, and also they are not contradicted by any other generation mechanisms of spiral waves (including those we described above). Thus, in [326] it is not shown that Lin's suggestion is the only one consistent with observations. In particular, as noted by the authors themselves [326], the instability mechanism could cause the spiral pattern of the M33 galaxy.

The comparison of the theory of the galactic density waves with observed data on a more comprehensive material (24 galaxies) was continued by Roberts and Shu [312]. Moreover, in this paper, an attempt is made to give a new classification of spiral galaxies based on a small number of observed characteristics.

The choice of these characteristics is prompted by the density wave theory. The linear theory of Lin and Shu involves one free parameter, Ω_p (or r_c, the corotation radius). The strength of the large-scale shock formed in the gaseous constituent of the galaxy is proportional to $(w_\perp/D)^2$, where D is the effective speed of sound, $w_\perp$ is the total (unperturbed + perturbed) component of the velocity of the gas perpendicular to the spiral arm. The latter quantity oscillates along a streamline due to the action of the spiral gravitational field, about the unperturbed value $w_{\perp 0}$. The shock[13] is formed in the case where the accelerating force acting from the side of the spiral arm is large enough in order that $w_\perp$ could reach supersonic values. For $w_{\perp 0} > D$, the larger part of the gas on the streamline moves at a supersonic speed. Strong shock waves formed in such a situation lead to the formation of *narrow* regions of *high* compression of the gas. Weak shocks formed for $w_{\perp 0} < D$ provide *broad* regions with relatively *low* compression of the gas. These two cases correspond to the observed *narrow* and *broad* arms, so that the formation of one or the other must critically depend on the value of $w_{\perp 0}/D$, or (since D probably does not change very much from galaxy to galaxy) simply on the value $w_{\perp 0} = r(\Omega - \Omega_p) \sin i$, where i is the angle of inclination of the wave to the azimuthal direction, which is found for a given Ω_p from the dispersion equation.

From the above follows the possibility of classification of spiral galaxies based on the two parameters, $w_{\perp 0}$ and i. The dimensional analysis shows that the typical values of $w_{\perp 0}$ and i may be expressed as

$$w_{\perp 0} = (GM/r_c)^{1/2} f(r_{0.5M}/r_c); \qquad \sin i = g(r_{0.5M}/r_c), \qquad M(r_{0.5M}) = M/2,$$

where f and g are the functions, the form of which should be determined from the equilibrium conditions, M is the total galactic mass. Hence the conclusion [312] follows that the main characteristics and the geometrical shape of the normal spirals should be determined by the two parameters: M/r_c (or equivalently, $M/r_{0.5M}$) and $r_{0.5M}/r_c$.

The principles stated above of galactic classification based on the density wave theory should attach to it, according to the opinion of the authors of the paper [312], a more objective character. They note a satisfying correlation between the galactic models in the proposed classification and Hubble's types.

---

[13] Note that the shock wave theory of Roberts was however subjected to criticism (see Section 1.6, Chapter VII, and in detail [52^{ad}]).

§ 6 Experimental Simulation of Spiral Structure Generation

6.1 In a Rotating Laboratory Plasma

6.1.1. Formulation of the Problem. In [100], the question of the analogy between the spiral arm formation process of galaxies and the density waves in a rotating laboratory plasma is considered. This question is not a new one: Bostick was the first to draw attention (about 20 years ago) to the external likelihood of the photographs of galactic spiral arms and the plasma clusters in laboratory experiment. Pictures taken at the time of collision of plasma clusters during the injection of these clusters from two or more injectors to one point of space, really very much resemble the pictures of galactic spirals. In Bostick's experiments, the analogy does not extend farther than a purely surface likelihood at the time of collision of the clusters, each being identified by Bostick with a "spiral" arm. Thus, the number of "arms" (according to Bostick) is exactly equal to the number of plasma injectors. Of course, such an analogy could not provoke a serious discussion. Nonetheless, if we turn to the question of the analogy between the variety of galactic spiral structures and more modestly sized objects, then among the latter attention is drawn by the rotating masses of gas and plasma: the very familiar satellite photographs of cyclones and anticyclones (Fig. 124), the "spiral structure" of the funnel of rotating liquid, and photographs

Figure 124. Space photo of a cyclone (negative) over the Pacific.

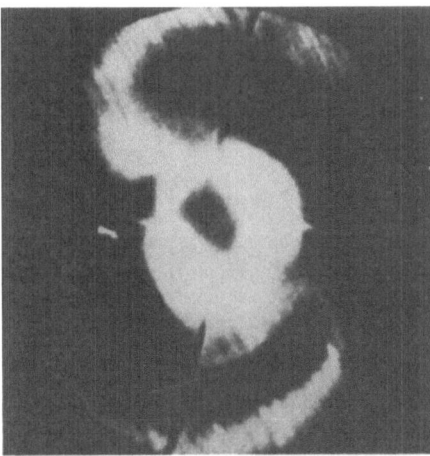

Figure 125. Two-armed spiral in a rotating plasma [34].

of density waves in rotating plasma [34] (Fig. 125). Figure 126 shows characteristic spiral structures of plasma density waves obtained on our plasma machine.

The similarity of the shapes of the galactic spiral structure and the rotating laboratory plasma, under certain conditions, may be a consequence of the available analogy between the mechanisms discussed above of the formation of the spiral structure in the two apparently quite different media. The scheme of the proof [100] of the existence of such an analogy is suggested to be the following.

First of all, the existence of such a mechanism among the different possible mechanisms of the galactic spiral arm formation, which turns out to be free from the influence of the gravitational effects associated with the presence of giant gravitating masses in the galaxy. The same instability must lead to large-scale waves of density in a rotating laboratory plasma. It is evident that such an "universal" instability responsible for the dynamics of the rotating continuous medium may be any of the hydrodynamic instabilities caused by the presence of the velocity gradients and density gradients in a gaseous disk of the flat subsystem of the spiral galaxy and in the rotating laboratory plasma. The possibility of the plasma experiment under discussion is provided owing to the fact that, as proved in [100], the dispersion equations describing the oscillations of the plasma and gravitating media, are similar in many interesting cases.

The elementary scheme of proof of the existence of such an analogy was proposed some time ago by one of the authors (A.M.F.)

Astronomical observations of recent years [315] have discovered a region of sharp decrease of the rotational velocity $V_\varphi(r)$ in the disks of flat galaxies. This fact can be explained by using the results of a calculation of a stationary model of a spiral galaxy in the form of a heterogeneous "disk + nucleus" system [93]. If the nucleus is chosen in the form of a sufficiently

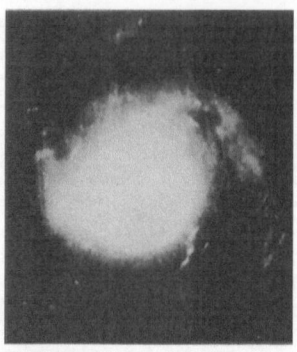

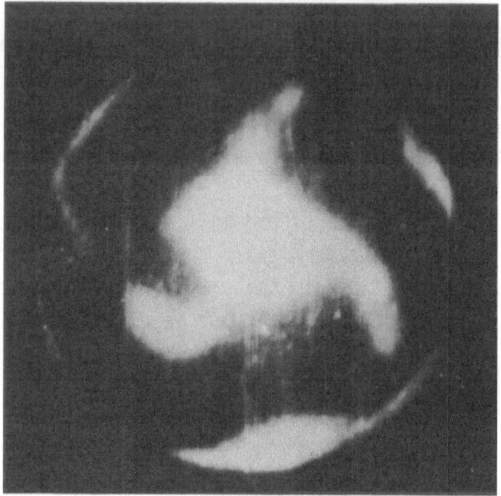

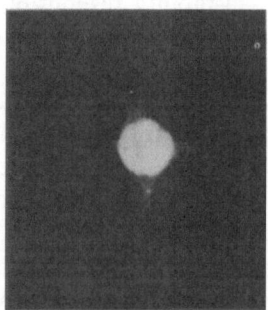

Figure 126. Plasma density waves in the special plasma machine modelling the formation of galactic spirals (photos a, b, c correspond to different conditions of the experiment) [100].

thin lens, which completely corresponds to the observed forms of the nuclei of spiral galaxies [37], then near the edge of the lens the gradient of the gravitational potential may change rather abruptly (almost discontinuously when the thickness of the lens tends to zero). It is not difficult to calculate the critical thickness of the lens at which $V_\varphi \sim 1/r$ from the equilibrium condition. If the thickness of the lens is less than the critical, then the Rayleigh instability criterion [233] is satisfied in the system; this is the necessary condition for the development of the Kelvin–Helmholtz instability. It so happens that the spiral arms extend over the region of radial growth of the surface density σ of atomic hydrogen, i.e., the region in which $g\nabla\sigma < 0$ (g is the acceleration of the force of gravity) [37]. In this connection, it has been conjectured [98a] (see also [24^{ad}]) that the spiral structure of flat galaxies is formed as a result of excitation of the Kelvin–Helmholtz instability in the region of rapid variation of $V_\varphi(r)$; a second observational fact determines the necessary condition for excitation of the flute instability (see subsections 6.1.2, 6.1.4). The growth rates of these instabilities may considerably exceed the Jeans growth rate, and the conditions of development of these instabilities are not related to a critical size.

In subsection 6.1.3 we show that in a gravitating medium for perturbations with wavelengths shorter than the Jeans length $\lambda_J = c_s/(4\pi G\rho_0)^{1/2}$ (in a galactic spiral structure $\lambda/\lambda_J \approx 0.2$–$0.4$) the relative influence of perturbations of the gravitational field on the dynamics of the Kelvin–Helmholtz instability is rather small—the corrections to the hydrodynamic effects are of order $(\lambda/\lambda_J)^2$ (see [98a]). With regard to the unperturbed gravitational field, it enters only into the condition of radial equilibrium and does not affect the dynamics of the perturbations [98a, 24^{ad}].

Because the gravitational effects are small, it is natural to consider verifying the hypothesis of A.M.F. under laboratory conditions. However, the use of a fluid or neutral gas as experimental medium does not enable one to specify independently the necessary gradients of the rotational velocity, especially if there is a large ratio of the velocity discontinuity Δv to the characteristic propagation velocity c_s of perturbations in the medium (for galaxies [116^{ad}] one usually has $\Delta v/c_s \gtrsim 5$). The fulfillment of these conditions is much simpler in a rotating (because of drift in crossed $E_r^{(0)}(r)$ and $B_z^{(0)}$ fields) plasma medium. Here, the role of the fields $E^{(0)}$ and $B^{(0)}$, like the gravitational field's, reduces merely to ensuring that the system is stationary (when $v_i \ll \omega_{Bi}$)^{14}.

Depending on the magnitudes of the characteristic particles of the process, the dynamics of the perturbations of such a plasma can be described either in the framework of magnetohydrodynamics ($\omega \ll v_i$) or in the framework of Chew–Goldberger–Low hydrodynamics [117^{ad}] ($v_i \ll \omega \ll \omega_{Bi}$).

The simplest models convenient for investigating the Kelvin–Helmholtz and flute instabilities are: (1) a plane-parallel flow of fluid with velocity and density that vary in the direction perpendicular to the flow velocity;

---

[14] v_i and ω_{Bi} are the collision and the Larmor frequencies of ions, respectively.

(2) differentially rotating cylindrical configurations of a fluid. Here it is appropriate to recall that the most general stability criterion of these models were obtained in [186, 129] in the approximation of an ideal incompressible fluid. The investigation of these instabilities in a gravitating medium of necessity requires allowance for compressibility, which significantly complicates the analysis and prevents one obtaining general stability criteria.

For this reason, for the original analysis of the problem (in a gravitational medium) we have chosen the simplest models: velocity and density shear layer of the gravitating medium (Subsection 6.1.2) and tangential shear between two gravitating cylinders rotating in opposite directions, their equilibrium being provided by the equality of centrifugal and gravitational forces (Subsection 6.1.4). The stability of a shear layer was investigated earlier in the approximation of an incompressible fluid in an external gravitational field [186, 67], or in a compressible fluid and in the magnetohydrodynamic approximation [126] in the absence of a gravitational field. In subsection 6.5.1 besides proving that the gravitational effects have little influence on the short-wave part of the spectrum in the framework of the Kelvin–Helmholtz instability that we investigate, we obtain estimates that characterize the important role of the Kelvin–Helmholtz instability in the formation of spiral galactic structure. In subsection 6.1.6, we consider the stability of a plasma flow with a tangential shear of the velocity in the Chew–Goldberger–Low approximation [117^{ad}]. In subsection 6.1.7, we compare the dispersion relations that describe the oscillation frequency ω as a function of the wave vector k and the characteristic parameters of the plasma and gravitating media. We show that under typical conditions of the plasma experiment, the corresponding dispersion relations are identical, which demonstrates that the similarity of the spiral patterns of the rotating gravitational and plasma media is not fortuitous but a consequence of the deep analogy between the process responsible for the formation of the spiral structure in these two very different but nevertheless "hydrodynamic" media.

6.1.2. Velocity and Density Shear Layer of a Gravitating Medium. 1. We consider the stability of a shear layer of the velocity and density in a compressible gravitating medium.

We begin with the effects due solely to the velocity shear (Kelvin–Helmholtz instability). Suppose[15]

$$\rho_{01} = \rho_{02}, \qquad c_1^2 = c_2^2, \qquad V_{01} = -V_{02} = V_0, \qquad g = 0.$$

We shall describe the solution by means of the dimensionless parameters

$$M = |V_0|/c, \qquad \beta = M \cos \alpha, \qquad \cos \alpha = (kV_0)/(|k||V_0|),$$

$$v = \omega_0/kc.$$

In the limit of short-wave perturbations, $\omega_0 \ll kc$, in the first approximation, we readily obtain from the dispersion relation the well-known result

---

[15] Formulae used below were derived in Section 3, Chapter VI.

of the theory of a compressible fluid [126]

$$\omega = ikc\beta\gamma, \qquad \gamma = \{(1 + 4\beta^2)^{1/2} - (1 + \beta^2)\}^{1/2}\beta. \tag{1}$$

In the following approximation in $v = \omega_0/kc$,

$$\omega = ikc\beta\gamma\{1 - v^2 A(\gamma)\}, \qquad A(\gamma) = \frac{(1 - \gamma)(1 + \gamma^2)[2\gamma + (1 + \gamma)^2]}{4\gamma(1 + \gamma)(3 - \gamma^2)}. \tag{2}$$

It is easy to see that $A(\gamma) > 0$ and A is a monotonically increasing function of β. Thus, perturbations of the shear surface are subject to an additional stabilization at longer wavelengths due to the gravitational properties of the medium. However, it must be borne in mind that (see Section 3.1, Chapter VI) the results (2) apply only for wavelengths $\lambda \ll \lambda_J = c/\omega_0$. At the same time, as can be seen from (2), the growth rate of the instability is much greater than the Jeans growth rate: $\text{Im}(\omega) \gg \omega_0$.

We now consider the effects associated with the change of the density. Assuming that $V_{01} = V_{02} = 0$ and that the magnitude of the change in the density is not too small compare with ρ_0, we obtain

$$\omega = \left(kg\frac{\rho_{02} - \rho_{01}}{\rho_{02} + \rho_{01}}\right)^{1/2} \left\{1 + \frac{\pi G}{kg}(\rho_{01} - \rho_{02})\right.$$
$$\left. + \frac{\rho_{01}\rho_{02}g(\rho_{01}c_1^2 + \rho_{02}c_2^2)}{kc_1^2 c_2^2(\rho_{01} - \rho_{02})(\rho_{01} + \rho_{02})^2}\right\}.$$

In the approximation $k \to \infty$, we obtain the well-known result of the theory of an incompressible fluid.

We give the expressions for the quantities $\chi_{1,2}$ that characterize the exponential decay of the perturbed pressure along the z-axis (which will be needed later):

$$\chi_{1,2}^2 = k^2 - \frac{\omega_{1,2}^2 + \omega_{01,2}^2}{c_{1,2}^2} + \frac{k^2 g^2}{\omega_{1,2}^2 c_{1,2}^2}, \qquad \text{Re}(\chi_{1,2}) \geq 0. \tag{3}$$

Here, the subscript 1 is appended to the variables of the region $z > 0$; the subscript 2, to those of the region $z < 0$; $\omega_{1,2} = \omega - kV_{01,2}$, $\omega_0^2 = 4\pi G\rho_0$.

6.1.3. Absence of Influence of Gravitational Forces on the Short-Wave Part of the Oscillation Spectrum. Thus, we have shown that hydrodynamic instabilities can develop in a gravitating medium. The Jeans instability characteristic of such a medium is stabilized by thermal spread in the region of short, $k^2 c^2 \gtrsim \omega_0^2$, wavelengths. The hydrodynamic instabilities, in contrast to the gravitational, are not stabilized by the thermal spread in the short-wave region.[16] Moreover, in accordance with (1) and (3), the growth rates of the

---

[16] This result is obvious if one recalls that the shear model considered in subsection 6.1.2 is also unstable in the approximation of an incompressible fluid, in which the thermal spread is by definition infinite.

hydrodynamic instabilities increase with decreasing wavelength of the perturbation.[17] This unique property of the Kelvin–Helmholtz and flute instabilities distinguishes them from the previously investigated hydrodynamic
instabilities of a gravitating medium.

If one assumes that the gravitating medium is in equilibrium, $\nabla p_0 + \rho_0 \nabla \psi_0$
$= 0$, then from the original system of equations with allowance for the
gradients of the unperturbed quantities one can readily see that $|\nabla p|$ is
greater than $|\rho \nabla \psi_0|$ by kL times (L is the characteristic inhomogeneity
scale, $kL \gg 1$) and $|\rho_0 \nabla \psi|$ is smaller than $|\rho \nabla \psi_0|$, also by kL times. Thus,
the influence of the "external" gravitational field can be regarded as a small
correction to the hydrodynamic effects. The influence of "self-gravitation"
is even smaller.

6.1.4. Cylindrical Tangential Shear Density of a Gravitating Medium. We
now investigate the possibility of exciting a flute instability in a gravitating
cylinder. For this we consider a model of an infinitely long cylinder with
abrupt change of the density ρ_0 at a distance R from the axis of the cylinder,
assuming that equilibrium is established by the resultant effect of the centrifugal and gravitational forces and the pressure force, so that $g = d\varphi_0/dr$
$- \Omega^2 r \neq 0$. Consider short-wave (compared with the Jeans length) oscillations, for which the influence of the perturbed gravitational potential is
negligibly small.

Since we are only interested in the basic possibility of exciting the flute
instability, we consider the case of a fairly hot ($c^2 \to \infty$) medium. Then for ρ_0
$= \rho_{01}(r > R) \neq \rho_{02} = \rho_0(r < R)$ we obtain the growth rate ($m^2 \ll k^2 R^2$,
see Section 3.2, Chapter VI):

$$\gamma \approx \left\{ kgA + \frac{(2 - A^2)m^2 \Omega^2}{k^2 R^2} \right\}^{1/2}, \qquad A = (\rho_{01} - \rho_{02})/(\rho_{01} + \rho_{02}). \quad (4)$$

As follows from the expression (4), the necessary condition for instability is

$$gA > 0. \quad (5)$$

This means that for $g = (\partial \Phi_0/\partial r) - \Omega^2 r > 0$ the flute instability develops
if $\rho_{01} > \rho_{02}$ while for $g < 0$ it develops if $\rho_{02} > \rho_{01}$.

Let us consider now a different limiting case: $\lambda \ll a$. For perturbations of
the type $\exp[i(kr + m\varphi - \omega t)]$, we obtain instead of (4) the following growth
rate of the flute instability:

$$\gamma = \left[g \frac{d \ln \rho_0}{dr} \frac{m^2}{k^2 r^2} \right]^{1/2}. \quad (6)$$

Naturally, the instability condition is analogous to (18). The growth rate (6)
is much greater than the Jeans growth rate when $m/kr \gg 1$.

---

[17] This assertion is true at least for wavelengths that are greater or of order of the thickness
of the transition layer.

6.1.5. Spiral Structure of Galaxies as a Possible Consequence of Hydrody-namic Instabilities. Hitherto, it has been assumed that the maximum growth rate of instabilities that can develop in gravitating systems is the Jeans growth rate $\gamma \approx (4\pi G \rho_0)^{1/2}$. The attempt to explain the formation of the spiral arams in our Galaxy by the Jeans instability led, as is well known [333], to a contradiction between the critical Jeans wavelength and the separation of spiral arms. All the remaining hitherto known instabilities of a gravi-tating medium have growth rates less than the Jeans growth rate.

The investigation in the preceding sections of hydrodynamic instabilities of gravitating systems with growth rates appreciably greater than the Jeans rate opens up a new possibility of explaining the origin of the spiral structure if the conditions observed in spiral galaxies correspond to the conditions of development of these instabilities. In the present section we bring forward arguments for the existence in spiral galaxies of the necessary conditions for the development of the hydrodynamic instabilities.

The recent investigations of the rotation curve of the nearest spiral galaxy —the Andromeda Nebula [315]—has revealed the presence of a region of abrupt change of the rotation velocity of the flat subsystem. In the region $0.4 \text{ kpc} \lesssim r \lesssim 2 \text{ kpc}$ there is a section of rapidly decreasing (from the center) rotation velocity $V_\varphi(r)$, in which $(d/dr)(\varkappa^2/2\Omega)$ changes sign. Such a distribu-tion of the rotational velocity is unstable in accordance with the Rayleigh criterion [233] in the approximation of an ideal incompressible fluid, and the finite compressibility of the medium evidently cannot signifi-cantly alter this result. According to the Rayleigh criterion [233], the Kelvin–Helmholtz instability can be excited in rotating systems if over a certain interval Δr the rotation velocity $V_\varphi(r)$ decreases faster than r^{-1}. A suf-ficiently detailed study of the rotation curves of the gaseous subsystems of flat galaxies has made it possible to find such regions in M31 (see [315]) and apparently in NGC 7436 (see [37]). The reasons for this behavior of $V_\varphi(r)$ are to be found in the strong oblateness of the dense nuclear regions of flat galaxies [37]. This may also be the case for barred galaxies. For example, in NGC 4027 (see [343]) the ratio of the semiaxes of the bar are $b/a \approx 0.6$, $c/a \approx 0.2$ (see [36]) (almost elliptical disk).

We now show how the number of spirals is determined in the case when a Kelvin–Helmholtz instability develops in the system. As follows from Sections 3.1 and 3.2, Chapter VI, the growth rates of the Kelvin–Helmholtz instability of gravitating systems with cylindrical and plane shears of the velocity for modes $m \geq 2$ have similar dependences on the wave numbers. Using, for simplicity, the results of subsection 6.1.2 and then making a transi-tion to cylindrical coordinates $(k_\parallel \to k_\varphi, k_\perp \to k_z)$, we can readily estimate the number of spiral arms. Indeed, let us set $k_\varphi = m/R_s$ (m is the number of spirals and R_s is the radius of the shear), $k_z = \pi/h$ (on the basis of the observa-tional data, we assume that approximately half a wavelength fits into the thickness h of the disk). For the Andromeda nebula, the magnitude of the discontinuity of the rotation velocity (see [315]) $\Delta v \approx 150 \text{ km/s}$ is much greater than the turbulent velocities of the gas and gas clouds, $v_T \approx 20 \text{ km/s}$, and therefore perturbations excited with the maximum growth rate (satisfying the relation $\Delta v k_{\parallel} \approx 3^{1/2} v_T (k_{\parallel}^2 + k_{\perp}^2)^{1/2}$) necessarily have $k_{\perp} \gg k_{\parallel}$ (see Section 3.1, Chapter VI).

Thus, the number m of spiral arms is

$$m \approx 3^{1/2} \pi (v_T/\Delta v)(R_s/h).$$

Since $R_s \approx 0.5$ kpc, $h \approx 0.1$ kpc (see [315]), the number of spirals is of the order of a few units.

It is also easy to establish the direction of winding of the spirals. For this (see subsection 6.1.2.) we go over to a frame of reference moving in the direction V_{01} with velocity $V_s > V_0$, so that $V_{01}' = V_0 + V_s, V_{02}' = V_s - V_0$. In such a reference system, the oscillation frequency is $\omega' = -k_{\parallel} V_s + ikc\beta\gamma$ (see (1)). Identifying $z > 0 \rightarrow r > R_s$ (R_s is the radius of the velocity shear $\Delta v = 2V_0$) and substituting ω into (3), we find that for perturbations with the maximum growth rate in the region $r > R_s$ the relation $\text{Im}(\chi_1) < 0$ necessarily holds. It can be seen from this that the equation of constant phase in the (r, φ) plane, $m\varphi - \text{Im}(\chi_1)r = \text{const}$, describes a trailing spiral ($m > 0$, $V_\varphi > 0$).

The distribution of the density in the gas disks of the flat subsystems of spiral galaxies has, as is known from observations, a bell-shaped form (with a point at which the density is maximum). Therefore, the presence of even a small radial gradient of the gas temperature (in the neighborhood of the extremum of the density) may lead to the development of the flute instability. Indeed, in this region $|(1/p_0)\nabla p_0/(1/\rho_0)\nabla \rho_0| > 1$ and, on either side of the extremum point of the density, $\nabla p_0/\nabla \rho_0 < 0$.

6.1.6. Tangential Velocity Shear of a Magnetized Plasma Medium.
In this case, we believe it is convenient to use the system of equations of Chew–Goldberger–Low hydrodynamics [117[ad]],

$$\frac{\partial \mathbf{V}}{\partial t} + (\mathbf{V}\nabla)\mathbf{V} = -\frac{1}{\rho} \text{div}\, \overleftrightarrow{P} + \frac{1}{4\pi\rho} [\text{rot}\, \mathbf{B} \times \mathbf{B}],$$

$$\frac{\partial \mathbf{B}}{\partial t} = \text{rot}[\mathbf{V} \times \mathbf{B}],$$

$$\frac{\partial p_{\parallel}}{\partial t} + \mathbf{V}\nabla p_{\parallel} + p_{\parallel}\, \text{div}\, \mathbf{V} + 2p_{\parallel} \tau(\tau\nabla)\mathbf{V} = 0, \qquad (7)$$

$$\frac{\partial p_{\perp}}{\partial t} + \mathbf{V}\nabla p_{\perp} + 2p_{\perp}\, \text{div}\, \mathbf{V} - p_{\perp}\tau(\tau\nabla)\mathbf{V} = 0,$$

$$\frac{\partial \rho}{\partial t} + \text{div}(\rho\mathbf{V}) = 0,$$

$$\tau = \mathbf{B}/|\mathbf{B}|, \quad \text{div}\, \overleftrightarrow{P} = \nabla p_{\perp} + (p_{\parallel} - p_{\perp})(\tau\nabla)\tau + \tau\, \text{div}(\tau(p_{\parallel} - p_{\perp}))$$

As was shown in subsection 6.1.3, for a gravitating system in equilibrium, the influence of the gravitational field on the stability of the system against short waves ($kL \gg 1$, k is the wave number and L is the characteristic in-homogeneity scale) is negligibly small.

On the other hand, it is known from the theory of plasma instabilities [86] that the maximum of the growth rate of the Kelvin–Helmholtz instability lies in the short-wavelength part of the spectrum, $kL \gg 1$. For a plasma cylinder of radius R, this condition corresponds to the condition $kR \gg 1$; in this case, to terms $\sim 1/kR$, a cylindrical plasma velocity shear can be replaced by a velocity shear layer in a plasma with homogeneous magnetic field B_{0y}. Assuming that the variation of V_{0x} near the plane $z = 0$ is smooth and linearizing Eqs. (7) for perturbations of the type $\exp\{i(k_x x + k_y y - \omega t)\}$, we obtain

$$(\omega_*^2 - k_y^2 V_{A\parallel}^2)\xi = c_{s\perp}^2 p' - k_y^2 c_\perp^2 (c_{s\perp}^2 - c_\perp^2)\left[\frac{p}{\omega_*^2 - k_y^2(c_{s\parallel}^2 - c_\perp^2)}\right]', \quad (8)$$

$$\xi' = p\left\{k_x^2 \frac{c_{s\perp}^2 - k_y^2 c_\perp^2 (c_{s\perp}^2 - c_\perp^2)/[\omega_*^2 - k_y^2(c_{s\parallel}^2 - c_\perp^2)]}{\omega_*^2 - k_y^2 V_{A\parallel}^2}\right.$$

$$\left. + k_y^2 \frac{c_\perp^2}{\omega_*^2 k_y^2(c_{s\parallel}^2 - c_\perp^2)} - 1\right\}, \quad (9)$$

where $\omega_* = \omega - k_x V_{0x}(z)$, $\xi = iV_{z1}/\omega_*$ is the displacement of the plasma in the z direction, $p = \rho_1/\rho_0$ is the ratio of the perturbed to the unperturbed density, $V_{A\parallel}^2 = V_A^2 + (p_{\perp 0} - p_{\parallel 0})/\rho_0$, $V_A^2 = B_0^2/4\pi\rho_0$, $c_\perp^2 = p_{\perp 0}/\rho_0$, $c_{s\perp}^2 = V_A^2 + 2c_\perp^2$, $c_{s\parallel}^2 = 3p_{\parallel 0}/\rho_0$, and the prime denotes differentiation with respect to z. Integrating Eqs. (8) and (9) over a narrow surface layer, we obtain the conditions for matching ξ and p at the plane $z = 0$:

$$c_{s\perp}^2 [p] = k_y^2 c_\perp^2 (c_{s\perp}^2 - c_\perp^2)\left[\frac{p}{\omega_*^2 - k_y^2(c_{s\parallel}^2 - c_\perp^2)}\right], \quad (10)$$

$$[\xi] = 0, \quad [A] \equiv A(z = +0) - A(z = -0). \quad (11)$$

Matching, in accordance with (10) and (11), the solutions of the system (8) and (9) which do not increase away from the velocity shear, we obtain the dispersion relation

$$\chi_1(\omega_{*2}^2 - k_y^2 V_{A\parallel}^2) + \chi_2(\omega_{*1}^2 - k_y^2 V_{A\parallel}^2) = 0; \quad (12)$$

the subscripts 1 and 2 are appended to the variables of the regions $z > 0$ and $z < 0$, respectively, and

$$\chi_{1,2}^2 = k^2 + \frac{(k_y^2 c_{s\parallel}^2 - \omega_{*1,2}^2)(\omega_{*1,2}^2 - k_y^2 V_{A\parallel}^2)}{c_{s\perp}^2 [\omega_{*1,2}^2 - k_y^2(c_{s\parallel}^2 - c_\perp^4/c_{s\perp}^2)]}. \quad (13)$$

It can be seen from this that in the investigated case we can expect a dependence $\omega = \omega(k_x, k_y)$ like the one observed in ordinary hydrodynamics [98a] when one of the following two conditions is satisfied:

$$V_{A\parallel}^2 \simeq c_{s\parallel}^2 - c_\perp^4/c_{s\perp}^2, \quad (14)$$

$$c_\perp^4 \ll c_{s\parallel}^2 c_{s\perp}^2. \quad (15)$$

The condition (14) corresponds to a plasma with $\beta \sim 1$. The solution of the dispersion relation (12) in this case,

$$\omega^2 \simeq k_x^2 V_0^2 + \tilde{k}^2 c_{s\perp}^2 - \{(k_x^2 c_{s\perp}^2 + k_y^2 c_\perp^4/c_{s\parallel}^2)^2 + 4k_x^2 V_0^2 k^2 c_{s\perp}^2\}^{1/2}, \quad (16)$$

where $\tilde{k}^2 c_{s\perp}^2 = k_x^2 c_{s\perp}^2 + k_y^2 c_{s\parallel}^2$, shows that instability ($\omega^2 < 0$) will occur for $V_0^2 < c_{s\perp}^2$ for perturbations with

$$\cos^2 \alpha = \frac{k_x^2}{k^2} > \cos^2 \alpha_1 = \frac{(c_{s\parallel}^2 c_{s\perp}^2 - c_\perp^4)}{(c_{s\parallel}^2 c_{s\perp}^2 - c_\perp^4) + V_0^2 c_{s\perp}^2},$$

and, for $V_0^2 > 2c_{s\perp}^2$, for perturbations with

$$\cos^2 \alpha_1 < \cos^2 \alpha < \cos^2 \alpha_2 = \frac{(c_{s\parallel}^2 c_{s\perp}^2 + c_\perp^4)}{(c_{s\parallel}^2 c_{s\perp}^2 + c_\perp^4) + c_{s\perp}^2(V_0^2 - 2c_{s\perp}^2)}.$$

The condition (15) corresponds to a plasma with $\beta \ll 1$. The solution of the dispersion relation (12) in this case:

$$\omega^2 = k_x^2 V_0^2 + k^2 V_A^2 - \{k_x^4 V_A^4 + 4k_x^2 V_0^2 k^2 V_A^2\}^{1/2}, \quad (17)$$

predicts instability when $V_0^2 < 2V_A^2$ for perturbations with

$$\cos^2 \alpha > \cos^2 \alpha_1 = V_A^2/(V_A^2 + V_0^2),$$

and when $V_0^2 > V_A^2$ for perturbations with

$$\cos^2 \alpha_1 < \cos^2 \alpha < \cos^2 \alpha_2 = V_A^2/(V_0^2 - V_A^2).$$

6.1.7. Analogy Between the Dispersion Relations Describing the Kelvin–Helmholtz and the Flute Instability in a Gravitating Medium and in a Plasma.

For the typical conditions of the plasma experiment, $E_0 \ll B_0^2/c(4\pi n_0 M)^{1/2}$, which corresponds to the inequality

$$V_0^2 \ll V_A^2 \sim c_{s\perp}^2. \quad (18)$$

The instability regions described by Eqs. (16) and (17) are shown in Fig. 127. It can be seen that, under the condition (18), $\gamma \to \gamma_{max}$ in a neighborhood of the straight line

$$k_y/k_x \approx 0. \quad (19)$$

Thus, under the conditions of the plasma experiment, the dispersion relations of nonelectrostatic oscillations of the plasma with tangential velocity shear have the form

$$\omega^2 = k_x^2 V_0^2 + k_x^2 c_{s\perp}^2 - (k_x^4 c_{s\perp}^4 + 4k_x^4 V_0^2 c_{s\perp}^2)^{1/2}, \quad \beta \sim 1, \quad (20)$$

$$\omega^2 = k_x^2 V_0^2 + k_x^2 V_A^2 - (k_x^4 V_A^4 + 4k_x^4 V_0^2 V_A^2)^{1/2}, \quad \beta \ll 1. \quad (21)$$

Since $c_{s\perp}^2 \sim V_A^2$, the two dispersion relations (20) and (21) are identical under the condition of instability (19).

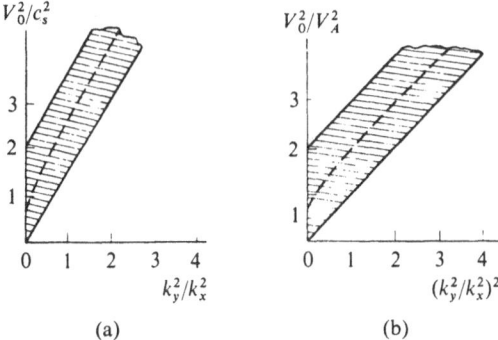

(a)   (b)

Figure 127. Region of the Kelvin–Helmholtz instability in a plasma (hatched). (a) The case $V_{A\parallel}^2 = c_{s\parallel}^2 - c_\perp^4/c_{s\perp}^2$; (b) $c_\perp^4 \ll c_{s\parallel}^2 c_{s\perp}^2$ ($\beta \ll 1$). The line of the maximal growth rate is the dashed curve.

In the case of a tangential velocity shear in a gravitating medium, the following dispersion relation was obtained for short wavelengths in subsection 6.1.2 (see (1)):

$$\omega^2 = k_x^2 V_0^2 + k^2 c_s^2 - (k^4 c_s^4 + 4k_x^2 V_0^2 k^2 c_s^2)^{1/2}. \tag{22}$$

In Fig. 127(a) the dashed curve shows the region where $\gamma \approx \gamma_{max}$. In this region

$$k_y \approx 2 \cdot 3^{-1/2} V_0 k_x/c_s. \tag{23}$$

In the case $V_0 \ll c_s$, $k_y \ll k_x$, Eq. (22) is identical with Eqs. (20) and (21).

In the case $V_0 \gg c_s$ (which corresponds to the observations of the galaxies)

$$\gamma_{max} \approx 3^{-1/2} k_x V_0. \tag{24}$$

We find a similar dependence of γ_{max} on k_x from Eqs. (20) and (21) under the condition (18).[18]

We now consider a plasma cylinder in which the electrons and ions drift in crossed electric E_r and magnetic B_z fields. Suppose that in a plasma with $\beta \ll 1$, $\omega_{pi} \gg \omega_{Bi}$, $V_0 \ll r\omega_{Bi}$ (ω_{pi} and ω_{Bi} are the plasma and cyclotron frequencies, respectively, and V_0 is the velocity of rotation of a particle about the axis) oscillations with $\omega \ll \omega_{Bi}$, $k_z = 0$ are excited. One can show that in the case of a radially decreasing plasma density these oscillations are unstable and that for $l \gg 1$ the growth rate is

$$\gamma = \left(\frac{\partial n_0}{n_0 \partial r} \frac{1}{r}\right)^{1/2} V_E \left(\frac{k_\varphi}{k}\right)^{1/2}, \tag{25}$$

$$V_E = cE_r/B_z, \qquad k_\varphi = l/r.$$

Denoting $V_E^2/r = g$, this growth rate is identical to the growth rate (6).

---

[18] In the framework of two-fluid hydrodynamics one can show that in a plasma with $\beta \ll 1$ and inhomogeneous velocity profile electrostatic oscillations can be excited. If certain conditions are satisfied, the equation for the perturbed potential is identical to the equation of the oscillations of a plane-parallel flow of an ideal fluid [89]. Therefore, in a plasma described by the equations of two-fluid hydrodynamics in the case of a tangential velocity shear, the growth rate of the instability associated with the excitation of electrostatic oscillations is $\gamma \approx k_x V_0$.

6.2 In Numerical Experiments

6.2.1. N-Body Simulations of Disk-like Systems. An immense number of papers have been devoted to computer experiments on the stability and evolution of flat stellar systems. Therefore, by referring the reader to the reviews [80[ad], 88[ad]] for more details, we shall consider briefly only the main results.

Probably, the principal unexpectedness of these experiments was the impossibility of obtaining rather long-lived spirals in purely stellar disks. The development of nonaxially-symmetrical instabilities leads to a rapid heating of the stellar disk, spiral-like structures arise for a short time only, and ultimately the system becomes a very hot elliptical disk. On the other hand, the experiments in question show that barlike disturbances develop very easy in rotating stellar disks; the bars generally are the most typical feature of evolution, and this, as noted by Toomre [88[ad]], is an important step toward understanding the nature of the formation of the SB-galaxy bars. The problem is, however, that normal spirals just do not have apparent bars. At the same time, N-body simulation shows that in order to stabilize the barlike instability in a stellar disk, one needs velocity dispersions far more than, for example, those observed in the solar vicinity of the Galaxy. This was noted in many works, but a specially detailed study was performed by Miller [80[ad]]. By carefully studying all the factors which could, in principle, bring about the heating of computer "stars" (including the causes due to the peculiarities of computation such as approximation, cut-off of the Newton potential at small distances, roughness of the integration scheme of equations of motion, etc.) he showed that heating is brought about just by physical causes. Thus, this fact may be considered as firmly established. What, however, may this mean? The first, most obvious, possibility is in the supposition about the existence of a massive but invisible halo, which, as we know, can effectively stabilize the barlike instability. Note also here that, according to some authors [74[ad], 88[ad]], the barlike instability can actually be not so dangerous, as ensues from the computer experiments as well as from theoretical studies on the linear theory of stability of disklike systems. The point is that the latter have so far been referred to the simplest models, while the former considered only those systems in which all the stars rotated in the same direction. The situation must be clarified upon completion of the very much delayed investigations of large-scale instabilities of the general models of stellar disks, and when computer experiments are performed for disks with stars rotating not only in the main but also in the opposite direction (reversely rotating stars, actively taking part in perturbations, exert a larger stabilizing action than the passive halo [74[ad]]).

Finally, to sum up it should be noted that the lame attempts to simulate spiral galaxies using purely stellar disk systems may be indicative of the essential role of the gaseous component of the galaxy (energy dissipation, shocks, etc.).

Here we do not give a review of papers on numerical modelling of spiral arms due to the tidal galaxy interaction. Except for the rare cases when the satellite-galaxy is a continuation of a spiral arm of the neighboring galaxy (as, for instance, in M51), the role of the tidal interaction in the spiral structure formation seems to be negligible. In detail, such a point of view on this problem is presented in [51[ad]].

6.2.2. On the Criterion of Applicability of Numerical Models of Interacting Galaxies [111[ad]], [50[ad]].

The first hypothesis about the decisive role of gravitational interaction in the process of the spiral structure formation seems to belong to Chamberlin [100[ad]]. True, he considered the spiral nebulae produced by a tidal force acting on the rotating star by a neighboring star. However, such considerations are similar in principle to the well-known outlines of the hypothetical origin of two galaxies suggested more than half a century later by Zwicky [107[ad]] in order to illustrate the process of birth of a spiral structure due to the formation of a bar between two galaxies.

The first numerical calculation to prove this hypothesis was carried out by Holmberg [101[ad]] by means of an unsophisticated procedure of graphical integration proposed by him, and later on, using new computers, by Pfleiderer and Siedentopf [102[ad], 103[ad]]. Then followed numerical experiments revealing details of the mechanism of the spiral structure formation for a large number of points: Tashpulatov [104[ad]], Toomre [110[ad]], Yabushita [106[ad]], Kozlov et al. [108[ad], 109[ad]] and, in particular, for ring structure: Toomre [105[ad]].

As is shown below, in the above models of flat galaxies the tidal effects analyzed are significantly weaker than the collective effects not regarded there. A simple criterion obtained in [40] by Ginzburg et al. is given, according to which the above case histories are not related in any way to any of the presently known spiral galaxies[19] and they may be related only to systems with anomalously large central mass of the type of Saturn's rings.

The model of interacting galaxies used in the above cited papers is presented schematically in Fig. 128(a). It is assumed that each point A interacts only with centers B and C and not with other points.

If the disk, as assumed by all the authors, is infinitely thin and there is no initial thermal scatter of points over velocities, such a model is absolutely unstable and must practically instantaneously[20] take the shape of a "cartwheel" (Fig. 128(b)).

Indeed, in this case the dispersion equation for the frequency of small oscillations of the disk with a central body of mass M, composed of gravitating points has the form [40] (see also next section):

$$(\omega - m\Omega_0)^2 = \varkappa_0^2 - 2\pi G\sigma_0 k, \tag{1}$$

---

[19] In some rare cases where the galaxy-satellite is a continuation of the spiral arm of a neighboring galaxy (such as M51), the perturbing center (galaxy-satellite) is responsible only for the spiral arm *orientation* of its giant neighbor but by no means for the fact of *their formation*.

[20] Time estimate see below.

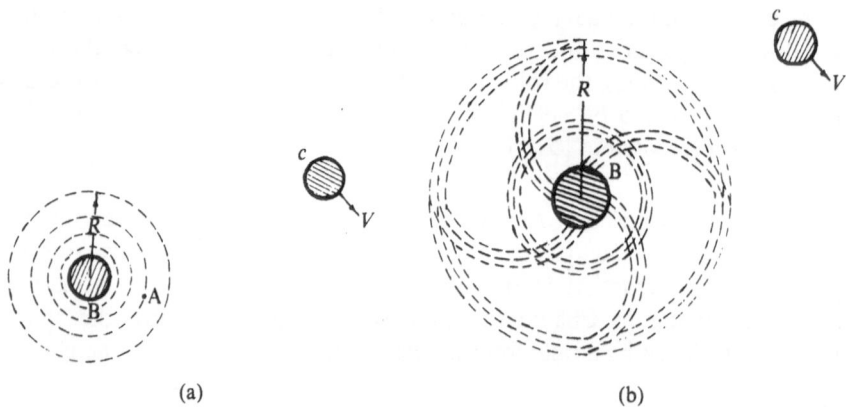

(a)   (b)

Figure 128. Interacting galaxies: (a) initial moment of interaction; (b) disk's instability.

where x_0^2, in the case where undisturbed orbits of points are Keplerian, is equal to

$$x_0^2 = GM/r^3. \tag{2}$$

The instability increment is independent of the mode number (in the highest order of the expansion over $1/kr$; in this case $k_{max} \sim 2\pi/\Delta$, where Δ is the distance between the points which, at a characteristic number of points ≈ 2000, is very small, i.e., the value $kr \gg 1$), therefore for the time

$$t \sim 1/\gamma_{max}, \quad \text{where } \gamma_{max} \sim \sqrt{4\pi^2 G\sigma_0/\Delta} \tag{3}$$

(G is the gravity constant, σ_0 is the surface density) the disk assumes the shape depicted in Fig. 128(b).

The instability time (3) is so small that any influence of the disturbed center C is out of the question.

Allow for the finite thickness of the disk, $h \neq 0$. In order that the model of interacting galaxies (Fig. 128(a) may be true, it is required that the perturbed motion of the test particle A be determined by the central mass B rather than by the remaining particles of the disk. In other words, it is necessary to write the condition of neglect of collective effects.

As is shown in the paper [40] (see also next Chapter) just that condition is realized for Saturn's rings. It has the form:

$$M/m \gtrsim 2R/h, \tag{4}$$

where $m = \sum m_i$, m_i is the mass of the ith particle of the disk, R is the radius of the disk.

It is easy to see that for spiral galaxies criterion (4) is not realized. Indeed, spiral galaxies are unknown for which the left-hand side of the inequality would be more than 2, while the characteristic values of the right-hand side of the inequality are 20 + 30.

If we introduce the initial velocity dispersion for particles such that the disk becomes stable in the linear approximation, then, as shown in [20] (see also Chapter VII), the presence of strong nonlinear effects will not permit the model in Fig. 128(a) to be used.

§ 7 The Hypothesis of the Origin of Spirals in the SB-Galaxies

According to de Vaucouleurs (see, e.g., in [344]), the structural properties of spiral galaxies change very smoothly along the sequence of types, and among them one can trace a smooth transition from the usual S-galaxies to the SB-galaxies. According to the opinion of Freeman and de Vaucouleurs, this fact suggests that the different structural features of spiral galaxies of different types are produced under the action of a certain single mechanism.

Such a mechanism, in particular, may be instability having a "barlike" shape at the center of the system and a spiral shape on the periphery (such a possibility was discussed in Sections 4.1–4.3, 4.5; see also Section 4.5, Chapter V). However, the bar can, of course, be produced also during the time of the initial collapse of the protogalaxy, as was shown in the paper by Lin *et al.* [268] and also by Lynden-Bell [282].

In some S- and SB-galaxies, large-scale radial motions of the gas have been discovered. As far as the SB-galaxies are concerned, it turns out that it is quite possible to reproduce the spiral structure, if one assumes [201, 344] that the matter flowing outwards along the bar takes on a necessary angular momentum due to twisting by the gravitational field of the bar and then, leaving the bar at its ends, forms the trailing spirals (such a scheme does not "work" in case of normal spirals).

The main component of the velocity of the gaseous clouds leaving the bar is the drift along the x-axis with a constant speed (see (14), §1, Chapter IV):

$$v_{cr} = yB_1^2/2\Omega. \tag{1}$$

Recall that for the stars constituting the quasistationary bar, this component of the velocity should be omitted. For the gaseous clouds, on the contrary, we should obviously omit the oscillative constituent of motion, since it must be damped by collisions. Provided that $A^2 = \Omega^2$, from (1) we get $v = 3.7\Omega y_0$, which for $y_0 = a/10$; $\Omega a = 150$ km/s yields $v \sim 60$ km/s, the value of the same order as the observed velocity of the gas flowing outwards along the bar. Thus, the ejection of gas may, in principle, occur under the action of the purely gravitational mechanism described.

The motion of gaseous clouds outside the bar occurs in a rather complex field and must therefore be calculated numerically. The clouds produce the spiral trailing arms, and, according to Freeman and de Vaucouleurs, the picture obtained is in reasonable agreement with observations of the SB(s)-spirals.

The matter flowing out of the bar, which carries out the mass and the angular momentum, causes a slow evolution of the bar. Freeman [202–204] (as well as Hunter [234]) have calculated the corresponding evolutionary pathways. It turns out that the evolution leads, in a typical case, to that the ellipsoid of the bar becomes still shorter and denser and rotates more rapidly, i.e., the evolution of the SB-systems in the scheme described proceeds in the direction SBc $\to$ SBa. The characteristic time of the evolution is of the same order as the galactic lifetime ($\sim 10^{10}$ years).

The hypothesis of the origin of spirals in the SB-systems suggested by Freeman, is not, of course, the sole possibility, although he provides some arguments which distinguishes it from the others (see also [344]). One may, for example, note the scheme of Antonov [5] (repeated later by Goldreich and Lynden-Bell [210]), according to which the matter is lost from the neutral points of the potential of the very much elongated bars.

CHAPTER XII

Other Applications

§ 1 On the Structure of Saturn's Rings

1.1 Introduction

Theoretical investigations of Saturn's rings (structure, composition, and stability) have a long history which abounds both in great names (Laplace, Maxwell, and others) and important results. However, it can be stated that the recent flights of Voyager I and II spacecrafts which transmitted to Earth the pictures of Saturn's rings (made from such small distances as several million kilometers and even less) have opened a new era in studying this (to use Maxwell's words) "great space arch." The rings revealed, quite unexpectedly, a much more complicated and interesting structure than had at first been thought. Particularly, they are actually divided into a huge number (thousands) of narrow concentric ringlets.[1]

In this connection the very statement of the problems associated, for instance, with investigations of the rings' system stability should now be

---

[1] The first *Voyager-Bulletin* [124^{ad}] began with the words "Rings within rings, within rings, within....", which fairly well describes the most striking features of the picture revealed to us. However, an insight into the history of observations of Saturn's rings shows that the very possibility of the rings' filamentation into narrower ringlets was, for the first time, discussed by Kant more than 200 years ago, in 1755. By the middle of the last century many astronomers (Vico in Rome, Bond in the United States, Struve in Russia, Dawes and Lassell in England) identified ten dark ringlets altogether. Proctor's drawing illustrates the plausible version on the fine structure of Saturn's rings (the drawing is taken from Flamarion's book *Les Terres du Ciel*, Paris, 1884 (Fig. 129)).

Figure 129. Structure of Saturn's rings according to Proctor (1882).

changed. Actually, theoreticians have so far stated the problem of determining a critical mass (or a critical density) of Saturn's rings assuming stability
of the distribution of matter in the system, and this distribution was considered to be sufficiently smooth. It actually looks so to the Earth's observer,
which is seen, for example, from graphs of dependence in the rings' surface
density on the distance from the center (these graphs are presented in a
great number of monographs, textbooks, and papers—see, for instance, the
monograph by Bobrov [30]). In these graphs, distinct gaps follow only the
few remarkable "divisions": that of Cassini, Encke, etc.[2] Observed smooth-
ness of the density distribution, as well as the fact that the rings' system has
existed for a very long time, has been thought to imply the stability of this
system.

Now, however, in a certain sense, a directly opposite problem should be
stated—on the reasons for the fine structure formation in Saturn's rings due,
possibly, to some instability.[3]

However, small disturbances induced by satellites (primarily, of course,
those due to Saturn's nearest satellite, Mimas) could well serve as a "seeding"
for subsequent apparent filamentation under the influence of some other
more powerful mechanism—that of instability. Particle–satellite interaction can determine the main type of disturbance which is growing

[2] Locations of these divisions are well correlated with resonant circles for which ratios of the
particles' rotation frequencies ω_0 to angular velocities of the nearest of Saturn's moons ω_s
are rational numbers, $\omega_0/\omega_s = p/q$, with not very large integers p and q. For example, for the
main (Cassini's) division one obtains $\omega_0/\omega_s \simeq 2$, for Encke's gap $\omega_0/\omega_s \simeq 3$, etc.

[3] Because of the information obtained by the Voyagers, we had to change completely the
section on Saturn's rings in the Russian edition of the book (1976) as the corresponding material
(with discussion of various estimations of the rings' parameters, which follow from the requirement of their stability) became rather archaic. It was necessary to change the very title of this
section because the old one, "One the critical mass of Saturn's rings," suggested a problem that
is no longer relevant.

due to instability. If growth rate of this instability exceeds the angular velocity of the satellite, initial disturbances may be regarded mainly as annular, since different points of the resonant circles have no distinction in this case.[4] Besides circular structures, spiral density perturbations may also increase due to instabilities. Given below are considerations for circular perturbations which are also valid for tightly wound spiral density waves.

Below we examine possible instabilities of Saturn's rings: Jeans (1.3), dissipative (1.4) and modulational (1.5).

Restricting ourselves to only purely gravitational aspects of explanation of the observable structure of Saturn's rings, we put aside, as a result, such phenomena as periodically appearing and disappearing dark transversal bands (so-called "spokes"). This phenomenon is very likely to be due to some electrodynamical processes in a gas of the smallest dust particles (this follows from the correlation between the period of evolution of spokes and that of Saturn's rotation).

1.2 Model. Basic Equations

Recall that the three "classical" rings of Saturn are traditionally denoted by the letters A, B, and C. Ring A is the outermost (its inner and outer edges are $R_i = 121.900$ km, $R_o = 136.600$ km). This ring includes a dark gap near the outer edge ($R = 133.400$ km), usually called Encke's gap (or division). The brightest ring B is separated from ring A by the well-known Cassini's division. $R_i = 91.000$ km and $R_o = 117.400$ km are the inner and outer radii of ring B. The faint ring C (so-called "crepe") is touching the inner side of the B ring.

The thickness of Saturn's rings is at any rate less than 1 km. So we use a rotating gravitational disk of finite thickness (which is very small in comparison with the disk's radial extension) in the field of the central body as a model of Saturn's rings. The most likely constituent of the ring's particles is probably (water) ice.

As the initial set we take the first-order perturbations to Euler's equation for a viscous fluid disk [67]:

$$\frac{\partial v_r}{\partial t} + \Omega \frac{\partial v_r}{\partial \varphi} - 2\Omega v_\varphi = -\frac{\partial \Phi}{\partial r} - \frac{1}{\rho_0} \frac{\partial P}{\partial r} + \frac{\rho}{\rho_0^2} \frac{\partial P_0}{\partial r}$$

$$+ \frac{1}{\rho_0} \left(\frac{1}{r} \frac{\partial S_{rr}}{\partial r} + \frac{1}{r} \frac{\partial S_{r\varphi}}{\partial \varphi} - \frac{S_{\varphi\varphi}}{r} \right), \qquad (1)$$

$$\frac{\partial v_\varphi}{\partial t} + \Omega \frac{\partial v_\varphi}{\partial \varphi} + \frac{\varkappa^2}{2\Omega} v_r = -\frac{\partial \Phi}{r \partial \varphi} - \frac{1}{\rho_0 r} \frac{\partial P}{\partial \varphi} + \frac{1}{\rho_0} \left(\frac{1}{r^2} \frac{\partial}{\partial r} (r^2 S_{r\varphi}) + \frac{1}{r} \frac{\partial S_{\varphi\varphi}}{\partial \varphi} \right)$$

$$(2)$$

---

[4] Annular perturbations are also selected due to the strong differentiality of rotation of Saturn's rings.

(where S_{ik} is the viscous stress tensor), the equation of continuity,

$$\frac{\partial \rho}{\partial t} + \text{div}(\rho \mathbf{v}) = 0, \tag{3}$$

the Poisson equation

$$\Delta \Phi = 4\pi G \rho, \tag{4}$$

and some state equation. The latter may differ in different situations; so we cannot write this equation. Also we do not write the z-component of Euler's equation as we shall consider only plane large-scale motions of a disk. Only the $S_{r\varphi}$ component of S_{ik} will be used below,

$$S_{r\varphi} = \mu \frac{d}{rdr}\left(\frac{v_\varphi}{r}\right), \tag{5}$$

μ being the dynamical viscosity.

1.3 Jeans Instability

Putting $\mu = 0$ in our set of equations and assuming that all perturbations are proportional to $\sim \exp(ikr)$, $kr \gg 1$, $\partial/\partial\varphi = 0$, we arrive at the usual Toomre dispersion relation

$$\omega^2 = \varkappa^2 - 2\pi G\sigma_0|k| + k^2 c^2, \tag{6}$$

where $c^2 = \partial P_0/\partial\rho_0$, c is the sound velocity. The local dispersion relation accounting for the stabilizing influence of the ring's finite thickness is

$$\omega^2 = \varkappa^2 + \frac{2\pi G\sigma_0|k|}{1+|k|h} + k^2 c^2, \tag{7}$$

where we used conventional notations: ω and k are the frequency and wave number of a disturbance; $\varkappa = (4\Omega^2 + r\,d\Omega^2/dr)^{1/2}$ is the epicyclic frequency; $\Omega = \Omega(r)$, $\sigma_0 = \sigma_0(r)$, and $c = c(r)$ are respectively the angular velocity, surface density, and dispersion of chaotic velocities of the ring's particles at a distance r from the system's center; $h \simeq c/\Omega$ is the half-thickness of the disk. The stability boundary ($\omega^2 = 0$, $\partial\omega^2/\partial k = 0$) corresponds to the value of Toomre's parameter ("stability margin") $Q = \varkappa c/\pi G\sigma_0 = Q_{cr} \simeq 0.55$; the system is stable for $Q > Q_{cr}$ and unstable for $Q < Q_{cr}$, the maximum growth rate corresponding to the wavelength

$$\lambda_0 = \frac{2\pi h}{0.65} \simeq 10h. \tag{8}$$

In this case of large central mass, $m/M \ll 1$ (m is the rings' mass, M is the planet's mass), we have the Keplerian rotation of particles in nearly circular orbits, so that $\varkappa = \Omega = \sqrt{GM}/r^{3/2}$. Assuming that $\sigma_0(r) \simeq m/\pi r^2$, the instability condition $Q < Q_{cr}$ can be rewritten in the form similar to that used by Ginzburg et al. [40]

$$H < H_{max} = m(r) \cdot r/M \qquad (9)$$

(here we already used $Q_{cr} \simeq 0.55$; $H = 2h$ is the full thickness of the rings).

An estimate of the maximum possible mass of Saturn's rings system has been made in the paper by Null et al. [122[ad]] by comparison of the real trajectory of the Pioneer 11 spacecraft with its prescribed trajectory (obtained without accounting for the influence of Saturn's rings). The authors conclude that the m/M ratio must be less than $1.7 \cdot 10^{-6}$ since the "heavier" rings would have led to appreciable (but actually not observed) deviations of the real trajectory from the prescribed trajectory. Assuming $(m/M)_{max} \simeq 1.7 \cdot 10^{-6}$, we obtain an estimate for the maximum possible thickness of a Jeans-unstable ring: $H < H_{max} = 1.7 \cdot 10^{-6} r$, i.e., for $r = 10^5$ km, $H_{max} = 170$ m. Saturn's rings are likely to have the same or even smaller thickness; so, generally speaking, they may be Jeans-unstable. The characteristic scale of this instability is $\lambda \simeq \lambda_0 \simeq 2$ km for $m/M = 1.7 \cdot 10^{-6}$ (and still less for $m/M < 1.7 \cdot 10^{-6}$), and only wavelengths close to λ_0 are unstable: both small-scale and large-scale disturbances are stable. Note that scales λ of the order of a few kilometers (or even a few hundred meters) correspond to the hyperfine structure of Saturn's rings discovered by Voyager II.

As the parameters of rings A, B, and C essentially differ from each other, it is natural to use the *local* Toomre criterion and, possibly, only the most dense ring B may be really unstable according to the Jeans mechanism. Indeed, some indirect estimates, obtained in recent papers (Lane et al. [120[ad]]), give $Q \simeq 2$ for ring B (and $Q \simeq 30$ for the Cassini division). However, an error in only two times can lead to the appearance of the most powerful (Jeans) instability giving both hyperfine (with scales $\lambda \sim H$) and, at the nonlinear stage, fine (with scales $\lambda \gg H$) structures of the rings. (Nonlinear theory of Jeans instability is discussed in Chapter VII).

1.4 Dissipative Instabilities

Aside from the Jeans solutions considered above (they may also be called "adiabatic"), Eqs. (1)–(5) for $\mu = 0$ have another solution corresponding to a new stationary state close to the initial one; this solution can be obtained from Eqs. (1)–(5) if we put in these equations $\omega = 0$:

$$v_r = 0; \qquad v_\varphi = \frac{ik}{2\Omega}\left(\Phi + \frac{P}{\rho_0}\right), \quad \text{etc.} \qquad (10)$$

Actually, solution (10) describes two essentially different types of perturbations [48a]: "entropic" (local density ρ is perturbed) and "whirl" (ρ is unperturbed). But in the long-wavelength limit ($kh \ll 1$) these perturbations are determined by one and the same formulae (10).

In the case when the adiabatic (Jeans) perturbations are stable, $Q \gtrsim 1$, a quite natural question arises: What is the character of evolution of perturbations described by Eq. (10) when dissipative factors (viscosity and heat conductivity) are operating? In other words, one should investigate whether dissipative instabilities are possible.

Let us consider the simplest model of Saturn's rings consisting of identical particles assumed to be indestructible—imperfectly elastic spheres. Then, following Goldreich and Tremaine [114[ad]], one can perform some simple estimates of the conditions when the heat balance is possible. On one hand, in any differentially rotating system, the viscous stress converts the kinetic energy or orbital rotation into the random kinematic energy ("thermal energy"). The velocity dispersion of particles v_T^2 increases, owing to this mechanism, according to:

$$\left(\frac{dv_T^2}{dt}\right)_1 \simeq c_1 v \left(r \frac{d\Omega}{dr}\right)^2, \tag{11}$$

c_1 being a constant of the order of unity [67]. On the other hand, the "thermal energy" is dissipated as the collisions of particles are not perfectly elastic,

$$\left(\frac{dv_T^2}{dt}\right)_2 \simeq -c_2 v_T^2 \omega_c (1 - \varepsilon^2), \tag{12}$$

where ω_c denotes the collision frequency, ε is the so-called coefficient of restitution (it is supposed that at each collision the tangential component of the relative velocity of particles is conserved while the normal component is reduced by a factor ε, where $\varepsilon \leq 1$), c_2 is a constant of the order of unity. Evidently, the stationary state is possible when

$$\left(\frac{dv_T^2}{dt}\right)_1 + \left(\frac{dv_T^2}{dt}\right)_2 = 0. \tag{13}$$

For a ring of thickness $h \gg a$ (a is the particle radius) the collision frequency $\omega_c \simeq v_T n \bar{\sigma}$, n being the volume particle density (cm^{-3}) and $\bar{\sigma} = 4\pi a^2$. It can readily be shown that $\omega_c \simeq \Omega \tau$, where $\tau = \Sigma \cdot \pi a^2$ is the optical depth, Σ being the surface density of particles (cm^{-2}). The "hydrodynamical" estimate for the kinematic viscosity v is

$$v \simeq l v_T, \tag{14}$$

$l \simeq 1/n\bar{\sigma} \simeq v_T/\omega_c$ being a mean free path; this estimate is valid for $\tau \gg 1$. A quite natural generalization of the formula (14) for arbitrary values of τ,

$$v \simeq \frac{v_T^2}{\omega_c} \left(1 + \frac{\Omega^2}{\omega_c^2}\right)^{-1}, \tag{15}$$

is analogous to the expression for the plasma conductivity in the magnetic field (see [138]). The approximate expression (14) is not very accurate but it gives correct asymptotics both for $\Omega/\omega_c \gg 1$ and $\Omega/\omega_c \ll 1$. Then the expression (15) can be also represented as

$$v \simeq \frac{v_T^2}{\Omega}\left(\tau + \frac{1}{\tau}\right)^{-1}. \tag{16}$$

By using the relations just derived, we can rewrite Eq. (13) in the form:

$$(1 - \varepsilon^2)(1 + \tau^2) = \text{const.} \tag{17}$$

Thus, the heat balance condition requires, for each τ, its own equilibrium value of ε. In particular, if $\tau^2 \gg 1$ we have, instead of (17), that

$$(1 - \varepsilon^2)\tau^2 \simeq \text{const}, \tag{18}$$

or

$$(1 - \varepsilon^2)\sigma^2 \simeq \text{const}, \tag{19}$$

where σ is the mass surface density (g cm^{-2}). The equilibrium condition (17) requires, in this case, ε being very close to unity, which means that the collisions should be almost absolutely elastic. If strong inelasticity actually takes place, the equilibrium of the kind considered is not possible: a dense layer with properties close to those of incompressible fluid must form. True, Goldreich and Tremaine [114^{ad}] showed the possibility of the interesting mechanism of automatic regulation. Indeed, generally speaking, the value $(1 - \varepsilon^2)$ in (17)–(19) is a monotonically increasing function of the thermal velocity v_T. Hence, when $(1 - \varepsilon^2)$ varies, with variation of v_T, within sufficiently broad limits, evolution of the disk may lead to the state in which v_T adjusts so that relation (19) is satisfied. However, possibilities of such self-regulation are probably restricted.[5]

It means that for the real situation the state of the problem itself on the dissipative instability of some *equilibrium* is not always correct. On the contrary, it may easily occur that any stationary state is quite absent; or before the system went to a stationary state, perturbations, increasing at the unsteady background, will completely break up the initial smooth distribution.

Let us, nevertheless, suppose that equilibrium on the (19)-type is present. Then, if we are interested only in long-wavelength motions in the ring, the equilibrium conditions across the ring's plane (along the z-axis) being

---

[5] It seems to be evident from the above discussion that the study of the dependence of the restitution coefficient ε on the impact velocities of (ice) particles at the temperature of the rings ($\simeq 90$ K) would be of great importance. Unfortunately, such data are not available at present.

remained, the relation (19) will play the role of the state equation. Indeed, assuming for simplicity the power dependence (Ward [125^{ad}]):

$$(1 - \varepsilon^2) \infty v_T^{2/\alpha} \sim T^{1/\alpha}, \qquad \alpha > 0, \tag{20}$$

we derive the following relationship for perturbations of σ and T:

$$2\frac{\sigma_1}{\sigma_0} + \frac{1}{\alpha}\frac{T_1}{T_0} = 0. \tag{21}$$

Equation (2) then gives:

$$v_r \simeq \frac{2\Omega}{\rho\varkappa^2}\frac{1}{r^2}\frac{\partial}{\partial r}r^3\eta_1\frac{d\Omega}{dr} \simeq ik\frac{2\Omega}{\rho\varkappa^2}\frac{d\Omega}{dr}\eta_1 r \tag{22}$$

In the right-hand side of the first equality (22) the term "proportional to v_φ" was neglected as much smaller at $kh \ll 1$ than the term remained. The second equality (22) is obtained for short-scale perturbations, $kr \gg 1$. According to (14), (21),

$$\frac{\eta_1}{\eta_0} = \frac{T_1}{T_0}. \tag{23}$$

Substituting (23) into the equation of continuity (3) gives

$$-i\omega\sigma_1 - ik\eta_0\frac{d\Omega^2/dr}{\varkappa^2}2\alpha\frac{\sigma_1}{\sigma_0}ik = 0, \tag{24}$$

i.e., the instability occurs with growth rate (Ward [125^{ad}]); Lin and Bodenheimer [121^{ad}]

$$\gamma = 6\alpha vk^2. \tag{25}$$

From the first equality (22) and the equation of continuity (3) we find the general criterion for the long-wavelength instability involved: the "state equation" of type (17)–(19) must lead to such a dependence between temperature T and surface density σ that the dynamical viscosity $\mu[\sigma, T(\sigma)]$ (i.e., in essence the viscous stress) was a decreasing function of σ:

$$\frac{d\mu}{d\sigma} = \frac{d(v\sigma)}{d\sigma} < 0. \tag{26}$$

Physics of this criterion is as follows. Suppose that some circular perturbations occur at any region of the rings. Then each ringlet of increased density (call it b) has neighboring ringlets of decreased density (a and c). The presence of the instability means that ringlet b will continue contracting. This feature of the evolution is a result of an anomalous diffusion: normal diffusion would tend to smooth density perturbations. Let us assume for definition that the ringlet c is inner relative to ringlet b while ringlet a is outer. In the case (opposite to (26)), when the viscous stress increases with growth of density, particles in the boundary area between b and c will mainly be carried by the more dense ringlet b; so these particles will slow down as the angular velocity of ringlet b is less than that of ringlet c). As a result, they will come "falling down" onto the center of the system. In other words, the boundary between b and c will move to the center. Similarly, particles in the boundary region between a and b will mainly be carried by the faster rotating ringlet b, moving away from the center. Thus, ringlet b will expand and eventually disappear: the "normal" viscous stress, increasing with the growth of density, leads to *normal* diffusion. The criterion (26) being satisfied, ringlet b will, on the contrary, contract (analogous considerations prove that). This means instability: anomalous viscous stress leads to *anomalous* diffusion.

Using the formulae above it is possible to study approximately the ring's stability at arbitrary values of τ (Ward [125[ad]]; Lin and Bodenheimer [121[ad]]). For each given value of the parameter α (in the model considered), which determines the velocity dependence of the restitution coefficient according to (20), the certain critical value of the optical depth τ_c exists, and the system is unstable only for $\tau > \tau_c$. An approximate estimation is $\alpha\tau_c^2 \simeq 1$. So it follows that rings with small τ may be unstable only at very large α. Factually, this obviously means that perturbations in optically thin rings must be damping.

We shall not go into the detail of this theory, as the model of the ring suggested by the above authors is too idealized, while conditions of this instability happen to be very sensitive to the choice of a model. For instance, principally different relations already occur for the model of a monolayer. However, it is more important to take into account the real distribution of the rings' particles over sizes (and, consequently, over masses). Observations (including those obtained by Voyagers I and II) show that the spectrum of particle sizes of Saturn's rings is very wide—from micrometers up to approximately a few tens of meters, maximum contribution into the optical depth belonging, probably, to centimeter-size particles. As we observe just a modulation of the rings' optical depth, studying the conditions of instability in the system of centimeter particles is of interest. But, as was noted in the review by Lane *et al.* [120[ad]], chaotization of these particles' velocities occurs mainly due to gravitation scattering (focusing) by large particles. Consequently, we should write, in this case, instead of (11):

$$\left(\frac{dv_T^2}{dt}\right)_1 \simeq \frac{v_T^2}{\tau_g}, \tag{27}$$

where (see formula (3), Section 2.1, Chapter VIII) $\tau_g \simeq v_T^3/n_0 M^2 G^2 \ln \Lambda$, where M and n_0 is the mass and density of large particles. So the condition of heat balance gives, instead of (17)–(19), quite a different state equation:

$$(1 - \varepsilon^2)v_T^3\tau \simeq \text{const}, \tag{28}$$

from where, particularly, the minimum optical depth, corresponding to the stability boundary, is $(\tau_c)_{min} = \sqrt{2} \simeq 1.4$, this being derived for ε independent of v_T ($\alpha = \infty$). Other cases[6] may be considered in similar fashion.

Now let us return to expression (25) for the growth rate of the instability; it predicts shorter wave perturbations to be more unstable, at least for $kh \ll 1$, when the theory is correct. Perhaps this instability vanishes for perturbations with $kh \lesssim 1$. Thus, the initial density perturbation being smooth, a set of narrow ringlets should appear for a time of the order of a few $t_0 = h^2/\nu$ in the regions where the instability criterion (26) is satisfied (i.e., in enough dense regions). Their widths and separations should be of the order of a few h. However, in Saturn's rings a hierarchy of scales of different circular structures is observed—from kilometers (or even a few hundreds of meters) up to hundreds and thousands of kilometers. An attempt to explain this hierarchy within the framework of the above instability meets substantial difficulties. (However, in the set of narrow ringlets an instability[7] could occur which in turn would lead to their "bunching"). Another embarrassment in the application of the theory of this instability to Saturn's rings is connected with the presence of τ_c, so that the instability considered may occur only for $\tau > \tau_c > 1$. This inequality contradicts conventional opinion on the value of the optical depth of rings: even the densest ring B is believed to have the optical depth $\tau \lesssim 1$. Nevertheless we should dwell more thoroughly upon the correctness of using the observable optical depth $\tau < 1$ for Saturn's rings. Let $\tau \gg 1$ in the homogeneous disk; then, as this disk breaks, as a result of some instability, into separate ringlets which are much narrower that their widths ($\sim H$), less than or comparable to a resolution of a device (the highest resolution $\Delta \sim 150$ m was obtained at photopolarimetry measurements (Lane et al. [120[ad]]); this value is not less than the usually admitted thickness H). The nonlinear stage of instability could lead to the appearance of clear zones between separate ringlets. Then a device with a resolution $\Delta > H$ would reveal $\tau < 1$ (Esposito et al. [113[ad]]).

Note also that the difficulty of this theory due to $\tau_c > 1$ might be principally removed when considering short wavelength perturbations, $\lambda \sim H$ (solely, they are interesting because of their rapid growth). However, for these short-scale perturbations, the theory should become much more sophisticated, firstly due to the necessity of solving the three-dimensional problem (instead of the plane problem as before) with detailed calculation of the z-dependence of perturbed quantities. Besides, one should include the perturbations of rotational velocity v_φ, pressure P, and gravitational potential

---

[6] For example, especially interesting are resonant regions (primarily, vicinities of the main resonances), for which effective state equations may be obtained, by equalization the rate of chaotization of the kinetic energy (due to "shaking" particles in these regions by satellites) to the rate of its annihilation in imperfectly elastic collisions. We note also that one would obtain the state equation, somewhat different from (28), by taking into account the thermal motion of large particles (for this case, $(\tau_c)_{min} \simeq 2$).

[7] Perhaps this instability belongs to the type discussed below. The question on the stability of the set of ringlets is very interesting, but a separate problem.

Φ (the latter for $Q \sim 1$). At last, the heat conductivity (both in the z-direction and in the plane of rings) becomes important, as well as viscosity, for finding the heat regime of such perturbations. Note that taking into account the heat conductivity along the vertical (z) direction is, strictly speaking, necessary even when considering the *equilibrium* heat balance in thin disks or rings. In the paper by Goldreich and Tremaine [114[ad]] the heat conductivity was omitted, which was incorrect. The statement of the problem being correct, the resulting "state equations" of the matter would link the density ρ, the temperature T, and the derivatives dT/dz, d^2T/dz^2, and not only ρ and T. The same inaccuracy remains in the stability investigations by Ward [125[ad]] and Lin and Bodenheimer [121[ad]]. This should not change considerably the main results of the theory at the long-wavelength limit, although even here a need arises to single out the sense of the quantities averaged in z.

If the system of particles determining the optical depth of the rings is thermostatized,[8] quite a new situation arises (Polyachenko and Fridman [123[ad]]). Now we begin to study this problem. Further, we restrict ourselves to long-wavelength perturbations of rings, $kh \ll 1$, and also assume that $\tau^2 \gg 1$.[9] The latter condition being fulfilled, the dynamical viscosity μ is the function only of the temperature. Since we consider the temperature to be constant, then, in Eq. (22), η_1 equals zero, so that the perturbation $v_{\varphi 1}$ of the Keplerian rotation must be taken into account in the expression (5) for the viscous stress (where $v_\varphi = \Omega r + v_{\varphi 1}$). So the set of linearized equations will be the following:

$$i\omega v_r + 2\Omega v_{\varphi 1} = \frac{ikc^2}{\sigma_0}\sigma_1 + ik\Phi_1 + vk^2 v_r, \qquad (29)$$

$$-i\omega v_{\varphi 1} + \frac{\varkappa^2}{2\Omega}v_r = -vk^2 v_{\varphi 1},$$

$$\omega\sigma_1 - \sigma_0 kv_r = 0, \qquad \Phi_1 = -\frac{2\pi G\sigma_1}{k}.$$

(c is the isothermic sound velocity). Equating the determinant of the set to zero, one can obtain a cubic dispersion equation—a simplified equation of Kumar [115[ad]].[10] Two roots of this equation correspond to the usual Jeans (Toomre) instability while the latter root describes the secular instability

---

[8] We do not discuss here the concrete nature of the thermostat. Note only that just taking into account the heat conductivity (which tends to smooth inhomogeneities of temperature) will further the "isothermalization" of perturbations. This is especially appreciable for short-wavelength perturbations.

[9] As can be readily shown, long-wavelength isothermic perturbations are damping if $\tau \ll 1$.

[10] The dispersion relation of Kumar [115[ad]] has five orders of frequency. In its derivation Kumar took into account temperature perturbations: $\delta T \neq 0$, adding the heat conduction equation to the initial set of equations. Besides that, he took the angular velocity in a more complicated form: $\Omega = \Omega(0, y, z)$. True, Kumar's dispersion relation corresponds to a viscous cylinder rather than a disk.

of the rings. Supposing $\omega \ll vk^2$, one can readily find, from the set (29), main relations of this solution [130[ad]]:

$$v_\varphi = -\frac{i}{2\Omega k}\frac{\sigma_1}{\sigma_0}(2\pi G\sigma_0|k| - k^2c^2); \tag{30}$$

$$v_r = \frac{i\sigma_1 vk^2(2\pi G\sigma_0|k| - k^2c^2)}{k\sigma_0/(\varkappa^2 + k^2c^2 - 2\pi G\sigma_0|k|)}; \tag{31}$$

$$F_g \equiv -\frac{\partial\Phi_1}{\partial r} = i\sigma_1 \cdot 2\pi G; \tag{32}$$

$$F_g \equiv \frac{1}{\sigma_0}\frac{\partial P_1}{\partial r} = -i\sigma_1\frac{kc^2}{\sigma_0}; \tag{33}$$

$$\omega = ivk^2\frac{2\pi G\sigma_0|k| - k^2c^2}{\varkappa^2 + k^2c^2 - 2\pi G\sigma_0|k|}. \tag{34}$$

From Eq. (34), we see that the secular instability takes place for $k < 2\pi G\sigma_0/c^2$ if the denominator in (34) is positive (i.e., the rings are stable according to Jeans).

To explain the physics of secular instability, let us, first of all, consider the case of uniform rotation: $\Omega \neq \Omega(r)$. The energy of perturbations in the disk is determined by the expression:

$$\delta E = \frac{1}{2}\int(\sigma_0 v^2 + \sigma\Phi + c^2\sigma^2/\sigma_0)\,dr, \tag{35}$$

where $c^2 = \partial P/\partial\sigma$; σ, Φ, and v are the perturbations of density, potential, and velocity, respectively. Expanding perturbed quantities into the Fourier integrals

$$(v, \sigma, \Phi) = \int d\mathbf{k}\, \exp(i\mathbf{k}\mathbf{r})(v_k, \sigma_k, \Phi_k), \tag{36}$$

and using here the connection $\Phi_k = -2\pi G\sigma_k/|k|$, we shall obtain

$$\delta E = (2\pi)^2\frac{\sigma_0}{2}\int d\mathbf{k}\left\{|v_k|^2 + \left|\frac{\sigma_k}{\sigma_0}\right|^2\left(c^2 - \frac{2\pi G\sigma_0}{|k|}\right)\right\}, \tag{37}$$

or for one harmonic

$$\delta E_k \infty |v_k|^2 + \left|\frac{\sigma_k}{\sigma_0}\right|^2\left(c^2 - \frac{2\pi G\sigma_0}{|k|}\right). \tag{38}$$

Among radial eigenoscillations of the disk with the wave number k, there is, apart from Jeans modes, another, neutral ($\omega = 0$) mode (10) which is of interest to us in this case. Since $\sigma/\sigma_0 = -k\Phi/2\pi G\sigma_0$, the wave energy δE is thus easily expressed through Φ. As a result, we have

$$\delta E \infty -\frac{k^2|\Phi|^2}{\varkappa^2}\left(1 - \frac{k^2c^2}{2\pi G\sigma_0|k|}\right)\left(\frac{k^2c^2 + \varkappa^2 - 2\pi G\sigma_0|k|}{2\pi G\sigma_0|k|}\right) \tag{39}$$

For a Jeans-stable disk, the second bracket in (39) is positive, so that $\delta E < 0$, if

$$|k| < \frac{2\pi G\sigma_0}{c^2} \qquad \left(\lambda > \lambda_{cr}^s \equiv \frac{c^2}{G\sigma_0}\right). \tag{40}$$

The disturbance under examination corresponds to a new, differentially rotating equilibrium disk, and from condition (40) this disk has a lower energy than the original one. Therefore, if some (any!) dissipative mechanism is at work in the system, for example, viscosity, the system indeed will go over into this new state at a rate proportional to the rate of energy release in a specific dissipative process. This is just the secular instability. It is quite analogous to the classical secular instability of viscous Maclaurin's ellipsoids (see Appendix, where we derive the instability condition in terms of energy).

For a differentially rotating disk ($\Omega = \Omega(r)$) the rate of change of the wave energy is determined, for $\omega \ll vk^2,$[11] by the equation (which is readily found from Eqs. (29)):

$$\frac{\partial}{\partial t} \frac{1}{2} \int \left(\sigma_0 v^2 + \sigma\Phi + \frac{c^2\sigma^2}{\sigma_0}\right) dr$$

$$= -\mu \int d\mathbf{r} \left\{\left(\frac{\partial v_r}{\partial r}\right)^2 + \left(\frac{\partial v_\varphi}{\partial r}\right)^2\right\} - \mu \int d\mathbf{r} \left[-\frac{r(d\Omega^2/dr)}{\varkappa^2}\right] \left(\frac{\partial v_\varphi}{\partial r}\right)^2 < 0. \tag{41}$$

In order to elucidate the physical mechanism that drives the instability ("in terms of forces") we return to formulas (30)–(34). Figure 130 visualizes the plots of disturbances (30)–(33). In the absence of viscosity (or if it is neglected) we now have the solution described above, i.e., a new, differentially rotating equilibrium disk. In this solution $v_r = 0$, $v_\varphi \neq 0$. This means that the particle from one radius *had* simply *passed to* another[12] (close) radius and remained there (but for the equilibrium not to be violated, the angular rotation rate, gravitational potential, and surface density should be correspondingly corrected).

Consider now formula (30) expressing the relationship between the surface density σ of disturbances and the azimuthal velocity v_φ. This formula does not involve viscosity, so that at $v = 0$ the relationship between σ and v_φ is given by the same formula (30) (see also Figs. 130(a) and (b)). Assume that at

---

[11] In the opposite case, $\omega \gg vk^2$ (Jeans modes), one can readily obtain:

$$\frac{\partial}{\partial t} \frac{1}{2} \int \left[\sigma_0 v^2 + \sigma\Phi + \frac{c^2\sigma^2}{\sigma_0} + \sigma_0 \frac{\varkappa^2}{4\Omega^2} r\left(-\frac{d\Omega^2}{dr}\right)\xi_r^2\right] dr$$

$$= -\mu \int d\mathbf{r} \left[\left(\frac{\partial v_r}{\partial r}\right)^2 + \left(\frac{\partial v_\varphi}{\partial r}\right)^2\right] < 0 \tag{41'}$$

Here, the additional term in the expression of the wave energy is due to the central body (see end of Introduction). These modes are damping owing to viscosity.

[12] This evidently requires that the angular momentum of each particle be correspondingly changed (unlike Jeans modes, where the angular momentum of each particle stays unchanged).

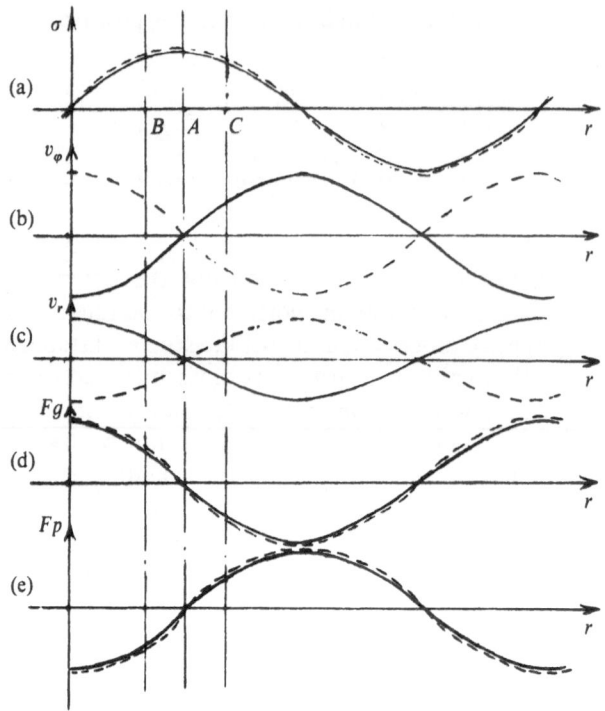

Figure 130. Profiles of perturbed quantities in the stable (by Jeans) gravitating disk. Solid lines correspond to the case of secular instability, dotted lines to the case of secular stability. (a) Perturbed surface density; (b) perturbed azimuthal velocity; (c) radial velocity; (d) perturbed radial gravitational force; (e) perturbed pressure force.

some moment there has been a density disturbance ($\sigma > 0$ at some point A). Then, according to Fig. 130(b), for $k^2 c^2 < 2\pi G\sigma_0 |k|$, $v_\varphi > 0$ at points C (at somewhat larger radii, $r_C > r_A$) and $v_\varphi < 0$ at points B (at somewhat smaller radii, $\tau_B < r_A$). If there were no viscosity ($v = 0$), the initial $\sigma(r)$, $v_\varphi(r)$, $F_g(r)$, $F_p(r)$, $v_r = 0$ would have remained at subsequent moments of time ($\omega = 0$, which corresponds to the new, close *equilibrium* state). The picture will, however, change when viscosity ($v \neq 0$) is taken into account. Viscous forces will evidently tend to decrease the velocity gradient so that v_φ will slightly increase at points B and slightly decrease at points C. Hence, both particles B and C will approach A, i.e., there will be such a radial motion of fluid (it corresponds to Fig. 130(c)) which will lead to a further increase in density σ, etc. Thus, it is clear that under condition (40) the (secular) instability must actually arise. For the inequality contrary to (40), it is also easy to make certain that this case (dotted curves in Fig. 130) corresponds to the stability.

The secular instability of the type just described is interesting for several reasons. Primarily, it works in a (Jeans–Toomre) stable disk. Further, it is universal in that it can be caused by any dissipative mechanism: for the system it is energetically advantageous to change to a new ("sliced") state, and the criterion for stability does not depend on a specific process of dissipation. Finally, it is essential that for a fairly cool disk the instability must manifest itself within a very broad wavelength range which is limited only on the side of very short wavelengths, according to (40).[13]

1.5 Modulational Instability

The aim of this section is to prove the following general statement: in an arbitrary gravitating disk rotating in the field of the central body with large mass, the sufficient condition of the modulational instability is fulfilled. As a consequence of this statement it can be concluded that Saturn's rings undergo modulational instability.

In the works by Petviashvili, Mikhailovskii, Fridman (see Section 1.1, Chapter VII) a nonlinear dispersion equation was obtained describing perturbation of the small but finite amplitude in a rotating gravitating disk provided this disk is near its stability boundary. This equation was generalized by Polyachenko, Churilov, and Shukhman (see in the same place), who did not use the latter condition. The dispersion equation obtained in the latter work has the form

$$\omega^2 = \omega_k^2 - \frac{k^2}{\gamma_p^2(\omega_{2k}^2 - 4\omega_k^2)} \cdot \{8[(2\gamma_p - 1)k^2c^2 - \gamma_p \pi G\sigma_0 |k|]^2$$

$$- [(2\gamma_p - 1)(5\gamma_p - 2)k^2c^2 - 4\gamma_p^2 \pi G\sigma_0 |k|](\omega_{2k}^2 - 4\omega_k^2)\} \, |\xi_k|^2. \quad (42)$$

Here γ_p is the "flat" adiabatic index, ξ_k is the element's displacement amplitude,

$$\omega_k^2 = \varkappa^2 + k^2c^2 - 2\pi G\sigma_0 |k|, \quad (43)$$

and

$$\omega_{2k}^2 = \varkappa^2 + 4k^2c^2 - 4\pi G\sigma_0 |k|. \quad (43')$$

Eq. (42) is valid for arbitrary k. Now we are only interested in a small vicinity of that wave number k_0 which corresponds to non convective oscillations. In other words we shall consider only those oscillations which remain in the disk. Their group velocity is zero, $d\omega_k/dk = 0$. The corresponding region of wave numbers is shaded in Fig. 131, where

$$k_0^m = k_{01}^J = \frac{\pi G\sigma_0}{c^2}. \quad (44)$$

---

[13] See also Section 1.5. We would like to mention some different (but gravitational too) instabilities which were used by Kadomtsev [126[ad]] and Igumenshev [127[ad]] to explain the wave-like appearance and "clumping" of F-ring (and of several other thin Saturn ringlets).

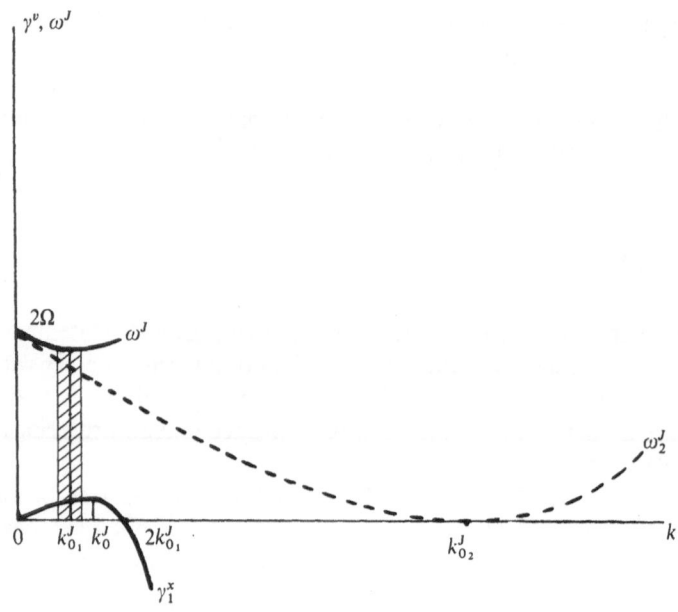

Figure 131. Dependence of the secular instability growth rate γ of a disk in the field of large central mass on the wave number k. Notation: $k_1 = 2\pi G\sigma_0/c^2$, $k_2 = \varkappa^2/\pi G\sigma_0$; $k_2/k_1 = Q^2/2 \gg 1$. Left: the dispersion curve in a very stable (by Jeans) gravitating disk. Nonconvective perturbations (having $v_g \equiv d\omega_k/dk = 0$) belong to the shaded area of wave numbers in the vicinity of the point $k = k_{01}^J \Rightarrow d\omega/dk = 0$. ω_2^J is the dispersion (marginal) curve corresponding to Jeans instability in the region of $k = k_0^J$. It is seen that $k_{02}^J \gg k_{01}$.

For comparison, we show there the dispersion curve for the case of the Jeans instability. It is seen that maximum Jeans instability is at $k_0^J \gg k_0^m$. As it follows from §1.3, $\lambda_0^J = 5H$; at the same time from the latter inequality it follows that $\lambda_0^m \gg H$. Besides, we are in a very stable (in the sense of Jeans) region where $\varkappa^2 \gg 2\pi G\sigma_0 |k_0^m|$.

Let us substitute into Eq. (42), instead of k, its value at the point $k = k_0$. Then we express the quantities entering into Eq. (42) in terms of Q and $\varkappa$:

$$k_0^2 c^2 = \frac{\varkappa^2}{Q^2}, \qquad k_0^2 c^2 \pi G\sigma_0 k_0 = \frac{\varkappa^4}{Q^4}, \qquad \pi G\sigma_0 k_0 = \frac{\varkappa^2}{Q^2}; \qquad (45)$$

a nonlinear dispersion equation for the vicinity of $k = k_0$ is obtained in the form

$$\omega^2 = \omega_{k_0}^2 - \frac{\varkappa^4}{\gamma_p^2 Q^4 (3Q^2 - 4)} \{16\gamma_p^2 - 20\gamma_p - 3Q^2(6\gamma_p^2 - 9\gamma_p + 2)\} |\xi_k|^2. \qquad (46)$$

Let us express the "surface" adiabatic index γ_p in terms of the "volume" adiabatic index γ_V. If this equilibrium is due to the balance between the pressure force and the external gravitational force ($Q \gg 1$), then (see Churilov and Shukhman [112^{ad}])

$$\gamma_p = 3 - 4/(\gamma_V + 1), \qquad Q \gg 1. \tag{47}$$

For $Q < 1$ the Jeans instability develops in the disk; as known this instability is an aperiodic one, i.e., $\text{Re}\,\omega_k = 0$. In this case, as we shall see below, the modulational instability is not possible. In the "very stable" (in the sense of Jeans) region, $Q \gg 1$, it is necessary to use Eq. (47). Then we obtain, instead of Eq. (46):

$$\omega^2 = \omega_{k_0}^2 - \frac{3\varkappa^4}{(3\gamma_V - 1)^2} \cdot \frac{|\xi_k|^2}{Q^2(3Q^2 - 4)}$$

$$\times \left[(50\gamma_V - 29\gamma_V^2 - 17) + \frac{84\gamma_V^2 - 136\gamma_V + 36}{3Q^2} \right]. \tag{48}$$

For large Q, the second term in the square brackets in Eq. (48) may be neglected, and then in the expression for the frequency

$$\omega = \omega_{k_0} - \beta|\xi_k|^2 \tag{49}$$

the nonlinear addition is negative, $\beta > 0$ (which, as we shall see below, corresponds to development of the modulational instability) if

$$1 \leq \gamma_V < 1.26. \tag{50}$$

Further we shall take $\gamma_V \simeq 1$ for the estimation of the wavelength of modulational instability. For $\gamma_V = \gamma_p = 1$, from Eq. (46) we obtain

$$\omega^2 = \omega_{k_0}^2 - \frac{\varkappa^4}{Q^4 c^2}|\xi_k|^2, \qquad \gamma_V = 1. \tag{51}$$

From the continuity equation

$$\frac{\partial \sigma_1}{\partial t} + \text{div}(\sigma_0 \mathbf{v}_1) = 0,$$

in the region of the minimum of the dispersion curve ($k \simeq k_0$) we have

$$\sigma_1/\sigma_0 = -i(\mathbf{k}_0 \boldsymbol{\xi}_0). \tag{52}$$

Hence

$$|\sigma_1/\sigma_0|^2 = k_0^2|\xi|^2.$$

Then, instead of Eq. (51) we obtain

$$\omega^2 = \omega_{k_0}^2 - \frac{\varkappa^4}{Q^4 k_0^2 c^2}\left|\frac{\sigma_1}{\sigma_0}\right|^2,$$

or, using the first formula of (45), we find

$$\omega^2 = \omega_{k_0}^2 - \frac{\varkappa^2}{Q^2}\left|\frac{\sigma_1}{\sigma_0}\right|^2. \tag{53}$$

Since

$$\omega_{k_0}^2 = \varkappa^2 \left(1 - \frac{1}{Q^2}\right), \tag{54}$$

then

$$\omega^2 = \varkappa^2 \left[1 - \frac{1}{Q^2}\left(1 + \left|\frac{\sigma_1}{\sigma_0}\right|^2\right)\right]. \tag{55}$$

In the stable (in the sense of Jeans) region, $Q^2 \gg 1$ (Q not necessarily being very large; it is sufficient for example to have $Q = 3$) we obtain

$$\omega = \varkappa \left[1 - \frac{1}{2Q^2}\left(1 + \left|\frac{\sigma_1}{\sigma_0}\right|^2\right)\right]. \tag{56}$$

Taking as before $Q^2 \gg 1$ we find the group velocity of the wave (in the vicinity of $k \approx k_0$)

$$v_g = \frac{d\omega_k}{dk} = \frac{kc^2 - 2\pi G\sigma_0}{\varkappa},$$

hence

$$\frac{dv_g}{dk} = \frac{c^2}{\varkappa} > 0. \tag{57}$$

It is seen from (56) that the nonlinear addition to the frequency α which is determined from the equation

$$\omega = \omega_k + \alpha \left|\frac{\sigma_1}{\sigma_0}\right|^2, \tag{58}$$

is negative:

$$\alpha = -\frac{\varkappa}{2Q^2} < 0; \tag{59}$$

consequently,

$$\alpha \frac{dv_g}{dk} < 0. \tag{60}$$

The last inequality coincides with the well-known Lighthill's condition [265a]—the sufficient condition of the modulational instability.

The boundary of the modulational instability in the wave vector q of the envelope is determined by the inequality

$$q^2 < q_{cr}^2 = -4\alpha \frac{|\sigma_1/\sigma_0|^2}{dv_g/dk}. \tag{61}$$

Using (57), (59), and also the equilibrium condition in z, $c \simeq \varkappa h$, we obtain from (61)

$$q_{cr} h = \frac{\sqrt{2}}{Q} \left| \frac{\sigma_1}{\sigma_0} \right| ,$$

or

$$\tilde{\lambda}^m_{cr} = \sqrt{2} \, \pi Q h \left| \frac{\sigma_1}{\sigma_0} \right|^{-1} . \tag{62}$$

Physics of the modulational instability is clearly described in the well-known books on plasma physics (see, e.g., [39]), and can be understood with the help of Fig. 132, where the modulated dependence of the perturbed density σ_1 on the coordinate x is represented. As seen from Fig. 132, the sign of $\partial |\sigma_1/\sigma_0|^2/\partial x$ in region 1 is negative, while in region 2 it is positive. According to Eq. (59), $\alpha < 0$, so from the definition (58) of the frequency ω it follows that

$$-\frac{\partial \omega}{\partial x} = \begin{cases} < 0, & \text{in region 1,} \\ > 0, & \text{in region 2.} \end{cases} \tag{63}$$

For short wavelengths which we are just considering the approximation of geometrical optics is valid in which equations for the connection between the frequency and the wave number have the form of Hamilton's equations (see, e.g., [70]). With the help of one of these equations,

$$\frac{\partial k}{\partial t} = -\frac{\partial \omega}{\partial x} ,$$

we determine the function $k = k(t)$ which, according to Eq. (63), decreases with the time t in region 1 and increases in region 2. Then, in accordance with Eq. (57), $\partial v_g/\partial k > 0$, the group velocity decreases in region 1 and increases in region 2. Thus the wave packet in region 1 will trail intensifying thereby the wave at the point A and decreasing its amplitude at the point B.

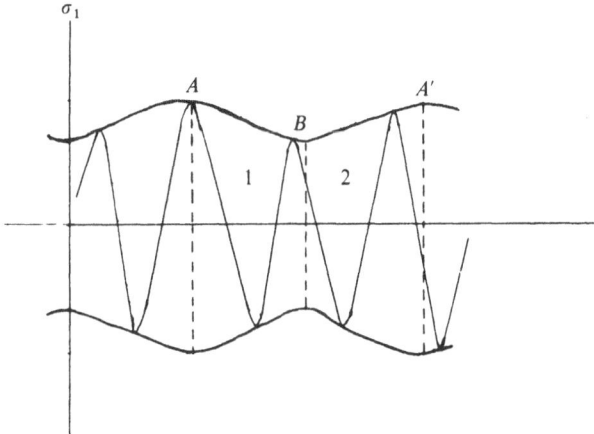

Figure 132. Initial stage of modulational instability.

In region 2 the packet will lead thereby deepening the minimum at the point B and increasing the maximum at the point A.

A similar situation occurs for the gravitational waves on the surface of a deep water, for Lengmoir solitons, and helikons in plasma, with the only specific feature that for the Lengmoir waves signs of nonlinear addition to the frequency α and $\partial v_g/\partial k$ coincide with signs of analogous expressions of nonlinear waves of the gravitation disk, while for the gravitation waves on the deep water and for the helikons the same terms have, respectively, opposite signs.

The necessary condition of the modulational instability (as it follows from the description of its mechanism) is the presence of the derivative of the group velocity $dv_g/dk \neq 0$ and consequently the presence of the real part of the frequency, $\mathrm{Re}\,\omega_k \neq 0$. So for $Q < 1$ when the aperiodic Jeans instability occurs, $\mathrm{Re}\,\omega_k = 0$, the modulational instability is impossible.

Let us estimate the characteristic time of the modulational instability as $1/\gamma^m_{\max}$ where γ^m is the growth rate of this instability determined by the expression (see, e.g., [39])

$$\gamma^m = \sqrt{-\alpha \left|\frac{\sigma_1}{\sigma_0}\right|^2 \frac{dv_g}{dk} \cdot q}. \tag{64}$$

Maximum growth rate of the modulational instability can be estimated if we substitute $q = q_{\mathrm{cr}}$ into Eq. (64):

$$\gamma^m_{\max} \simeq \Omega Q^{-2} \left|\frac{\sigma_1}{\sigma_0}\right|^2. \tag{65}$$

Appendix

Derivation of the Expression for the Perturbation Energy of Maclaurin's Ellipsoid

Let us restrict ourselves to the case of radial perturbations which have the frequency $\omega = 0$ in the reference system rotating with the angular velocity Ω (where the ellipsoid is at rest). Then the perturbation energy δE may be written in the form

$$\delta E = T + W, \tag{A1}$$

where the kinetic energy

$$T = \frac{1}{2}\int \rho v_\varphi^2 \, dV, \tag{A2}$$

and the potential energy W can be calculated (similarly to Chandrasekhar [186]) by the formula

$$W = -\frac{1}{2}\int \nabla\left(\Phi - \frac{\Omega^2 r^2}{2}\right)\xi \, dV, \tag{A3}$$

where ξ is a Lagrange displacement of a fluid element, Φ is a potential.

First we calculate W. Representing Φ as the sum $\Phi = \Phi_0 + \Phi_1$ of the unperturbed $\Phi_0 = \frac{1}{2}Ar^2 + \frac{1}{2}Bz^2$ and perturbed Φ_1 potentials, we divide W into two parts

$$W = -\frac{1}{2}\int \nabla\left(\Phi_0 - \frac{\Omega^2 r^2}{2}\right)\xi \, dV - \frac{1}{2}\int \nabla\Phi_1\xi \, dV$$

$$= \frac{1}{2}\int\left(\Phi_0 - \frac{\Omega^2 r^2}{2}\right)\xi \, dS + \frac{1}{2}\int \Phi_1\xi \, dS, \tag{A4}$$

where we pass to integration over the ellipsoid's surface (generally, a perturbed one). Reducing integration in (A4) to the unperturbed surface of the ellipsoid and using oblate spheroidal coordinates μ, $\zeta (r^2 = R^2(1 - \mu^2) \times (1 + \zeta^2), z = R\zeta\mu)$ we obtain

$$W = \pi \int \delta\zeta_0 h_\zeta h_\mu h_\varphi \left[\Phi_1 + \delta\zeta_0 \frac{\partial}{\partial \zeta}\left(\Phi_0 - \frac{\Omega^2 r^2}{2}\right)\Big|_{\zeta = \zeta_0}\right], \tag{A5}$$

where $\zeta = \zeta_0$ is the equation of ellipsoid's surface; h_ζ, h_μ, h_φ are the Lame coefficients:

$$h_\zeta = R\sqrt{(\zeta^2 + \mu^2)(\zeta^2 + 1)}, \qquad h_\mu = R\sqrt{(\zeta^2 + \mu^2)(\mu^2 + 1)},$$

$$h_\varphi = R\sqrt{(1 - \mu^2)(1 + \zeta^2)}.$$

The perturbed potential of Maclaurin's ellipsoid for the certain mode, characterized by the index n, may be written as

$$\Phi_1 = \begin{cases} \Phi_1 p_n(\zeta)P_n(\mu)q_n(\zeta_0), & \zeta < \zeta_0, \\ \Phi_1 q_n(\zeta)P_n(\mu)p_n(\zeta_0), & \zeta > \zeta_0, \end{cases} \tag{A6}$$

where the continuity of Φ_1 at $\zeta = \zeta_0$ is already taken into account; $P_n(z)$ is the Legendre polynomial,

$$p_n(z) = i^{-n}P_n(iz), \quad q_n(z) = p_n(z)\int_z^\infty (1 + t^2)^{-1}[p_n(t)]^{-2}\, dt.$$

We shall not seek to express δE in terms of $\overline{\Phi}_1$. Connection between $\overline{\Phi}_1$ and $\delta\zeta_0$ may be found from the boundary condition

$$\frac{1}{h_\zeta}\frac{\partial \Phi}{\partial \zeta}\Big|_{\zeta = \zeta_0 + 0} - \frac{1}{h_\zeta}\frac{\partial \Phi}{\partial \zeta}\Big|_{\zeta = \zeta_0 - 0} = 4\pi G\rho h_\zeta \delta\zeta_0, \tag{A7}$$

so that

$$\delta\zeta_0 = -\frac{P_n(\mu)}{4\pi G\rho h_\zeta^2(1 + \zeta_0^2)}\cdot \overline{\Phi}_1. \tag{A8}$$

Taking into account also that

$$\frac{\partial}{\partial \zeta}(\Phi_0 - \Omega^2 r^2/2) = BR^2\zeta_0(\zeta_0^2 + \mu^2)/(\zeta^2 + 1),$$

we obtain after simple calculations:

$$W = \frac{\pi R^2}{4\pi G\rho} K_n(\zeta_0) \frac{2}{2n + 1} \bar{\Phi}_1^2, \tag{A9}$$

where Bryan's [172] notation $K_n(z) = p_1(z)q_1(z) - p_n(z)q_n(z)$ was introduced.

Let us now calculate T. To do this, it is necessary to find a connection between v_φ and $\bar{\Phi}_1$. From the equilibrium condition in r we have

$$v_\varphi = \frac{1}{2\Omega} \frac{\partial \psi}{\partial r} = \frac{1}{2\Omega} \frac{\partial \psi_1}{\partial r}, \tag{A10}$$

where $\psi = \Phi + P/\rho - \Omega^2 r^2/2$ (P is the pressure). On the other hand, from the motion equation in z-axis it follows that $\partial \psi/\partial z = 0$, i.e., $\psi = \psi(r)$. It is sufficient to calculate ψ at the boundary $\zeta = \zeta_0$, where $P = 0$, $\psi = \text{const.}$ We obtain

$$\psi_1(r) = - \bar{\Phi}_1 K_n(\zeta_0) P_n(\sqrt{1 - r^2/a^2}). \tag{A11}$$

From the expression

$$\psi_1(\zeta_0) = \Phi_1(\zeta_0) + \frac{\partial}{\partial \zeta}(\Phi_0 - \Omega^2 r^2/2)\delta\zeta_0 = - \bar{\Phi}_1 K_n(\zeta_0) P_n(\mu), \tag{A12}$$

since $\mu|_{\zeta_0} = \sqrt{1 - r^2/a^2}$. Respectively,

$$v_\varphi = \frac{1}{2\Omega} \bar{\Phi}_1 K_n(\zeta_0) \frac{\partial}{\partial r} P_n(\sqrt{1 - r^2/a^2}). \tag{A13}$$

Calculating the integral in (A2) we obtain

$$T = \frac{1}{8\Omega^2} \bar{\Phi}_1^2 K_n^2(\zeta_0) 2\pi R\zeta_0 n(n + 1) \frac{2}{2n + 1}. \tag{A14}$$

Taking into account that $\Omega^2 = 4\pi G\rho q_2(\zeta_0)\zeta_0$, then from (A9) and (A14) we find the following final expression for the energy of the radial mode (A6):

$$\delta E = \bar{\Phi}_1^2 \frac{2\pi n(n + 1)}{2n + 1} \cdot \frac{R\zeta_0}{4\Omega^2} \cdot K_n(\zeta_0) \left\{ K_n(\zeta_0) + \frac{4q_2(\zeta_0)}{n(n + 1)} \right\}. \tag{A15}$$

The dispersion relation of Bryan [172] for the same mode (A6) is

$$F_n^{(0)}(\zeta_0, \omega) \equiv K_n(\zeta_0) - \frac{\omega^2}{\Omega^2} \cdot \frac{q_2(\zeta_0) P_n(v_0)}{v_0(1 + 1/\zeta^2)(dP_n(v_0)/dv_0)} = 0, \tag{A16}$$

$$v_0 \equiv \frac{\omega\zeta_0}{\sqrt{4\Omega^2(1 + \zeta_0^2) - \omega^2}}.$$

It is easy to see that the expression in curly brackets in (A15) and the left side of the dispersion relation (A16) are identical for $\omega = 0$.

§ 2 On the Law of Planetary Distances

Another interesting application of the theory of the stability of flat gravitating systems in the field of a large central mass is the problem of the evolution of the protoplanetary cloud. This topic[14] is dealt with in a number of original papers, principally in the papers by Schmidt [151], Fesenkov [137], Weizsäcker [347, 348], Kuiper [262], Berlaga [164, 165]. Polyachenko and Fridman [110] investigate the possibility of the explanation of the law of planetary distances by the gravitational instability in sufficiently flat systems.

As is well known, the distances of the planets from the sun are well enough described with the help of the following empirical rule formulated as long ago as the eighteenth century by Bode and Titius (see, e.g., [3]):

$$r_n = r_0 + c \cdot 2^n, \tag{1}$$

where $r_0 = 0.4$; $c = 0.3$; $n = -\infty, 0, 1, \ldots, 6$; r_n are the distances (in a.u.), respectively, to Mercury, Venus, ..., Pluto.

In accordance with this point of view, assume that the planetary system has been produced from a gas–dust disk rotating in a field of the central mass [110]. Let the instability condition of such a disk be satisfied. Then if the disk is nonhomogeneous in its density σ_0, the perturbation wavelength is a function of the radius-vector r. Thus, the problem is reduced to the determination of function $\sigma_0 = \sigma_0(r)$ from the function $\lambda = \lambda(r)$ given from (1).

Above all, note that the Titius–Bode law (1) can be written in the following form:

$$a \ln \frac{r_n - r_0}{c} = 2\pi n, \tag{2}$$

where

$$a = \frac{2\pi}{\ln 2} \approx 9.1. \tag{3}$$

Represent the density σ in the form of the sum $\sigma_0(r) + \sigma_1(r)$. Let

$$\sigma_1(r) = \tilde{\sigma}_1(r) \exp\left[ia \ln \frac{r - r_0}{c} \right], \tag{4}$$

where $\tilde{\sigma}_1(r)$ is the amplitude of the perturbed density varying slowly with radius, and $a \ln[(r - r_0)/c]$ is the phase.

From Eq. (2), it follows that the maxima of the perturbed density σ_1 in (4) coincide with true locations of all the planets, except for the last two— Neptune and Pluto. This exception is easily removed if one assumes that up to $r \sim 14$ a.u. $\sigma_0(r) \sim 1/r^2$, and farther it is $\sigma_0 \sim 1/r^3$ (see Fig. 133).

With due regard for the pressure of the medium, the dispersion equation for the frequencies of small radial oscillations of a flat disk rotating about

---

[14] A detailed review of relevant papers is contained in Safronov's book [119].

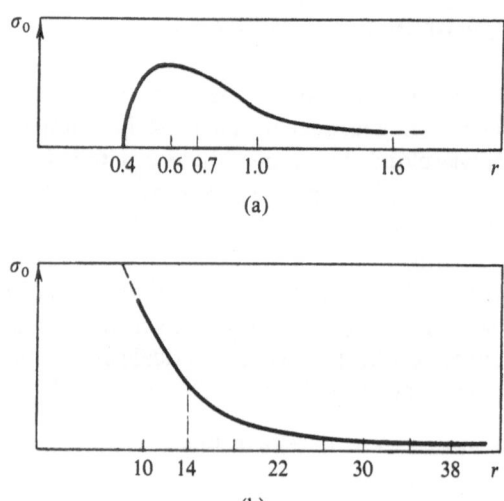

Figure 133. Dependence of the unperturbed density of the gas–dust protoplanetary cloud $\sigma_0(r)$ on the radius r (in astronomical units): (a) $0.4 \le r \le 1.6$; (b) $10 \le r \le 40$.

a central mass M, is Eq. (6) of §1. (The quantity c may denote also the mean velocity of turbulent motions of the gas.)

(a) $c = 0$, *the dust disk model.* For perturbations of the type of (4) it is easy to find

$$\sigma_0(r) = \frac{\Omega^2 - \omega^2}{2\pi Ga}(r - r_0),$$ (5)

where $\Omega^2(r) = GM/r^3$. By determining the mass of the nth protoplanet from the formula $m_n = \int_{r_n}^{r_{n+1}} \sigma_0(r)r \, dr$, we obtain that the masses of the proto-planets in the region where $\sigma_0(r)$ is determined by formula (5), are of the same order of magnitude. At the stability boundary $\omega^2 = 0$, and from (5) we have (taking into account that $a = 9.1, c = 0.3$) $M/m_n \simeq 28$.

If the ratio of masses M/m_n is more than 28, then the gravitational insta-bility of a dust disk with the parameters above is absent, and the density waves do not increase.

(b) $c \ne 0$, *the gas–dust disk.* As follows from the dispersion equation, at $k = \pi G\sigma_0/c^2$ the instability is maximum. Substituting this value of k again into the dispersion equation in (1), Section 4.1, we find the relation between the stationary parameters of the system at the stability boundary ($\omega^2 = 0$): $\Omega = \pi G\sigma_0/c$, hence

$$\sigma_0(r) = \frac{M(r - r_0)}{\pi ar^3}.$$ (6)

Comparing the obtained density of the stationary state of the gas–dust disk with a similar expression in (5) for the dust disk, we see that the difference is only twofold.

Thus, in the case where the disk density is dependent on radius according to the law of (5) or (6), the time-increasing ring-shaped disk density perturbations are arranged in it according to the law of Titius–Bode (1). If one does not assume any definite dependence $\sigma_0(r)$, the condition of "quantization" of planetary orbits will appear in the following way (on the stability boundary $\omega^2 = 0$):

$$\Phi_1(r_n) \equiv \frac{M}{2\pi} \int^{r_n} \frac{dr}{r^3 \sigma_0(r)} = 2\pi n. \tag{7}$$

The Titius–Bode rule is obtained from this for

$$\sigma_0(r) = \frac{M \ln 2}{4\pi^2} \frac{r - r_0}{r^3}.$$

It is easy to see that for $\sigma_0 = c/r^3$ ($c = \text{const}$) we shall obtain from (7) equal distances between the planetary orbits. The constant may be expressed through the distance Δr between these planets ($\Delta r \approx 10$ a.u.): $c = M\Delta r/4\pi^2$. From the continuity condition of the surface density we have

$$\sigma_0(r) = \begin{cases} \dfrac{M \ln 2}{4\pi^2} \dfrac{r - r_0}{r^3}, & r \lesssim 14 \text{ a.u.} \\[2mm] \dfrac{M\Delta r}{4\pi^2} \dfrac{1}{r^3}, & r > 14 \text{ a.u.} \end{cases} \tag{8}$$

From (8), it follows that the Titius–Bode law is valid up to $r \approx 14$ a.u. Farther (according to observations), the separations between the orbits remain unaltered.

Thus, if the space dependence of the surface density of a gas–dust disk rotating in a field of the central mass M, has the form (8), the increasing ring-shaped perturbations are located in places of the solar system planets (Fig. 134). From Fig. 133, it is seen that approximately from 0.6 to 14 a.u., the surface density drops $\sim 1/r^2$. Within this interval, the masses of protoplanets prove to be of the same order of magnitude. Starting with 14 a.u., $\sigma_0 \sim 1/r^3$, therefore the masses of the last two protoplanets should decrease with distance.

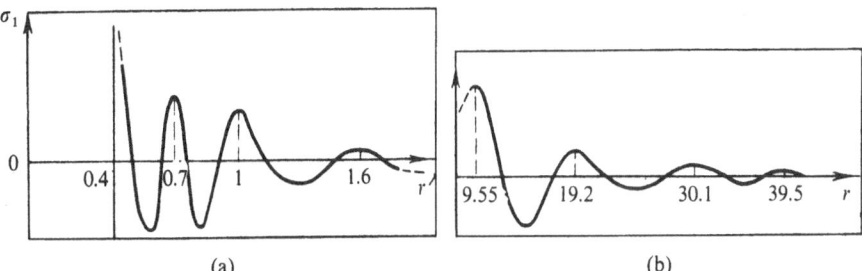

Figure 134. Dependence of the perturbed density of the gas–dust protoplanetary cloud $\sigma_1(r)$ on the radius: (a) $0 < r \le 1.8$ a.u.; (b) $g \le r \le 40$ a.u.

For the development of the instability in the dust disk model, it is necessary that $M/m_n \lesssim 28$, while in the gas–dust model $M/m_n \lesssim 14$. This means that in the present planets there is contained not more than $\approx 3\%$ ($\approx 1.5\%$) of the mass of the protoplanet cloud.

According to Kuiper's hypothesis [262], the mass of the present planets constitutes 1–10% of the protoplanetary cloud, and Hoyle postulates that the residual mass makes up 9% [223].

It is suggested that the larger part of the initial mass of protoplanets was blown away due to intensive corpuscular emission of the sun. The difference between the masses of the giant planets and planets of the Earth's group is ascribed to the action of thermal emission of the sun, which has burned away from the latter nearly all the light elements.

Such a point of view, according to Fesenkov [137], is not unnatural since the planets of the Earth's type consist, as is known, mainly of elements with a high melting temperature and are almost lacking light elements such as, for example, hydrogen, which is the main material of which stars and giant planets are built.

By adding to the present masses of planets of the Earth's group the amount of light gases which is necessary to restore the chemical composition of giant planets, we obtain masses larger by a factor of several hundreds, coincident in order of magnitude with the masses of giant planets [137]. Therefore, it is quite logical to suggest that the original diffuse medium that has served for planetary formation, has had the same composition as in the sun.

Of course, the above linear treatment may serve only as a very approximate idea for a future theory, which must be essentially nonlinear. If an originally stable, slowly evolving system loses at some moment its stability, then it would be natural to assume that the growth rate of the corresponding perturbations will be rather small. Then a question arises of the further fate of these perturbations (already in the nonlinear stage). Practically, one is able to investigate only perturbations with values relatively small as compared to equilibrium, though with finite amplitudes. But even such a treatment allows one to answer a number of interesting questions. Taking account of nonlinearity can lead either to the stabilizing effect or, on the contrary, to an increase in the growth rate of perturbations. Which case will be realized in reality is determined by the physical properties of the system. In the simplified treatment described in Section 1.2, Chapter VII, these properties are generally characterized by the gaseous adiabatic exponent γ whose value, as is well known, is dependent on many factors: temperature, molecular or dissociated state of the gas, ionization degree, etc. It turns out that the behavior of the evolving perturbations is critically dependent on the value of γ: with the adiabatic exponents larger than some critical value γ_c, perturbations increase only to some small finite value. Therefore, only with such physical states of the gas in the system as ensure sufficiently low values of the adiabatic index $\gamma < \gamma_c$, perturbations can increase strongly.

The critical value of the adiabatic index for a rotating gaseous layer turn out to be equal to $\gamma_c = 1.404$. It is interesting that γ_c turned out to be close to the adiabatic index of the two-atomic gas ($\gamma = \frac{7}{5}$ under normal conditions). In the one-atomic gas of neutral atoms, for which $\gamma = \frac{5}{3}$, perturbations must be stabilized with small amplitudes. But this also refers to the molecular hydrogen gas under the conditions typical for dense gas–dust complexes of the Galaxy ($T \sim 10 \div 50$ K). For the hydrogen molecules, the rotational degrees of freedom are "frozen in" due to an anomalously large value of the rotational quantum of energy [68]: $\hbar\omega \approx 8 \cdot 10^{-3}$ eV, which corresponds to the temperature $\hbar\omega/k \simeq 85$ K.

In what cases may perturbations of the gas be assumed to be approximately isothermic ($\gamma < \gamma_c$)? Note two possibilities. The adiabatic index may be close to 1 for the gaseous layer in the state of partial ionization. Such a state could be established in a gaseous layer surrounding the young sun if the gas is heated by its radiation up to $T \sim 8000$ K. The same may be applicable to the case if, in the layer of the gas (even neutral) or the dust, losses for radiation are large, so that under compression the temperature changes negligibly.[15]

§ 3 Galactic Plane Bending

Hunter and Toomre [230] applied the membrane oscillation theory of an infinitely thin cold gravitating disk which they developed (cf. Section 2.1, Chapter V) to test the different hypotheses proposed for the explanation of the observed bending of the plane of the Galaxy (Burke [177]).

Data obtained from radio observations at $\lambda = 21$ cm (see [230]) come to the following. The outer edge of the layer of the interstellar atomic hydrogen is lifted upwards (by about $h = 1$ kps, or ($6 \div 7\%$ of radius) near one (solar) edge of the Galaxy, while on the opposite side it is lowered downwards by approximately the same amount. At the same time, a part of the disk of the Galaxy, internal relative to the sun, appears to be flat. Thus, in the cross-section, the hydrogen layer resembles in shape the integral sign.

At least four different explanations of the described phenomenon are suggested.

(1) The possibility was considered [230] that this is simply a tidal distortion due to the Magellanic Clouds at their present distance from the Galaxy (in order of magnitude).

(2) Kahn and Woltjer (see [230]) suggested that the phenomenon is due to the intergalactic gas flow round the Galaxy and its halo.

---

[15] In conclusion, we should note that the Jeans instability is apparently not only interesting in the problem discussed. For example, the secular instability of the type considered in the preceding section could also play an essential role.

(3) The same authors, as well as Lynden-Bell [283], considered also the possibility of free oscillations of the plane of the Galaxy (of the membrane type) which might have been excited at the epoch of the formation of the Galaxy.

(4) The orbit of the Magellanic Clouds might in the past have been located much closer to the Galaxy than it is now [230]. Then the amount of the effect may be essentially larger than that according to the first of the versions listed. For the first time such a possibility, but in a somewhat different connection, was considered by Idlis [51].

To the four possibilities listed above, one may add still another implying an *a priori* nonprohibited instability. However, this possibility may be excluded due to the general theorem about the perturbation stability with $m = 1$ (in a cold disk) proved by Hunter and Toomre, which just corresponds to the observable angular dependence $\sim e^{i\varphi}$. The available thickness of the Galaxy is likely to be enough to stabilize the "fire-hose" instability.

Let us consider the remaining possibilities.

3.1 Quasistationary Tidal Deformation

To make an estimate of the present tidal effect of the Magellanic Clouds, these latter may be represented in the form of a point (with a mass $M_c \simeq 10^{10} M_\odot$, where $M_\odot$ is the solar mass), located at a distance $r_c \simeq 55$ kps at the latitude $b = 35°$. It is convenient to assume the longitude of this point (from the center of the Galaxy disk) to be $\varphi = 0$. Then, in the first order of magnitude with respect to r/r_c, the vertical tidal force F_* from the clouds is approximately equal to

$$F_*(r, \varphi) \approx \left(\frac{3GM}{2r_c^2}\right)(r/r_c) \sin 2b \cos \varphi. \tag{1}$$

Hunter and Toomre [230] calculated the response to the stationary external perturbation of the form

$$F_* = (\pi GM/R^2)(r/R) \cos \varphi. \tag{2}$$

It is represented in the following form:

$$h(r, \varphi, t) = S\tau r \sin \varphi + RH_{st}(r/R) \cos \varphi, \tag{3}$$

where $S \approx$ const is the constant velocity of precession ($S \sim 1/2\Omega$) while H is the function of radius, which is presented by Hunter and Toomre [230] in the form of plots (Fig. 135). In order that they might be used, it is necessary in this case to take into account that F_* in Eqs. (1) and (2) differ by a factor of $g_{st} = (3/2\pi)(M_c R^3/Mr_c^3) \sin 2b$. Assume for the Galaxy that

$$M = 1.2 \cdot 10^{11} M_\odot, R = 16 \text{ kps};$$

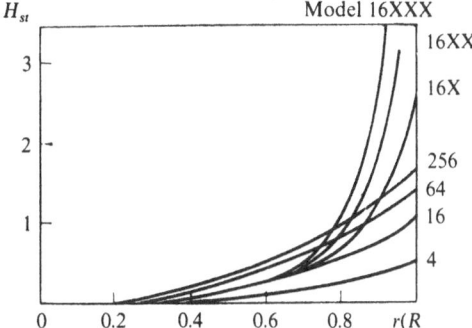

Figure 135. Perturbations with $m = 1$ of various disks under the action of external force [230]; amplitudes at the disk's edge are 2.72, 5.31, and 7.94 for models 16X, 16XX, and 16XXX, respectively.

the last value yields for the model $16X^{16}$ a reasonable surface density: $\sigma = 87\ M_\odot/\text{ps}^2$ and the rate of rotation $V = 255$ km/s at $r = 10$ kps. Then the factor $g_{st} = 9.2 \cdot 10^{-4}$, and from Fig. 135 it follows that for the 16X model, for example, we have a maximum deviation of the plane by only $2.72g_1R \simeq 40$ ps (if the tidal force of such an amount is constantly acting). For the 16XX model, a similar displacement will be 78 ps and even for the 16XXX model with a very low density on the periphery it is only 117 ps on the very edge. Of course, these quasistationary estimates may easily contain an error by a factor of 2 or 3 due to uncertainties in the models of the Galaxy and of clouds. Nevertheless, because of the fact that the observed deformation is ~ 1 kps, Hunter and Toomre [230] arrive at the conclusion that the possibility considered here should be rejected, in any event it cannot be the main cause of the observed bending of the Galaxy plane.

3.2 Free Modes of Oscillations

The disk with discrete spectra are not typical since the real disks have no sharp boundary. In such disks (with a continuous frequency spectrum) simple initial deformations cannot remain for a sufficiently long time. One of the reasons common for perturbations of any form, was noted earlier, in Section 2.1, Chapter V—this is the drift of the initial perturbation toward the disk edge (where it must *damp* rapidly). The other reason, which is essential, at least in the case of perturbations with $m = 1$, lies in the fact that different ringlets simply tend to rotate at different rates under the action of the gravitational torque from the side of a massive internal disk.

---

[16] These models were considered in §1, Chapter V.

3.3 Close Passage

According to the opinions of Hunter and Toomre [230], this is a single promising suggestion. The observed radial velocity of the Large Magellanic Cloud (LMC) corrected for the movement of the sun and for galactic rotation (250 km/s) shows that the LMC is at present moving away not only from us but also from the galactic center. This velocity is of the order of 50 km/s [230]. The time scale of $T \sim (4 \div 6) \cdot 10^8$ years was determined thus: at that era, the Galaxy and LMC were separated from each other by a distance essentially less than at present. Numerical estimates of the Galaxy plane bending in this case, which are similar to those made above (Section 3.1), yield a reasonable amount of deviation provided that the LMC (about $5 \cdot 10^8$ years ago) had passed at a distance within 20 kps from the center of the Galaxy (for the mass of the LMC $\gtrsim 2 \cdot 10^{10} \, M_\odot$).

From this point of view, the presently observed picture of the Galaxy plane bending is the evolved tidal perturbation from the LMC, which had been excited several hundred million years ago.

§ 4 Instabilities in Collisions of Elementary Particles

In conclusion, let us briefly consider[17] one somewhat unexpected possibility of applying the idea of the gravitational instability to the problem of multiple production of particles (see [136]).

Let us now turn our attention to collisions of hadrons (leading to the formation of a very much flattened disk and its subsequent scattering). In the most consistent way, this scattering is described, as is well known, in the hydrodynamic Landau theory [66]. This theory, however, does not describe the "fireball" (or "parton") structure of scattering. In reality, such a structure may probably be obtained with the preservation of the main idea of the Landau theory, i.e., of the hydrodynamics itself. Only the suggestion in it that the motion of the matter in the course of an essential part of scattering is one-dimensional is possibly inconsistent. The one-dimensional motion, as follows from elementary estimates, is unstable relative to perturbations lying in the disk plane.[18] The instability must lead to the formation of "quasi-independent" clusters. As we are aware, the dispersion equation for plane perturbations of a gravitating disk has, roughly speaking, the form

$$\omega^2 \approx k^2 c^2 - 2\pi G \sigma_0 k \, (+\varkappa^2).$$

In the case of interest, the role of gravitation is played by the attraction between "elements" of nuclear liquid of the disk. Since rather large distances are

[17] Presented by V. L. Polyachenko at the All-Union Conference on Plasma Astrophysics (Irkutsk, 1976).

[18] Other types of perturbations may also be unstable. It is generally interesting to investigate the problem of the role of "collective phenomena" in the problems under consideration.

essential here, we have, for the Yukawa (for example) attraction between the nuclear charges q and Q: $U = -g\,(qQ/r)e^{-\mu r}$ (g is the constant of interaction, $\hbar = c = 1$). Since the size of the disk is less than $1/\mu$, one may, for the estimate, assume that $e^{-\mu r} \approx 1$: the Yukawa attraction becomes long-range (and $\infty -1/r$), similar to gravitation. Accordingly, we have the following approximate dispersion equation: $\omega^2 = k^2 c^2 - 2\pi(g/m)\sigma k$, where m is the mean mass of nuclear particles, σ is their surface density (cm^{-2}). Hence, as one can easily confirm, it follows that $\omega^2 < 0$ (i.e. the instability takes place) already for multiplicity of the order of 1 (and when c is the velocity of light).

Of course, we have stated above only a "rough substantiation" of the idea. A more detailed discussion would be irrelevant in this book.

Appendix

§1 Collisionless Kinetic Equation and Poisson Equation in Different Coordinate Systems

(a) *Kinetic equation.* In Cartesian coordinates x, y, z the collisionless kinetic equation is written in the form

$$\frac{\partial f}{\partial t} + v_x \frac{\partial f}{\partial x} + v_y \frac{\partial f}{\partial y} + v_z \frac{\partial f}{\partial z} - \frac{\partial \Phi}{\partial x} \frac{\partial f}{\partial v_x} - \frac{\partial \Phi}{\partial y} \frac{\partial f}{\partial v_y} - \frac{\partial \Phi}{\partial z} \frac{\partial f}{\partial v_z} = 0. \tag{1}$$

It is easy to see that the transition to the curvilinear coordinates or into a rotating system may be realized via direct transformation of Eq. (1). This procedure, however, is normally rather cumbersome. A simpler method of obtaining the kinetic equation immediately in the required coordinate system is as follows. Note that the collisionless kinetic equation may be presented in the form

$$\frac{df}{dt} = \frac{\partial f}{\partial t} + [f, H] = 0. \tag{2}$$

Here H is the Hamiltonian of a particle in the self-consistent field

$$H = \tfrac{1}{2}\mathbf{v}^2 + \Phi(\mathbf{r}, t), \tag{3}$$

while $[f, H]$ is the Poisson bracket:

$$[f, H] \equiv \sum_i \left(\frac{\partial f}{\partial q_i} \frac{\partial H}{\partial p_i} - \frac{\partial f}{\partial p_i} \frac{\partial H}{\partial q_i} \right), \tag{4}$$

where q_i and p_i are the generalized coordinates and conjugate impulses of the particle. Recall that according to the definition $p_i = \partial L/\partial \dot{q}_i$, where L is the Lagrange function:

$$L = \tfrac{1}{2}v^2 - \Phi(\mathbf{r}, t). \tag{5}$$

Generally, it is easy to write the expression for the Hamiltonian in any coordinate system.

In the most usable (including those in this book) coordinates, the Hamilton function has the following form. In the cylindrical coordinates r, φ, z

$$H = \frac{1}{2}\left(p_r^2 + \frac{p_\varphi^2}{r^2} + p_z^2\right) + \Phi(r, \varphi, z), \tag{6}$$

where $p_r = v_r$, $p_\varphi = rv_\varphi$, $p_z = v_z$.

In the spherical coordinates r, θ, φ

$$H = \frac{1}{2}\left(p_r^2 + \frac{p_\theta^2}{r^2} + \frac{p_\varphi^2}{r^2 \sin^2 \theta}\right) + \Phi(r, \theta, \varphi). \tag{7}$$

In this case, $p_r = v_r$, $p_\theta = rv_\theta$, $p_\varphi = r \sin \theta\, v_\varphi$.

In accordance with these expressions for the Hamilton function, it is easy to obtain, using (2)–(4), the kinetic equation in the coordinate systems under consideration.

In *cylindrical* coordinates, it happens to be

$$\frac{\partial f}{\partial t} + v_r \frac{\partial f}{\partial r} + \frac{v_\varphi}{r}\frac{\partial f}{\partial \varphi} + v_z \frac{\partial f}{\partial z} + \left(\frac{v_\varphi^2}{r} - \frac{\partial \Phi}{\partial r}\right)\frac{\partial f}{\partial v_r}$$

$$-\left(\frac{v_r v_\varphi}{r} + \frac{\partial \Phi}{r\,\partial \varphi}\right)\frac{\partial f}{\partial v_\varphi} - \frac{\partial \Phi}{\partial z}\frac{\partial f}{\partial v_z} = 0. \tag{8}$$

In *spherical* coordinates, respectively, the kinetic equation is of the form

$$\frac{\partial f}{\partial t} + v_r \frac{\partial f}{\partial r} + \frac{v_\theta}{r}\frac{\partial f}{\partial \theta} + \frac{v_\varphi}{r \sin \theta}\frac{\partial f}{\partial \varphi} + \left(\frac{v_\theta^2 + v_\varphi^2}{r} - \frac{\partial \Phi}{\partial r}\right)\frac{\partial f}{\partial v_r}$$

$$-\left(\frac{v_r v_\theta}{r} - \cot \theta \frac{v_\varphi^2}{r} + \frac{\partial \Phi}{r\,\partial \theta}\right)\frac{\partial f}{\partial v_\theta}$$

$$-\left(\frac{v_r v_\varphi}{r} + \cot \theta \frac{v_\theta v_\varphi}{r} + \frac{1}{r \sin \theta}\frac{\partial \Phi}{\partial \varphi}\right)\frac{\partial f}{\partial v_\varphi} = 0. \tag{9}$$

In the spherically symmetrical case, it is frequently more convenient (see, e.g., §3, Chapter III) to use somewhat different coordinates in the velocity space: instead of v_r, v_θ, and v_φ, v_r, $v_\perp = (v_\theta^2 + v_\varphi^2)^{1/2}$, and $\alpha = \arctan v_\varphi/v_\theta$.

In these coordinates, the kinetic equation is written in the form

$$\frac{\partial f}{\partial t} + \frac{v_\perp}{r}\left(\cos\alpha\,\frac{\partial f}{\partial\theta} + \frac{\sin\alpha}{\sin\theta}\frac{\partial f}{\partial\varphi} - \cot\theta\sin\alpha\,\frac{\partial f}{\partial\alpha}\right)$$

$$+ v_r\left(\frac{\partial f}{\partial r} - \frac{v_\perp}{r}\frac{\partial f}{\partial v_\perp}\right) + \left(\frac{v_\perp^2}{r} - \frac{\partial\Phi}{\partial r}\right)\frac{\partial f}{\partial v_r}$$

$$- \frac{1}{r}\left(\cos\alpha\,\frac{\partial\Phi}{\partial\theta} + \frac{\sin\alpha}{\sin\theta}\frac{\partial\Phi}{\partial\varphi}\right)\frac{\partial f}{\partial v_\perp} - \frac{1}{rv_\perp}\left(\sin\alpha\,\frac{\partial\Phi}{\partial\theta} - \frac{\cos\alpha}{\sin\theta}\frac{\partial\Phi}{\partial\varphi}\right)\frac{\partial f}{\partial\alpha} = 0.$$

$$(10)$$

One may write the kinetic equation also in an arbitrary orthogonal coordinate system q_1, q_2, q_3, with the square of the line element of the form

$$dl^2 = h_1^2\,dq_1^2 + h_2^2\,dq_2^2 + h_3^2\,dq_3^2, \tag{11}$$

where the h_i are some coordinate functions. The velocity components are

$$v_1 = h_1\dot{q}_1, \qquad v_2 = h_2\dot{q}_2, \qquad v_3 = h_3\dot{q}_3 \tag{12}$$

so that the Lagrangian is

$$L = \tfrac{1}{2}(h_1^2\dot{q}_1^2 + h_2^2\dot{q}_2^2 + h_3^2\dot{q}_3^2) - \Phi(q_1, q_2, q_3). \tag{13}$$

Hence we find the generalized impulses p_1, p_2, p_3, corresponding to the coordinates q_1, q_2, q_3:

$$p_1 = \frac{\partial L}{\partial\dot{q}_1} = h_1^2\dot{q}_1, \dots. \tag{14}$$

We express, from (14), $\dot{q}_1, \dot{q}_2, \dot{q}_3$ through p_1, p_2, p_3 and the Lamé coefficients h_1, h_2, h_3:

$$\dot{q}_1 = p_1/h_1^2, \dots. \tag{15}$$

Therefore, the velocity components are

$$v_1 = p_1/h_1, \dots. \tag{16}$$

The Hamiltonian of the particle is

$$H = \tfrac{1}{2}(v_1^2 + v_2^2 + v_3^2) + \Phi = \frac{1}{2}\left(\frac{p_1^2}{h_1^2} + \frac{p_2^2}{h_2^2} + \frac{p_3^2}{h_3^2}\right) + \Phi(q_1, q_2, q_3). \tag{17}$$

Hence according to (2)–(4) we derive the general form of the kinetic equation

$$\frac{\partial f}{\partial t} + \sum_{i=1}^{3}\left[\frac{p_i}{h_i^2}\frac{\partial f}{\partial q_i} - \frac{\partial\Phi}{\partial q_i}\frac{\partial f}{\partial p_i} - \frac{1}{2}\frac{\partial}{\partial q_i}\left(\frac{p_1^2}{h_1^2} + \frac{p_2^2}{h_2^2} + \frac{p_3^2}{h_3^2}\right)\frac{\partial f}{\partial p_i}\right] = 0. \tag{18}$$

Then one has to turn again from the impulses p_i to the velocities v_i [by formulae (16)].

Write now the kinetic equation in a rectangular coordinate system x, y, z, *rotating* with an angular rate Ω. If it is written in the form

$$\frac{\partial f}{\partial t} + \mathbf{v}\frac{\partial f}{\partial \mathbf{r}} + \dot{\mathbf{v}}\frac{\partial f}{\partial \mathbf{v}} = 0, \tag{19}$$

then, instead of the acceleration $\dot{\mathbf{v}}$, in this case one evidently has to substitute the force

$$\dot{\mathbf{v}} = \mathbf{F} = -\frac{\partial \Phi}{\partial \mathbf{r}} + 2[\mathbf{v}\Omega] + [\Omega[\mathbf{r}\Omega]], \tag{20}$$

where the first term on the right-hand side provides the ordinary gravitational force, while the remaining two terms provide the inertia forces (respectively, the Coriolis and centrifugal force). If the rotation axis is chosen as the z-axis of the coordinate system, then the required kinetic equation will take the form

$$\frac{\partial f}{\partial t} + v_x\frac{\partial f}{\partial x} + v_y\frac{\partial f}{\partial y} + v_z\frac{\partial f}{\partial z} + \left(\Omega^2 x + 2\Omega v_y - \frac{\partial \Phi}{\partial x}\right)\frac{\partial f}{\partial v_x}$$

$$+ \left(\Omega^2 y - 2\Omega v_x - \frac{\partial \Phi}{\partial y}\right)\frac{\partial f}{\partial v_y} - \frac{\partial \Phi}{\partial z}\frac{\partial f}{\partial v_z} = 0. \tag{21}$$

Let us further give this equation in the cylindrical (also rotating) coordinate system:

$$\frac{\partial f}{\partial t} + v_r\frac{\partial f}{\partial r} + \frac{v_\varphi}{r}\frac{\partial f}{\partial \varphi} + v_z\frac{\partial f}{\partial z} + \left(-\frac{\partial \Phi}{\partial r} + \Omega^2 r + 2\Omega v_\varphi + \frac{v_\varphi^2}{r}\right)\frac{\partial f}{\partial v_r}$$

$$- \left(\frac{\partial \Phi}{r\,\partial \varphi} + 2\Omega v_r + \frac{v_r v_\varphi}{r}\right)\frac{\partial f}{\partial v_\varphi} - \frac{\partial \Phi}{\partial z}\frac{\partial f}{\partial v_z} = 0. \tag{22}$$

(b) *Poisson equation.* The second basic equation of the theory, the Poisson equation, has, in the arbitrary orthogonal coordinate system (11), the following symmetrical form:

$$\Delta\Phi = \frac{1}{h_1 h_2 h_3}\sum\frac{\partial}{\partial q_1}\left(\frac{h_2 h_3}{h_1}\frac{\partial f}{\partial q_1}\right) = 4\pi G\rho, \tag{23}$$

where the summing is carried out with circular interchanges of the indices 1, 2, 3. In *cylindrical* coordinates

$$\Delta\Phi = \frac{1}{r}\frac{\partial}{\partial r}\left(r\frac{\partial \Phi}{\partial r}\right) + \frac{1}{r^2}\frac{\partial^2 \Phi}{\partial \varphi^2} + \frac{\partial^2 \Phi}{\partial z^2} = 4\pi G\rho. \tag{24}$$

In *spherical* coordinates

$$\Delta\Phi = \frac{1}{r^2}\frac{\partial}{\partial r}\left(r^2\frac{\partial \Phi}{\partial r}\right) + \frac{1}{r^2 \sin\theta}\frac{\partial}{\partial \theta}\left(\sin\theta\frac{\partial \Phi}{\partial \theta}\right) + \frac{1}{r^2 \sin^2\theta}\frac{\partial^2 \Phi}{\partial \varphi^2} = 4\pi G\rho. \tag{25}$$

§2 Separation of Angular Variables in the Problem of Small Perturbations of Spherically Symmetrical Collisionless Systems

For the arbitrary distribution function of the form $f_0 = f_0(E, L^2)$ in the spherically symmetrical case, the angular part of the spatial dependence of perturbation may always be separated in the form proportional to the spherical harmonics: $\Phi_1 \sim Y_l^m(\theta, \varphi)$.

This natural statement may be formally substantiated with the aid of the following calculations. Simultaneously, we shall obtain, seemingly, the most natural representation of the equations describing the perturbations of spherically symmetrical stellar systems which takes into account completely the symmetry of the problem.

We represent the perturbation of the distribution function in the form of the expansion

$$f_1 = \sum_s \lambda_{ms}^l(v_\perp, v_r, r) T_{ms}^l(\varphi, \theta, \alpha), \tag{1}$$

where the functions

$$T_{ms}^l(\varphi_1, \theta, \varphi_2) = e^{-im\varphi_1 - is\varphi_2} P_{ms}^l(\cos \theta) \tag{2}$$

are introduced, and $P_{ms}^l(\cos \theta)$ are the three-index functions [33], in particular, the $P_{m0}^l(\cos \theta)$ functions, within an accuracy of coefficients coincident with the associated Legendre functions. Therefore, it is convenient to write that

$$\Phi_1 = \chi(r, t) T_{m0}^l(\varphi, \theta, \alpha). \tag{3}$$

Let us substitute (1) and (3) into the linearized equation in the form (see §1, Appendix)

$$\frac{\partial f_1}{\partial t} + \frac{v_\perp}{r} \hat{L} f_1 + \hat{D} f_1 = \frac{\partial \Phi_1}{\partial r} \frac{\partial f_0}{\partial E} v_r + \frac{1}{r} \hat{L} \Phi_1 \left(\frac{\partial f_0}{\partial E} v_\perp + \frac{\partial f_0}{\partial L^2} 2 v_\perp r^2 \right), \tag{4}$$

where

$$\hat{L} \equiv \cos \alpha \frac{\partial}{\partial \theta} + \frac{\sin \alpha}{\sin \theta} \frac{\partial}{\partial \varphi} - \sin \alpha \cot \theta \frac{\partial}{\partial \alpha},$$

$$\hat{D} = v_r \frac{\partial}{\partial r} - \frac{v_r v_\perp}{r} \frac{\partial}{\partial v_\perp} + \left(\frac{v_\perp^2}{r} - \frac{\partial \Phi_0}{\partial r} \right) \frac{\partial}{\partial v_r}.$$

As a result, we have

$$\frac{\partial}{\partial t} \sum_s \lambda_{ms}^l T_{ms}^l + \frac{v_\perp}{r} \sum_s \lambda_{ms}^l \hat{L} T_{ms}^l + \sum_s \hat{D} \lambda_{ms}^l T_{ms}^l$$

$$= \frac{\partial \chi}{\partial r} T_{m0}^l v_r \frac{\partial f_0}{\partial E} + \frac{\chi}{r} \left(\frac{\partial f_0}{\partial E} v_\perp + 2 \frac{\partial f_0}{\partial L^2} v_\perp r^2 \right) \hat{L} T_{m0}^l. \tag{5}$$

Further, the operator L will be represented in the form (33)

$$\hat{L} = \frac{1}{2i}(\hat{H}_+ + \hat{H}_-),\tag{6}$$

where the operators $\hat{H}_+$ and $\hat{H}_-$ act on the T_{ms}^l functions according to the formulae (33):

$$\hat{H}_+ T_{ms}^l = \alpha_{s+1} T_{m,\,s+1}^l,\tag{7}$$

$$\hat{H}_- T_{ms}^l = \alpha_s T_{m,\,s-1}^l,\tag{8}$$

where $\alpha_s = \sqrt{(l+s)(l-s+1)}$. Therefore,

$$\hat{L}T_{ms}^l = \frac{1}{2i}(\alpha_{s+1} T_{m,\,s+1}^l + \alpha_s T_{m,\,s-1}^l).\tag{9}$$

Further, the spherical harmonic $Y_m^l(\theta, \varphi)$ is rewritten through $T_{m0}^l(\varphi_1, \theta, \varphi_2)$ in the following way:

$$e^{im\varphi_1}P_{m0}^l(\cos\theta) = e^{2im\varphi_1}T_{m0}^l(\varphi_1, \theta, \varphi_2) = Y_m^l\tag{10}$$

[if all normal coefficients are included in $\chi(r)$]. Through T_{ms}^l one may write also the expression $\hat{L}T_{m0}^l(\theta, \varphi)$, on the right-hand side of (5). For that purpose, make use of the equality

$$LT_{m0}^l(\varphi, \theta, \alpha) = \frac{1}{2i}(\alpha_1 T_{m,\,1}^l + \alpha_0 T_{m,\,-1}^l).\tag{11}$$

Thus, the angular variables have been separated, and the equations for the function $\lambda_{\pm s}$

$$\lambda_s + \frac{v_\perp}{2ir}(\alpha_s\lambda_{(s-1)} + \alpha_{s+1}\lambda_{(s+1)}) + \hat{D}\lambda_s$$

$$= \chi'\delta_{s0}v_r\frac{\partial f_0}{\partial E} + \frac{\chi}{2ir}\left(\frac{\partial f_0}{\partial E}v_\perp + \frac{\partial f_0}{\partial L^2}2v_\perp r^2\right)(\alpha_s\delta_{s,\,1} + \alpha_{s+1}\delta_{s,\,-1}),\tag{12}$$

$$s = -l, \ldots, l,$$

have been obtained.

Since $\lambda_{-s} = \lambda_{+s}$ (as may be easily shown), we can restrict ourselves to only nonnegative $s = 0, 1, 2, \ldots, l$. The density $\rho_1 = \int f_1 \, dv$ needs to be calculated, according to (1) and (2), by the formula

$$\rho_1 = 2\pi \int \lambda_0 v_\perp \, dv_r \, dv_\perp.\tag{13}$$

The further investigation depends already on a specific form of the distribution function under consideration, and is unlikely to be carried out in the general form. In particular, in the general case $f_0 = f_0(E, L^2)$, it is impossible to construct the variational principle, except for the isotropic case $f_0 = f_0(E)$ (or for radial perturbations).

However, for sufficiently simple systems, possessing some additional properties of symmetry, the problem is solvable. For example, in case of a homogeneous sphere with circular orbits of particles, from derived Eqs. (15) and (16) it is easy to obtain all eigenfrequencies of oscillations (cf. Section 3.1, Chapter III).

In Section 3.4, Chapter III, the above method (for $l = 2$) is applied to derivation of the dispersion equation describing the local disturbances of the spherically symmetrical systems with nearly circular orbits.

§ 3 Statistical Simulation of Stellar Systems

In this section we consider some details of the method of statistical simulation of the stellar systems with different geometries. Description of the method was begun in Section 5.2, Chapter III.

3.1 Simulation of Stellar Spheres of the First Camm Series

1. First of all let us switch from the distribution function in (7), Section 6.2, Chapter III:

$$f_0(\varepsilon, \mu) = C\mu^\beta [\varphi_0(R) - \varepsilon]^2, \qquad dM = f_0 \, d\mathbf{r} \, d\mathbf{v}, \qquad \int dM = M, \quad (1)$$

which is normalized for the system's mass M to the function $F(\varepsilon, \mu)$ necessary for us which is normalized for 1; the latter expresses the density of probability that the particle has the energy and the angular momentum within the range $(\varepsilon, \varepsilon + d\varepsilon; \mu, \mu + d\mu)$

$$dw = F(\varepsilon, \mu) \, d\varepsilon \, d\mu, \qquad \int dw = 1. \qquad (2)$$

Let us take the general case $f_0 = f_0(\varepsilon, \mu, \mu_z)$, where μ_z is the z component of the angular momentum, and then in the formula

$$1 = \frac{1}{M} \int f_0(\varepsilon, \mu, \mu_z) \, d\mathbf{r} \, d\mathbf{v} \qquad (3)$$

switch to the variables action-angle I_i and w_i. Since

$$d\mathbf{r} \, d\mathbf{v} = dI_1 \, dI_2 \, dI_3 \, dw_1 \, dw_2 \, dw_3,$$

and since the equilibrium distribution function is not dependent of the angular variables w_i, we obtain

$$1 = \frac{8\pi^3}{M} \int f_0[\varepsilon(I_i), \mu(I_i), \mu_z(I_i)] \, dI_1 \, dI_2 \, dI_3. \qquad (4)$$

Thus, in the general case the probability density in the variables I_i is

$$F_I(I_1, I_2, I_3) = \frac{8\pi^3}{M} f_0[\varepsilon(I_i), \mu(I_i), \mu_z(I_i)]. \tag{5}$$

When the function f_0 is independent of $\mu_z = I_3$, $f_0 = f_0(\varepsilon, \mu)$, one may integrate in (4) over I_3 (from $-\mu$ to $+\mu$), so that we have

$$1 = \frac{16\pi^3}{M} \int \mu f_0[\varepsilon(I_i), \mu(I_i)] \, dI_1 \, dI_2. \tag{6}$$

Thus

$$F_I(I_1, I_2) = \frac{16\pi^3}{M} \mu(I_1, I_2) f_0[\varepsilon(I_1, I_2), \mu(I_1, I_2)]. \tag{7}$$

If one switches in (6) from I_1, I_2 to ε, μ, one finds

$$1 = \frac{16\pi^3}{M} \int f_0(\varepsilon, \mu)\mu \frac{D(I_1, I_2)}{D(\varepsilon, \mu)} \, d\varepsilon \, d\mu, \tag{8}$$

that is, the desired probability density equals

$$F(\varepsilon, \mu) = \frac{16\pi^3}{M} \mu f_0(\varepsilon, \mu) \frac{D(I_1, I_2)}{D(\varepsilon, \mu)}. \tag{9}$$

Since (see §4, Appendix)

$$I_1 = I_1(\varepsilon, \mu), \qquad I_2 = \mu - |\mu_z|, \tag{10}$$

the Jacobian on the right side of (9) is equal to $\partial I_1/\partial \varepsilon = 1/v_1$ so that we, indeed, arrive at formula (5), Section 5.2, Chapter III:

$$F(\varepsilon, \mu) = \frac{16\pi^3}{M} \mu f_0(\varepsilon, \mu)/v_1(\varepsilon, \mu). \tag{11}$$

This formula is valid for arbitrary systems with the distribution functions of the form $f_0(E, L)$. In the given case we must substitute into (11) $f_0(\varepsilon, \mu)$ from (1), so that, finally, for the models of the first Camm series

$$F(\varepsilon, \mu) = \frac{16\pi^3}{M} \frac{C\mu^{\beta+1}\tilde{\varepsilon}^\alpha}{v_1(\tilde{\varepsilon}, \mu)}, \qquad \tilde{\varepsilon} \equiv \varphi(R) - \varepsilon. \tag{12}$$

The normalized constant C is expressed by the formula (see §6)

$$C = (2\pi)^{-3/2} \cdot 2^{-\beta/2} \frac{\Gamma(\alpha + (\beta + 5)/2)}{\Gamma(\alpha + 1)\Gamma(\beta/2 + 1)}. \tag{13}$$

2. *"Playing"* of energies ε_i and of angular momenta μ_i, which correspond to (12), may be performed by one of the standard methods in the statistical simulation. Here we recall briefly two normally used methods (details may be found in the special courses of the statistical simulation [3ad, 9ad]). For the simulations of different systems, results of which are discussed in the main text of the book (§§5 and 6, Chapter III; §3, Chapter IV) we used both these methods. Since in all the cases we must first of all "play" two-dimensional random vectors [for example, (ε, μ) for Camm's models], we shall describe these methods just for some two-dimensional distribution function $f(x, y)$ which is determined in a region Δ:

$$\iint_\Delta f(x, y)\, dx\, dy = 1.$$

First method (Neumann's method). Let us suppose that $f(x, y) \leq M_0$ everywhere in Δ. We introduce also the auxiliary probability density

$$f_1(x, y) = \begin{cases} 1/S, & \text{if } (x, y) \in \Delta, \\ 0, & \text{if } (x, y) \notin \Delta, \end{cases} \tag{14}$$

where S is the area of the Δ region: $S = \iint_\Delta dx\, dy$. Then, the following algorithm of the playing of the vector (x, y) is valid $[3^{ad}, 9^{ad}]$. At first, we play (x, y) according to $f_1(x, y)$ (it is usually a very simple task—see, for example, below); let us assume that we obtain some vector (x_i, y_i). Then we play the value α_1 distributed uniformly in $(0, 1)$, and, if it turns out that $\alpha_1 M_0 < f(x_i, y_i)$, then the first pair (x_i, y_i) is found: but if $\alpha_1 M_0 > f(x_i, y_i)$ then this pair must be discarded, and we play (x, y) again according to f_1, compare $f(x, y)$ with $\alpha_2 M_0$, and so on.

The algorithm described is the most simple one. Though this algorithm is not the fastest of the known algorithms, we used just one in most cases since, as a rule, the time of playing of an initial state is only a negligible part of the total computer time for any task.

For the distribution function (12) with $\beta > -1$ and $\alpha > 0$, one can choose, as M_0, the maximum of the function $F(\varepsilon, \mu)$ in the region Δ: $M_0 = \max_\Delta (F(\varepsilon, \mu))$. In this case the region Δ is the "triangle" in Fig. 25(a) (§6, Chapter III), so that the playing algorithm of (ε_i, μ_i) in accordance with $f_1(\varepsilon, \mu) = 1/S_\Delta$ is especially simple: it reduces to the playing of coordinates of the point which has the uniform distribution within the rectangular region $(0 < \tilde{\varepsilon} < 1, 0 < \mu < \mu_{\max})$, if one will simultaneously discard those points which have for a given ε the ordinate $\mu > \mu_k(\tilde{\varepsilon})$, where $\mu_k = \mu_k(\tilde{\varepsilon})$ is the line of circular orbits.

For $\beta < -1$ (we assume α to be always positive) $F(\varepsilon, \mu) \to \infty$ as $\mu \to 0$; so in this case for the application of Neumann's method one must cut off $F(\varepsilon, \mu)$ at some $\mu = \mu_{\min}$, assuming, for example, that $F(\varepsilon, \mu < \mu_{\min}) = F(\varepsilon, \mu_{\min})$. It is clear that if we choose $\mu_{\min} \ll \bar{\mu}$, then an error due to such a cutoff must not influence the results of computations. It is possible, however, that in this case the playing methods of (ε, μ) connected with piece-constant approximation of the distribution function $[3^{ad}]$ are more profitable.

Second method (inversion). Using this method of playing the vector (x, y), one should, for each given case of the random number α_i having a uniform distribution on the segment $(0, 1)$, "invert," i.e., resolve with respect to y, the equation

$$\alpha_i = \int_{y_{\min}}^{y} \tilde{F}(y') \, dy' \tag{15}$$

[$\tilde{F}(y)$ is the one-dimensional distribution of the value y]. As a result, for y, we find some value $y = y_i$. Then, the conditional distribution of the random value x is considered:

$$g(x \mid y) = \frac{f(x, y_i)}{\tilde{F}(y_i)}. \tag{16}$$

We select again, within the segment $(0, 1)$, the random number α_{i+1} and determine with it the random number x_i with the distribution $g(x \mid y_i)$ by solving, with respect to x_i, the "equation of inversion" similar to (15):

$$\alpha_{i+1} = \int_{x_{\min}}^{x_i} g(x' \mid y_i) \, dx' = \frac{1}{\tilde{F}(y_i)} \int_{x_{\min}}^{x_i} f(x', y_i) \, dx'. \tag{17}$$

The sequence of the pairs of numbers (x_i, y_i) resulting in this way has, as may be shown [3^{ad}], the required distribution $f(x, y)$.

Some combined methods of playing are also possible, of course. For example, in the second stage of the inversion method just described [determination of x_i from α_{i+1} in accordance with the probability density $g(x \mid y_i)$] it may prove to be easy to make use of the Neumann method, etc.

In [41^{ad}] the problem was solved by both an N-body method and a matrix method. The latter requires the computation of different orbit characteristics of moving particles $x_{\min}$, $x_{\max}$, v_1, v_2, ... on the sufficiently "dense" network (in order to obtain the necessary exactness). So in fact we played [41^{ad}] just these "set" values $(\tilde{\varepsilon}_i, \mu_i)$ in most cases.

Playing of the coordinates and the velocities of the particles with the given ε, μ, μ_z may be performed, for example, in the way described in Section 5.2, Chapter III (or in [41^{ad}]).

As to the solution methods of the large sets of differential equations, then, in this connection, one can say that different high-order methods (of those normally used) are practically almost identical.

3.2 Simulation of Homogeneous Nonrotating Ellipsoids

From the distribution function of the ellipsoid at rest in the form (64), §1, Chapter IV, which was normalized by density ρ_0, we go to the probability density

$$F(u, v) = \frac{3}{4\sqrt{v}}, \tag{1}$$

which explicitly depends only on v and vanishes out of the triangular region shaded in Fig. 47(b).

The playing of the values u, v according to (1) may easily be executed by both Neumann's method [replacing in (1) $v \to v + \varepsilon$, where $\varepsilon \ll v$] and the method of inversion. According to the latter, we calculate the one-dimensional distribution function

$$f(v) = \int F(u, v) \, du = \frac{3}{4} \frac{(1 - v)}{\sqrt{v}}, \qquad 0 \le v \le 1; \tag{2}$$

after that we calculate the integral

$$x = \int_0^v f(v) \, dv = \frac{3}{2}\sqrt{v} - \frac{1}{2}(\sqrt{v})^3. \tag{3}$$

Playing the value x uniformly within the interval $(0, 1)$, $x = x_i$, then we solve the cubic equation following from (3):

$$y_i^3 - 3y_i + 2x_i = 0 \qquad (y_i \equiv \sqrt{v_i}), \tag{4}$$

thus determining y_i and $v = v_i = y_i^2$. With given $v = v_i$ we finally find $u = u_i$, playing this value uniformly within the interval $(v_i, 1)$. Simultaneously, we also know the following quantities: the energy of the particle at the plane (x, y), $E_\perp = (u + v)/2$; the modulus of the z-component of the angular momentum $|L_z| = \sqrt{uv}$; the radial action $I_1 = (E_\perp - |L_z|)/2$. The sign of L_z we play according to random law.

After this we display [uniformly within the interval $(0, 2\pi)$] the radial angular variable w_1, determining the radius of the particle in the plane (x, y) as

$$r^2 = \frac{u + v}{2} - \frac{(u - v)}{2} \cos w_1. \tag{5}$$

Similarly, playing uniformly in the interval $(0, 2\pi)$ the angular variable w_2 corresponding to the azimuthal motion of the particle at the plane (x, y), we determine the angle φ according to the following formula (which may be easily derived):

$$\varphi = w_2 - \frac{w_1}{2} + \frac{\pi}{4} + \frac{1}{2} \arcsin \left[\frac{(2I_1 - |L_z|)r^2 - L_z^2}{2r^2\sqrt{I_1^2 + |L_z|I_1}} \right]. \tag{6}$$

Then we calculate components of the particle velocity at the plane (x, y): $v_\varphi = L_z/r, |v_r| = \sqrt{2E_\perp - r^2 - v_\varphi^2}$; signs of v_r must be defined in accordance with the value w_1: these signs for $w_1 > \pi$ and $w_1 < \pi$ are opposite to each other.

Now we must only determine z and v_z. Due to the presence of the δ function in the distribution function of the system (64), §1, Chapter IV, under investigation, we have

$$\rho^2 \equiv \frac{2E_z}{\omega_0^2} = c^2(1 - u). \tag{7}$$

Let us play [again uniformly in the interval $(0, 2\pi)$] the angular variable w_3 which corresponds to the z-motion of the particle. Then

$$v_z = \omega_0 \rho \cos w_3, \qquad z = \rho \sin w_3 \qquad (8)$$

and playing of coordinates and velocities of the particle is completed. Repeating this procedure N times ($N = 100$–300), we obtain the realization of the system considered.

§ 4 The Matrix Formulation of the Problem of Eigenoscillations of a Spherically-Symmetrical Collisionless System

Introduce the variables action-angle [69] for the particle moving in a centrally-symmetrical potential $\Phi_0(r)$. The action variable:

$$I_1 = \frac{1}{2\pi} \oint p_r \, dr = \frac{2}{2\pi} \int_{r_{min}}^{r_{max}} dr \, \sqrt{2E - 2\Phi_0(r) - \frac{L^2}{r^2}}, \qquad (1)$$

$$I_2 = \frac{1}{2\pi} \oint p_\theta \, d\theta = \frac{2}{2\pi} \int_{\theta_0}^{\pi - \theta_0} d\theta \, \sqrt{L^2 - \frac{L_z^2}{\sin^2 \theta}} = L - |L_z|, \qquad (2)$$

$$I_3 = \frac{1}{2\pi} \oint p_\varphi \, d\varphi = L_z. \qquad (3)$$

Here r, θ, φ are the spherical coordinates; E is the energy of the particle; L is the module of the angular momentum; L_z is the projection of the angular momentum onto the z-axis; the angle θ_0 is determined from the equality $\sin^2 \theta_0 = L_z^2/L^2$; p_r, p_θ, p_φ are the corresponding generalized impulses:

$$p_r^2 = 2E - 2\Phi_0 - \frac{L^2}{r^2},$$

$$p_\theta^2 = L^2 - \frac{L_z^2}{\sin^2 \theta},$$

$$p_\varphi = L_z.[1]$$

The angular variables, conjugated to the action variables, are defined by the relations:

$$w_i = \frac{\partial S}{\partial I_i}, \qquad (4)$$

---

[1] Indeed, the Lagrange function of a particle in this case is [69]: $\mathscr{L} = \dot{r}^2/2 + r^2\dot{\theta}^2/2 + r^2 \sin^2 \theta \, \dot{\varphi}^2/2 - \Phi_0(r)$; therefore, $p_r = \partial\mathscr{L}/\partial\dot{r} = \dot{r}$, $p_\theta = \partial\mathscr{L}/\partial\dot{\theta} = r^2\dot{\theta}$, $p_\varphi = \partial\mathscr{L}/\partial\dot{\varphi} = r^2 \sin^2 \theta \, \dot{\varphi}$. Since $L_z^2 = r^2 \sin^2 \theta \, \dot{\varphi}$, $L^2 = r^2(r^2\dot{\theta}^2 + r^2 \sin^2 \theta \, \dot{\varphi}^2)$, $E = \dot{r}^2/2 + r^2\dot{\theta}^2/2 + r^2 \sin^2 \theta \, \dot{\varphi}^2/2 + \Phi_0(r)$, one can readily make sure that the written expressions for p_r, p_θ, p_φ are correct.

where

$$S = S(I_1, I_2, I_3, r, \theta, \varphi) = S_1 + S_2 + S_3,$$

$$S_1 = \int_{r_{min}}^{r_{max}} dr' \sqrt{2E(I_j) - 2\Phi_0(r') - \frac{(I_2 + |I_3|)^2}{r'^2}},$$

$$S_2 = \int_{\theta_0}^{\theta} d\theta' \sqrt{(I_2 + |I_3|)^2 - \frac{I_3^2}{\sin^2 \theta'}}, \tag{5}$$

$$S_3 = \int_0^{\varphi} d\varphi' \, I_3.$$

From (4), (5) we find

$$w_1 = \frac{\partial S_1}{\partial I_1}, \qquad w_2 = \frac{\partial S_1}{\partial I_2} + \arccos\left(\frac{\cos \theta}{\cos \theta_0}\right), \qquad w_3 = \frac{\partial S_1}{\partial I_3} + \frac{\partial S_2}{\partial I_3} + \varphi. \tag{6}$$

The linearized kinetic equation in the variables action-angle has the form

$$\frac{df_1}{dt} \equiv \frac{\partial f_1}{\partial t} + \Omega_i \frac{\partial f_1}{\partial w_i} = \frac{\partial f_0}{\partial I_i} \frac{\partial \Phi_1}{\partial w_i}, \tag{7}$$

where $\Omega_i = \partial E(I_1, I_2 + |I_3|)/\partial I_i$ are the oscillation frequencies of the particles, $f_0(E, L) = f_0(I_1, I_2, I_3)$ is the unperturbed distribution function depending on the action variables, and f_1 and Φ_1 are the perturbations of the distribution function and the potential, respectively, $\Omega_3 = \Omega_2 \operatorname{sgn}(I_3)$. Owing to periodicity of the movement of the particles with respect to w_1, w_2, w_3 (periods 2π), the potential along the trajectory may be written in the form

$$\Phi_1 = \frac{1}{(2\pi)^3} \sum_{l_1 l_2 l_3} \Phi_{l_1 l_2 l_3} \, e^{-i(\omega t - l_1 w_1 - l_2 w_2 - l_3 w_3)} + \text{c.c.}, \tag{8}$$

where the symbol c.c. denotes complex conjugation, and

$$\Phi_{l_1 l_2 l_3}(I_i) = \int_0^{2\pi} \int_0^{2\pi} \int_0^{2\pi} dw_1 \, dw_2 \, dw_3$$

$$\times \Phi_1(I_i, w_i) \exp[-i(l_1 w_1 + l_2 w_2 + l_3 w_3)]. \tag{9}$$

Integrating (7) along the trajectory of the particle, we find the perturbation of the distribution function

$$f_1 = -\frac{1}{(2\pi)^3} \sum_{l_1 l_2 l_3} \Phi_{l_1 l_2 l_3} \frac{e^{i(l_i w_i - \omega t)}}{\omega - l_i \Omega_i} \cdot l_j \frac{\partial f_0}{\partial I_j} + \text{c.c.} \tag{10}$$

Thus, the first part of the problem of the search for the eigenfrequencies of the system—the determination of the response of the system to a given perturbation of the potential—is solved in the variables action-angle in the simplest way. More cumbersome is the second part of the problem—the solution of the Poisson equation in the same variables.

Due to the spherical symmetry of the problem it is clear that the oscillation frequencies are independent of the azimuthal quantum number m. Therefore, it is possible to choose for the sake of simplicity the axially symmetrical perturbed potential

$$\Phi_1 = \Phi_1(r, \theta) = \chi(r)P_l(\cos \theta), \qquad (11)$$

where P_l is the Legendre polynomial.

Assuming, according to (6), that $l_3 = 0$, we obtain from (9) and (11)

$$\Phi_{l_1, l_2} \equiv \Phi_{l_1, l_2, l_3 = 0} = \int_0^{2\pi} dw_1 \int_0^{2\pi} dw_2\, \chi[r(I_1, I_2 + |I_3|, w_1]$$

$$\times P_l\left[\cos \theta_0 \cos\left(\frac{\partial S_1}{\partial I_z} - w_2\right)\right] \exp[-i(l_1 w_1 + l_2 w_2)]. \qquad (12)$$

Use the addition theorem for the Legendre polynomials [33]:

$$P_l(\cos \theta_1 \cos \theta_2 - \sin \theta_1 \sin \theta_2 \cos \varphi_2) = \sum_{k=-l}^{l} e^{-ik\varphi_2} P_l^k(\cos \theta_1)P_l^{-k}(\cos \theta_2).$$

Assume in this formula that

$$\theta_2 = \frac{\pi}{2}, \qquad \theta_1 = \frac{\pi}{2} - \theta_0, \qquad \varphi_2 = w_2 - \frac{\partial S_1}{\partial I_2} + \pi;$$

then it is possible to write

$$P_l\left[\cos \theta_0 \cos\left(w_2 - \frac{\partial S_1}{\partial I_2}\right)\right]$$

$$= \sum_{k=-l}^{l} \exp\left[-ik\left(w_2 - \frac{\partial S_1}{\partial I_2}\right)\right] P_l^k(\sin \theta_0)P_l^{-k}(0) \times e^{-ik\pi}. \qquad (13)$$

From (12) and (13) we obtain ($k = -l_2$):

$$\Phi_{l_1 l_2} = 2\pi P_l^{l_2}(0)P_l^{-l_2}(\sin \theta_0)\chi_{l_1 l_2}(E, L)e^{il_2\pi}, \qquad (14)$$

where the notation

$$\chi_{l_1 l_2}(E, L) = \int_0^{2\pi} dw_1 \times \exp\left[-i\left(l_1 w_1 + l_2 \frac{\partial S_1}{\partial I_2}\right)\right] \times \chi[r(E, L, w_1)] \qquad (15)$$

is introduced.

Similarly to [59^{ad}, 73^{ad}] where the oscillations of the disk systems are treated, consider the set of the functions $\chi_\alpha(r)$ and $\rho_\alpha(r)$ ($\alpha = 1, 2, 3, \ldots$), such that the density perturbation $\rho_1(r)$ of the form

$$\rho_1(\mathbf{r}) = P_l(\cos \theta) \sum_\alpha a_\alpha \rho_\alpha(r) \qquad (16)$$

corresponds to the potential perturbation

$$\Phi_1(\mathbf{r}) = P_l(\cos \theta) \sum_\alpha a_\alpha \chi_\alpha(r). \qquad (17)$$

Assume the density perturbation $\int f_1 \, dv$ and the potential perturbation in the form of (16) and (17). Then we obtain

$$\rho_1(r, \theta) = P_l(\cos \theta) \sum_\alpha a_\alpha \rho_\alpha(r)$$

$$= \frac{1}{2\pi} \sum_{l_1 l_2} \int dv \, P_l^{l_2}(0) \times P_l^{-l_2}(\sin \theta_0)$$

$$\times \frac{\exp(i l_j w_j)}{\omega - l_j w_j} l_j \frac{\partial f_0}{\partial I_j} \sum_\beta a_\beta \int_0^{2\pi} \chi_\beta[r(\lambda)] e^{-i(l_1 \lambda + l_2 \, \partial S_1/\partial I_2)} \, d\lambda. \quad (18)$$

Multiplying (18) by $P_l(\cos \theta)\chi_\gamma^*(r)$ and integrating over r, θ with the weight $r^2 \sin \theta$, after some transformations, we obtain the following infinite system of equations:

$$\sum_\alpha a_\alpha M_{\alpha\beta}^{(l)}(\omega) = \sum_\alpha \langle \rho_\alpha, \chi_\beta^* \rangle a_\alpha, \quad (19)$$

where

$$\langle A, B \rangle \equiv \int AB r^2 \, dr,$$

$$M_{\alpha\beta}^{(l)} = \sum_{l_1 = -\infty}^{\infty} \sum_{l_2 = -l}^{l} D_{ll_2} \int \frac{L \, dL \, dE}{\Omega_1} \frac{(\chi_\alpha)_{l_1 l_2}(\chi_\beta)_{l_1 l_2}^*}{\omega - (l_1 \Omega_1 + l_2 \Omega_2)}$$

$$\times \left[l_1 \Omega_1 \frac{\partial}{\partial E} + l_2 \left(\Omega_2 \frac{\partial}{\partial E} + \frac{\partial}{\partial L} \right) \right] F(E, L), \quad (20)$$

$$D_{ll_2} = \begin{cases} \dfrac{1}{2^{2l}} \dfrac{(l - l_2)!(l + l_2)!}{\{[(l - l_2)/2]![(l + l_2)/2]!\}^2} & \text{for even } |l - l_2|, \\ 0 & \text{for odd } |l - l_2|, \end{cases}$$

$$(\chi_\alpha)_{l_1 l_2} = \int_0^{2\pi} \chi_\alpha[r(E, L, \lambda)] \exp\left[-i \left(l_1 \lambda + l_2 \frac{\partial S_1}{\partial I_2} \right) \right] d\lambda.$$

In the derivation of these equations, we have used the following formulae (see [33, 42]):

$$P_l^{l_2}(z) P_l^{-l_2}(z) P_l^{l_2}(0) P_l^{-l_2}(0) = [P_l^{-l_2}(z)]^2 [P_l^{l_2}(0)]^2,$$

$$\int_{-1}^{1} dz [P_l^{-l_2}(z)]^2 = \left[\frac{(l - l_2)!}{(l + l_2)!} \right]^2 \int_{-1}^{1} [P_l^{l_2}(z)]^2 \, dz = \frac{2}{2l + 1} \frac{(l - l_2)!}{(l + l_2)!},$$

$$P_l^{l_2}(0) = \begin{cases} (-1)^{(l+l_2)/2} \dfrac{1}{2^l} \dfrac{(l + l_2)!}{[(l - l_2)/2]![(l + l_2)/2]!}, & |l - l_2| \text{ even}, \\ 0, & |l - l_2| \text{ odd}. \end{cases}$$

The other, equivalent form of the matrix element $M_{\alpha\beta}$ (it ensues naturally in the derivation of equations by the "Lagrange method"—cf. such a derivation in §5, Appendix, where a similar problem for disk systems is solved):

$$M_{mn}^{(l)} = - \sum_{l_1 l_2} D_{l l_2} \int \frac{dE\, dL}{\Omega_1} F(E, L)$$

$$\times \left[(l_1\Omega_1 + l_2\Omega_2) \frac{\partial}{\partial E} + l_2 \frac{\partial}{\partial L} \right] \left(L \cdot \frac{(\chi_m)_{l_1 l_2}(\chi_n)_{l_1 l_2}}{\omega - l_1\Omega_1 - l_2\Omega_2} \right). \quad (21)$$

By equating to zero the determinant of the system (19), we obtain the dispersion equation for the determination of eigenfrequencies ω.

It is convenient as a set of the χ_α and ρ_α functions, to choose some biortho-normalized (similar to [59^{ad}, 74^{ad}]) system, i.e., a set of the χ_α and ρ_α functions, for which

$$\langle \rho_\alpha \chi_\beta^* \rangle = \delta_{\alpha\beta}. \quad (22)$$

In this case, the dispersion equation takes an especially simple form:

$$\det \| M_{\alpha\beta}^{(l)}(\omega) - \delta_{\alpha\beta} \| = 0. \quad (23)$$

Examples of some biorthonormalized systems, convenient for the numerical computations, will be given below.

Equation (23) must be solved numerically for a concrete distribution function $F(E, L)$.

Note some transforms useful in practical computations. Switch in (15) to integration over the radius r (instead of integration over the angular variable w_1 corresponding to r). Since $dw_1 = dt\Omega_1 = \Omega_1\, dr/v_r$, then we have, by separating the integral into two parts in r (corresponding to the "direct" movement from r_{min} to r_{max} and to "inverse" movement from r_{max} to r_{min}):

$$\chi_{l_1 l_2}(E, L) = \Omega_1 \int_{r_{min}}^{r_{max}} \frac{dr\, \chi(r)}{\sqrt{2[E - L^2/2r^2 - \Phi_0(r)]}} e^{-i(l_1 w_1^+ + l_2 \partial S_1^+/\partial I_2)}$$

$$+ \Omega_1 \int_{r_{min}}^{r_{max}} \frac{dr\, \chi(r)}{\sqrt{2[E - L^2/2r^2 - \Phi_0(r)]}} e^{-i(l_1 w_1^- + l_2 \partial S_1^-/\partial I_2)}$$

$$(24)$$

In Eq. (24) w_1^+, S_1^+ and w_1^-, S_1^- correspond to the direct ($w_1 \in (0, \pi)$) and inverse ($w_1 \in (\pi, 2\pi)$) movement. It can be written as

$$S_1^+ = \tilde{I}(r, E, L) = \int_{r_{min}}^{r_{max}} dr' [2E - 2\Phi_0(r') - L^2/r'^2]^{1/2},$$

$$S_1^- = 2I - \tilde{I}, \qquad I = \tilde{I}(r_{max}, E, L), \quad (25)$$

$$w_1^+ = \Omega_1 \frac{\partial \tilde{I}}{\partial E}, \qquad w_1^- = 2\pi - \Omega_1 \frac{\partial \tilde{I}}{\partial E} = 2\pi - w_1^+;$$

therefore,

$$l_1 w_1^- + l_2 \frac{\partial S_1^-}{\partial I_2} = l_1(2\pi - w_1^+) + l_2 \frac{\partial}{\partial I_2}(2I - \tilde{I})$$

$$= 2\pi l_1 - l_1 w_1^+ - l_2 \frac{\partial S_1^+}{\partial I_2} + 2l_2 \left(\frac{\partial I}{\partial E}\Omega_2 + \frac{\partial I}{\partial L}\right).$$

But the expression in the parentheses in the last line of this equality is zero; therefore,

$$l_1 w_1^- + l_2 \frac{\partial S_1^-}{\partial I_2} = 2\pi l_1 - \left(l_1 w_1^+ + l_2 \frac{\partial S_1^+}{\partial I_2}\right).$$

Denote

$$\psi_{l_1 l_2}(r, E, L) = l_1 w_1^+ + l_2 \frac{\partial S_1^+}{\partial I_2} = \left[(l_1\Omega_1 + l_2\Omega_2)\frac{\partial}{\partial E} + l_2 \frac{\partial}{\partial L}\right]\tilde{I}(r, E, L).$$

Then we obtain the following final expression

$$\chi_{l_1, l_2} = \chi_{-l_1, -l_2}(E, L) = 2\Omega_1 \int_{r_{\min}}^{r_{\max}} \frac{dr\, \chi(r)}{\sqrt{2E - 2\Phi_0(r) - L^2/r^2}} \cos \psi_{l_1 l_2}(r, E, L).$$

$$(26)$$

In the investigation of the stability of the distribution functions of Camm series (Section 6.2, Chapter III) with the help of the matrix formulation developed here there arise some peculiarities. For the first Camm series

$$F(\varepsilon, \mu) = C(\alpha, \beta)\mu^\beta \tilde{\varepsilon}^\alpha,$$

where $\varepsilon = \varphi(R) - \tilde{\varepsilon}$ and μ are the dimensionless energy and the angular momentum, defined in Section 6.2, Chapter III and φ is the dimensionless potential. It is easy to see that, for the values of the parameter $\beta < -1$, the matrix elements M_{mn}, if calculated, for example, by formula (20), are divergent as $L \sim \mu \to 0$. In this case, one has to perform regularization. For that purpose, we introduce the sequence of the functions

$$F_k(\varepsilon, \mu) = C_k(\alpha, \beta)(a_k + \mu)^\beta \tilde{\varepsilon}^\alpha, \qquad k = 1, 2, \ldots,$$

$a_k \to 0$ as $k \to \infty$. Substituting these F_k into (20), transform M_{mn} in the following manner:

$$M_{mn} = \sum_{l_1 l_2} \iint \frac{d\varepsilon\, \mu\, d\mu}{v_1} D_{ll_2} \frac{\partial F_k}{\partial \tilde{\varepsilon}} + \sum_{l_1 l_2} \iint \frac{d\varepsilon\, \mu\, d\mu}{v_1}$$

$$\times D_{ll_2}\left[-\omega \frac{\partial}{\partial \tilde{\varepsilon}} + l_2 \frac{\partial}{\partial \mu}\right] F_k \frac{(\chi_m)_{l_1 l_2}(\chi_n)_{l_1 l_2}}{\omega - l_1 v_1 - l_2 v_2}, \qquad (27)$$

where v_1 and v_2 are the dimensionless frequencies. We transform now the second summand in (27): We multiply and divide the integrand by $\mu/(a_k + \mu)^\beta$; we get

$$\sum_{l_1 l_2} D_{ll_2} \iint \frac{d\varepsilon\, d\mu}{v_1} (a_k + \mu)^\beta \left[\frac{\mu}{(a_k + \mu)^\beta} \left(-\omega \frac{\partial F_k}{\partial \tilde{\varepsilon}} + l_2 \frac{\partial F_k}{\partial \mu} \right) \right] \frac{(\chi_m)_{l_1 l_2}(\chi_n)_{l_1 l_2}}{\omega - l_1 v_1 - l_2 v_2}$$

$$= \sum_{l_1 l_2} \frac{D_{ll_2}}{(\beta + 1)} \int d\varepsilon \left\{ \mu(a_k + \mu) \left[-\omega \frac{\partial F_k}{\partial \tilde{\varepsilon}} + l_2 \frac{\partial F_k}{\partial \mu} \right] \frac{(\chi_m)_{l_1 l_2}(\chi_n)_{l_1 l_2}}{v_1(\omega - l_1 v_1 - l_2 v_2)} \right\}_{C_1}^{C_2}$$

$$- \sum_{l_1 l_2} D_{ll_2} \iint d\varepsilon\, d\mu \frac{(a_k + \mu)^{\beta + 1}}{\beta + 1}$$

$$\times \frac{\partial}{\partial \mu} \left[\frac{\mu}{(a_k + \mu)^\beta} \left(-\omega \frac{\partial F_k}{\partial \tilde{\varepsilon}} + l_2 \frac{\partial F_k}{\partial \mu} \right) \frac{(\chi_m)_{l_1 l_2}(\chi_n)_{l_1 l_2}}{(\omega - l_1 v_1 - l_2 v_2)v_1} \right].$$

The two-dimensional integration of $(d\varepsilon\, d\mu)$ is here (as above) performed over the "triangle" phase region of the system limited by the straight lines $\tilde{\varepsilon} = 0$, $\mu = 0$ (the line C_1) and by the line (C_2) of circular orbits (cf. Section 6.2, Chapter III); the expression $\{\ \}_{C_1}^{C_2}$ means the difference of two expressions taken on the lines C_1 and C_2. But, on C_1, the expression in the braces is zero. On the line of circular orbits (C_2) $(\chi_m)_{l_1 l_2} = 0$, if $l_1 \neq 0$, and there remains only the terms with $l_1 = 0$. Since now as $k \to \infty$ $(a_k \to 0)$ all the integrals written are convergent, we can proceed to the limit $k \to \infty$; finally we get

$$M_{mn}^{(l)} = \left(\alpha + \frac{\beta + 3}{2} \right) \int_0^R dx\, \chi_m \chi_n x^{\beta + 2} y^{\alpha + (\beta + 1)/2}$$

$$+ \frac{1}{\beta + 1} \sum_{l_2 = -l}^{l} \int_{C_2} d\varepsilon\, \mu^2(\varepsilon) \left(-\omega \frac{\partial F}{\partial \tilde{\varepsilon}} + l_2 \frac{\partial F}{\partial \mu} \right) \frac{(\chi_m)_{0 l_2}(\chi_n)_{0 l_2}}{v_1(\omega - l_2 v_2)}$$

$$- \sum_{l_1 = -\infty}^{\infty} \sum_{l_2 = -l}^{l} D_{ll_2} \iint d\varepsilon\, d\mu \frac{\partial}{\partial \mu} \left\{ \mu^{-\beta + 1} \left(-\omega \frac{\partial F}{\partial \tilde{\varepsilon}} + l_2 \frac{\partial F}{\partial \mu} \right) \right.$$

$$\left. \times \frac{(\chi_m)_{l_1 l_2}(\chi_n)_{l_1 l_2}}{v_1(\omega - l_1 v_1 - l_2 v_2)} \right\} \frac{1}{\beta + 1} \mu^{\beta + 1}. \tag{28}$$

Some biorthonormal systems. Consider the (dimensionless) Poisson equation for the lth harmonic:

$$\frac{d^2 \varphi}{dx^2} + \frac{2}{x} \frac{d\varphi}{dx} - \frac{l(l + 1)}{x^2} \varphi = \sigma, \tag{29}$$

where φ, σ, x are the dimensionless potential, density, and radius. On the boundary of the system $(x = R)$, the condition

$$\left(\frac{d\varphi}{dx} + \frac{l + 1}{a} \varphi \right) \Bigg|_{x = a} = 0 \tag{30}$$

must be satisfied. Make the substitution $y = \varphi(x)\sqrt{x}$:

$$\left(\frac{d^2}{dx^2} + \frac{1}{x}\frac{d}{dx} - \frac{(l+\frac{1}{2})^2}{x^2}\right)y = \sqrt{x}\sigma(x). \tag{31}$$

The operator on the right-hand side of (31) is the "Bessel" operator; therefore, it is natural to test the Bessel function as a probable candidate for the biorthogonal pair:

$$y_n(x) = J_{l+1/2}\left(\alpha_n \frac{x}{a}\right), \qquad n = 1, 2, 3, \ldots, \tag{32}$$

and, consequently,

$$\sigma_n(x) = -\frac{1}{\sqrt{x}}\left(\frac{\alpha_n}{a}\right)^2 J_{l+1/2}\left(\alpha_n \frac{x}{a}\right), \tag{33}$$

where α_n must be determined from the boundary condition (30). Indeed, we prove the biorthogonality of (32) and (33):

$$\int_0^a \varphi_m(x)\sigma_n(x)x^2 \, dx = -\alpha_n^2 \int_0^1 z\, dz J_{l+1/2}(\alpha_n z)J_{l+1/2}(\alpha_m z)$$

$$= -\alpha_n^2 \,\delta_{mn}\cdot \tfrac{1}{2}[J_{l+1/2}(\alpha_n)]^2.$$

Therefore, the desired biorthonormalized system is:

$$\chi_n^l(x) = \frac{\sqrt{2}}{\alpha_n}\frac{1}{|J_{l+1/2}(\alpha_n)|}\frac{1}{\sqrt{x}} J_{l+1/2}\left(\alpha_n \frac{x}{a}\right), \tag{34}$$

$$\sigma_n^{(l)}(x) = -\frac{\sqrt{2}}{a^2}\alpha_n \frac{1}{|J_{l+1/2}(\alpha_n)|}\cdot\frac{1}{\sqrt{x}} J_{l+1/2}\left(\alpha_n \frac{x}{a}\right). \tag{35}$$

One has to define α_n; from condition (30) we have

$$\frac{\partial\varphi_n}{\partial x} + \frac{(l+1)}{a}\varphi_n\bigg|_{x=a} = \frac{1}{a^{3/2}}\left[\alpha_n J'_{l+1/2}(\alpha_n) + (l+\tfrac{1}{2})J_{l+1/2}(\alpha_n)\right] = 0.$$

Since [42], for the Bessel function, the identity

$$zJ'_\nu(z) + \nu J_\nu(z) = zJ_{\nu-1}(z)$$

is valid, then we obtain the following condition for α_n:

$$J_{l-1/2}(\alpha_n) = 0. \tag{36}$$

The system (34), (35) is the particular case of a more general biorthonormal set, which can be obtained in the following manner. We make in Eq. (29) the substitution $x \to z = x^\nu$:

$$\frac{d^2\varphi}{dz^2} + \frac{(\nu+1)}{\nu}\frac{1}{z}\frac{d\varphi}{dz} - \frac{l(l+1)}{\nu^2}\frac{\varphi}{z^2} = \frac{\sigma}{\nu^2 z^{2(1-1/\nu)}}. \tag{37}$$

Substitute now $\varphi \to y = \varphi \cdot z^{1/2\nu}$:

$$\frac{d^2y}{dz^2} + \frac{1}{z}\frac{dy}{dz} - \frac{\mu^2}{z^2}y = \frac{\sigma}{\nu^2 z^{2-5/2\nu}}, \qquad \mu \equiv \frac{(l + \frac{1}{2})}{\nu}. \tag{38}$$

Thus, we must consider the system:

$$y_n(z) = J_\mu\left(\alpha_n \frac{z}{a^\nu}\right) = J_\mu\left[\alpha_n\left(\frac{x}{a}\right)^\nu\right], \tag{39}$$

or

$$\varphi_n(z) = \frac{1}{z^{1/2\nu}} J_\mu\left(\alpha_n \frac{z}{a^\nu}\right) = \frac{1}{\sqrt{x}} J_\mu\left[\alpha_n\left(\frac{x}{a}\right)^\nu\right], \tag{40}$$

and, consequently,

$$\sigma_n(z) = -\left(\nu\frac{\alpha_n}{a^\nu}\right)^2 z^{2-5/2\nu} J_\mu\left(\alpha_n \frac{z}{a^\nu}\right) = -\left(\nu\frac{\alpha_n}{a^\nu}\right)^2 x^{2\nu-5/2} J_\mu\left(\alpha_n \frac{x^\nu}{a^\nu}\right). \tag{41}$$

Since

$$\int_0^a x^2 \varphi_n(x)\sigma_m(x)\, dx = -\nu\alpha_n^2\, \delta_{nm}\tfrac{1}{2}[J_\mu(\alpha_n)]^2,$$

then the desired biorthogonal system is

$$\chi_n(x) = \sqrt{\frac{2}{\nu\alpha_n^2 x}}\, \frac{1}{|J_\mu(\alpha_n)|}\, J_\mu\left[\alpha_n\left(\frac{x}{\alpha}\right)^\nu\right],$$

$$\sigma_n(x) = -\sqrt{2}\,\nu^{3/2}\alpha_n \frac{1}{|J_\mu(\alpha_n)|}\, J_\mu\left[\alpha_n\left(\frac{x}{a}\right)^\nu\right] x^{2\nu-5/2}, \tag{42}$$

where α_n must be defined, as is easily proved, from the equation

$$J_{\mu-1}(\alpha_n) = 0 \qquad \left(\mu = \frac{l + \frac{1}{2}}{\nu}\right). \tag{43}$$

From (42), it is easy to see that $\varphi_n(x)$ expands in series of the form

$$\varphi_n(x) \infty x^l(1 + ax^{2\nu} + bx^{4\nu} + \cdots). \tag{44}$$

In the particular case, when $\nu = 1$, we return to the first biorthonormal system of (34), (35). According to (44), all $\varphi_n(x)$ for $\nu = 1$ are the analytical functions of x at $x \to 0$. Therefore, it is natural to use the system (34), (35) in the numerical investigation for the stability of the systems with fairly "good," smooth equilibrium distributions of potential, density, etc. (as, e.g., for the Idlis models, considered in detail in Section 6.1, Chapter III). At the same time, in the study of the stability of the models possessing some peculiarities at center (for example, a volumetric density softly divergent as $r \to 0$, etc.), the second biorthonormal system (42) may be of more use. Indeed, for the first Camm series, the unperturbed potential is expanded in the series

$$\varphi_0 \infty 1 + ax^{\beta+2} + bx^{2(\beta+2)} + \cdots. \tag{45}$$

Comparing (45) with (44), we see that, in this case, it is natural to choose $v = (\beta + 2)/2$. Just this biorthogonal system was used in obtaining the results described in Section 6.2, Chapter III.

§ 5 The Matrix Formulation of the Problem of Eigenoscillations of Collisionless Disk Systems

We shall give here, following Kalnajs [74[ad]], the derivation of the matrix equation, the solutions of which give the eigenfrequencies of oscillations of the disk stellar systems of a rather general form. We recall that a similar problem for the spherically symmetrical systems was considered in §4, Appendix [34[ad]]. The way of derivation of the matrix equation that is used therein (conventionally, it may be called the "Euler" method) fits, of course, also in the case of stellar disks. Below, however, this equation is derived by the method used in the original work [74[ad]]; as will be clear from further treatment, this method can logically be called the "Lagrange" one. It distinguishes itself by its somewhat greater generality: For example, it at once leads to reasonable expressions not only for smooth but also for a certain class of singular distribution functions (where in the "Euler" approach there may arise the formally divergent integrals that require regularization).

5.1 The Main Ideas of the Derivation of the Matrix Equation

If one imposes on the equilibrium state of the disk a small potential perturbation, this will lead to a change in the original orbits of stars, resulting in the surface density perturbation σ_{orb}. At the same time, the Poisson equation also defines the density perturbation σ_p, which is required to produce potential perturbation. For the eigenoscillations, the self-consistency condition

$$\sigma_{\text{orb}} = \sigma_p \tag{1}$$

which we have repeatedly used (for example, in the derivation of the integral equation in Section 4.4, Chapter V), must be satisfied. It is convenient to replace (1) by the other initial requirement—by the equality of the scalar products of the form

$$\iint g^* \sigma_{\text{orb}} r \, dr \, d\varphi = \iint g^* \sigma_p r \, dr \, d\varphi, \tag{2}$$

where g is the arbitrary potential,[2] r and φ are the polar coordinates. The

---

[2] Belonging to the perturbation space $\mathscr{P}$ with a finite potential energy introduced by Kalnajs in [74[ad]]; in this space, the perturbation potential energy may be used as a norm, and the energy of interaction between the two perturbations as a scalar product.

convenience is due to the fact that the "orbital" scalar product on the left-hand side of (2) is nothing other than the value of the potential g^*, taken in perturbed orbits, averaged with the weight $f_0(r, v)$ over the phase volume of the system, without its value in equilibrium orbits. Just such a value will be calculated below; this is the first idea of the considered method of obtaining the matrix equation.

The second idea concerns the use of the action-angle variables. They are ideally appropriate in this case, since there is no need here for a separate calculation of the perturbed surface density and the complicated transformation to the coordinates r, v of phase space.

The third and last idea consists of using biorthonormal systems of the density-potential functions (cf. §4, Appendix, and below, in Section 5.2). The perturbed potential and the surface density are written in the form of a sum of the terms of the biorthonormal system and the calculation of the "Poisson" scalar product [of the right-hand side of Eq. (2)] then becomes a trivial problem. This yields the matrix equation for the coefficients of these expansions.[3] From the condition of zero equality of the (infinitely dimensional) determinant of the system, the characteristic equation for eigenfrequencies ensues. Examples of biorthonormal systems of functions for gravitating disks are given in papers by Clutton–Brock [59^{ad}] and Kalnajs [73^{ad}]; however, they are too complicated to be given here.

The matrix formulation of the problem of eigenvalues of the stellar disk obtained by the described method is seemingly the most appropriate for numerical calculations.

5.2 "Lagrange" Derivation of the Matrix Equation

Thus, we introduce the action-angle variables: I_1 is the radial action, I_2 is the angular momentum of the particle, w_1 and w_2 are the corresponding angles (for a detailed description see, e.g., §2, Chapter XI).

To begin with, consider the calculation of the orbital scalar product. In the presence of the perturbed gravitational field, actions I_i and angles w_i will get the increments ΔI_i, Δw_i, and (see §2, Chapter XI) in the first order of magnitude in the perturbations:

$$\Delta I_i = \frac{\partial \chi}{\partial w_i}, \qquad \Delta w_i = -\frac{\partial \chi}{\partial I_i}, \tag{3}$$

where the generating function

$$\chi = -\frac{1}{4\pi^2} \sum_{lm} \frac{\Phi_{lm}(I_i) \exp[i(lw_1 + mw_2 - \omega t)]}{i(l_1\Omega_1 + m\Omega_2 - \omega)}, \tag{4}$$

---

[3] If the biorthonormal system is countable. If, however, this system is nonenumerable, the integral equation ensues.

and $\Phi_{l,m}$ are the Fourier coefficients of the expansion of the perturbed potential $\Phi_1(I_i, w_i)$:

$$\Phi_{lm}(I_i) = \int_0^{2\pi} \int_0^{2\pi} \Phi_1(I_i, w_i) \exp[-i(lw_1 + mw_2)]. \tag{5}$$

The potential g in the star orbit is $g(I_i + \Delta I_i, w_i + \Delta w_i)$, where I_i and w_i are the stationary actions and angles. Expanding g in the Taylor series near the unperturbed orbit and using Eq. (3), we find

$$g(I_i + \Delta I_i, w_i + \Delta w_i, t) = g(I_i, w_i, t) + \sum_{i=1}^{2} \frac{\partial g}{\partial I_i} \frac{\partial \chi}{\partial w_i} - \frac{\partial g}{\partial w_i} \frac{\partial \chi}{\partial I_i} + \cdots. \tag{6}$$

Let us integrate (6) over phase space: $\iiint dI_1\, dI_2\, dw_1\, dw_2\, F(I_i)$. To begin with, we perform integration over the angular variables $dw_1\, dw_2$; on the right-hand side we obtain the expression

$$\int_0^{2\pi} \int_0^{2\pi} g(I_i, w_i)\, dw_1\, dw_2 + \sum_{i=1}^{2} \int_0^{2\pi} \int_0^{2\pi} \left[\frac{\partial g}{\partial I_i} \frac{\partial \chi}{\partial w_i} - \frac{\partial g}{\partial w_i} \frac{\partial \chi}{\partial I_i} \right] dw_1\, dw_2. \tag{7}$$

The first term is the potential energy of the unperturbed stars with the given I_1 and I_2, in the field Φ_1. We omit it (see Section 5.1) and integrate the last term in (7) by parts and combine with the middle term; as a result, we obtain

$$\sum_{i=1}^{2} \frac{\partial}{\partial I_i} \int_0^{2\pi} \int_0^{2\pi} g \frac{\partial \chi}{\partial w_i}\, dw_1\, dw_2. \tag{8}$$

The boundary terms are cancelled due to periodicity of all values with respect to w_i with the period 2π. We now integrate (8) over I_1 and I_2 with the weight $F(I_i)$. Substituting instead of the value χ its expression in (4) and integrating over the angular variables, we transform expression (8) for the orbital scalar product to the final form

$$\frac{1}{4\pi^2} \sum_{l,m} \iint F(I_i) \left\{ \left(l \frac{\partial}{\partial I_1} + m \frac{\partial}{\partial I_2} \right) \left[\frac{g_{lm}(I_i)\Phi_{lm}(I_i)}{l\Omega_1 + m\Omega_2 - \omega} \right] \right\} dI_1\, dI_2, \tag{9}$$

where g_{lm} are the Fourier coefficients of the function g, defined as in (5). Note that since the orbital scalar product (9) is calculated by the Lagrange method, it automatically involves all the contributions which may arise due to the displacement of the system boundaries and, besides, is suitable for any $F(I_i)$ functions with integrated singularities.

We turn to the calculation of the Poisson scalar product. Consider any biorthonormal system of surface densities $\{\sigma_i\}$ and the potentials $\{\Phi_i\}$

corresponding to them, such that the scalar product

$$-\frac{1}{2\pi G} \iint \Phi_i^* \sigma_j r\, dr\, d\varphi = \delta_{ij} = \begin{cases} 1, & i = j, \\ 0, & i \neq j. \end{cases} \tag{10}$$

We assume that the system is complete, i.e., assume that any potential $\Phi_1{}^4$ can be expanded in a series with respect to Φ_i:

$$\Phi_1 = \sum_{j=0}^{\infty} a_j \Phi_j. \tag{11}$$

The surface density which corresponds to Φ_1 is

$$\sigma_p = \sum_{j=0}^{\infty} a_j \sigma_j, \tag{12}$$

and the coefficients a_j are calculated by means of the biorthonormalization condition (10):

$$a_j = -\frac{1}{2\pi G} \iint \Phi_j^* \sigma_p r\, dr\, d\varphi = -\frac{1}{2\pi G} \iint \Phi_1^* \sigma_j r\, dr\, d\varphi. \tag{13}$$

Equality (2) must be satisfied for each potential of the biorthonormal system. Substituting into (2) the expansion (11) for Φ_1 and (12) for σ_p and assuming in (9) that $g = \Phi_i$, $i = 0, 1, 2, \ldots$, we arrive at an infinite system of linear equations which must be satisfied by the coefficients a_i:

$$\sum_{j=0}^{\infty} [M_{ij}(\omega) - \delta_{ij}] a_j = 0, \tag{14}$$

where the matrix elements

$$M_{ij}(\omega) = \frac{1}{8\pi^3 G} \sum_{l,m} \iint F(I_1, I_2)$$
$$\times \left\{ \left(l_1 \frac{\partial}{\partial I_1} + m \frac{\partial}{\partial I_2} \right) \left[\frac{(\Phi_i)_{lm}^*(\Phi_j)_{lm}}{l\Omega_1 + m\Omega_2 - \omega} \right] \right\} dI_1\, dI_2. \tag{15}$$

The characteristic equation for the oscillation eigenfrequencies ω is

$$\det \| M_{ij}(\omega) - \delta_{ij} \| = 0.$$

For a disk system in the linear approximation in question, the azimuthal number m may be considered as fixed, in investigating separately the stability of radial ($m = 0$), "barlike" ($m = 2$) and other perturbations. To calculate on computer, it is useful to turn, in expressions for matrix elements (15),

---

[4] Belonging to the space $\mathscr{P}$ mentioned above.

from the action variables I_1 and I_2 to the energy E and the angular momentum L. The derivatives are transformed as

$$\frac{\partial}{\partial I_1} = \Omega_1 \frac{\partial}{\partial E}, \qquad \frac{\partial}{\partial I_2} = \Omega_2 \frac{\partial}{\partial E} + \frac{\partial}{\partial L}, \qquad dI_1\, dI_2 = \frac{dE\, dL}{\Omega_1(E, L)}. \qquad (16)$$

§6 Derivation of the Dispersion Equation for Perturbations of the Three-Axial Freeman Ellipsoid

Restricting ourselves to the largest-scale oscillations of the ellipsoid–ellipsoid type retaining the direction of the main axis z, one may assume that

$$\Phi_1 = \alpha_1 x^2 - 2i\alpha_2 xy + \alpha_3 y^2 + \alpha_4 z^2. \qquad (1)$$

Then from Eq. (9) (§2, Chapter IV) we obtain for the B function the following expression,

$$B = \frac{2i\alpha_1}{\omega}\left(x - \frac{2\Omega v_y}{\beta^2}\right) + \frac{1}{\omega^2 - \beta^2}(-4\alpha_1\Omega y - 2i\alpha_2 v_y)$$
$$+ \frac{2\omega}{\omega^2 - \beta^2}\left(\frac{4\alpha_1\Omega v_y}{\beta^2} - 2i\alpha_2 y\right), \qquad (2)$$

and from Eq. (8) we find the function C (which in this case is a constant)

$$C = -\frac{2\alpha_1}{\omega^2} + \left[\frac{2\alpha_1}{\omega^2}\frac{4\Omega^2}{\beta^2} + \frac{4\Omega\alpha_2}{\omega(\omega^2 - \beta^2)} - \frac{8\alpha_1\Omega^2}{\beta^2(\omega^2 - \beta^2)}\right]. \qquad (3)$$

The cumbersome equation for the function A ensuing from (7) has a solution which we shall represent in the form:

$$A = D_1 x^2 + 2iD_2 xy + D_3 y^2 + D_4 z^2 + D_5 c_y^2 + D_6 v_z^2 + R, \qquad (4)$$

where

$$D_1 = d_{11}\alpha_1 + d_{12}\alpha_2 + d_{13}\alpha_3,$$

$$D_2 = \frac{1}{b^2}d_{21}\alpha_1 - \frac{1}{b^2}d_{22}\alpha_2 + \frac{1}{b^2}d_{23}\alpha_3,$$

$$D_3 = \frac{1}{b^2}d_{31}\alpha_1 + \frac{1}{b^2}d_{23}\alpha_2 + \frac{1}{b^2}d_{33}\alpha_3,$$

$$D_4 = \frac{1}{c^2}d_{44}\alpha_4, \qquad (5)$$

$$D_5 = \frac{1}{b^2\mu^2}d_{51}\alpha_1 + \frac{1}{b^2\mu^2}d_{52}\alpha_2 + \frac{1}{b^2\mu^2}d_{53}\alpha_3,$$

$$D_6 = \frac{d_{64}\alpha_4}{\gamma^2 c^2}.$$

R contains all the terms, nonessential further on, that contain odd degrees v_z and c_y. The coefficients d_{ik} are given by the following expressions:

$$d_{11} = \frac{4\mu^2(4\Omega^2 - 1)}{\omega^2} + \frac{16\Omega^2(1 - 2\mu^2) - 4(1 - \mu^2)}{\omega^2 - 1} - \frac{16\Omega^2(1 - \mu^2)}{\omega^2 - 4}$$

$$+ \frac{8(1 - \mu^2)(4\Omega^2 - 1)}{(\omega^2 - 1)^2},$$

$$d_{12} = \frac{1}{\omega\Omega}\left[\frac{4(1 - \mu^2) + 8\Omega^2(4\mu^2 - 3)}{\omega^2 - 1} + \frac{32\Omega^2(1 - \mu^2)}{\omega^2 - 4}\right.$$

$$\left. - \frac{4(1 - \mu^2)(4\Omega^2 - 1)}{(\omega^2 - 1)^2}\right],$$

$$d_{13} = 4(1 - \mu^2)\left(\frac{1}{\omega^2 - 4} - \frac{1}{\omega^2 - 1}\right), \qquad d_{21} = -d_{12},$$

$$d_{22} = \frac{1}{\Omega^2}\left[\frac{4\Omega^2(3 - 4\mu^2) - (1 - \mu^2)}{\omega^2 - 1} - \frac{16\Omega^2(1 - \mu^2)}{\omega^2 - 4}\right.$$

$$\left. + \frac{2(1 - \mu^2)(4\Omega^2 - 1)}{(\omega^2 - 1)^2}\right],$$

$$d_{23} = \frac{2(1 - \mu^2)}{\Omega}\omega\left(\frac{1}{\omega^2 - 4} - \frac{1}{\omega^2 - 1}\right), \qquad (6)$$

$$d_{31} = 16\Omega^2\left(\frac{1}{\omega^2 - 4} - \frac{1}{\omega^2 - 1}\right),$$

$$d_{32} = -\frac{8\Omega}{\omega}\left(\frac{4}{\omega^2 - 4} - \frac{1}{\omega^2 - 1}\right), \qquad d_{33} = -\frac{4}{\omega^2 - 4},$$

$$d_{51} = \frac{4(1 - \mu^2)(4\Omega^2 - 1)}{\omega^2} - \frac{16\Omega^2(1 - 2\mu^2) - 4(1 - \mu^2)}{\omega^2 - 1}$$

$$- \frac{16\Omega^2\mu^2}{\omega^2 - 4} - \frac{8(1 - \mu^2)(4\Omega^2 - 1)}{(\omega^2 - 1)^2},$$

$$d_{52} = \frac{1}{\omega\Omega}\left[-\frac{4(1 - \mu^2) + 8\Omega^2(4\mu^2 - 3)}{\omega^2 - 1} + \frac{32\Omega^2\mu^2}{\omega^2 - 4}\right.$$

$$\left. + \frac{4(1 - \mu^2)(4\Omega^2 - 1)}{(\omega^2 - 1)^2}\right],$$

$$d_{53} = 4\left(\frac{\mu^2}{\omega^2 - 4} + \frac{1 - \mu^2}{\omega^2 - 1}\right), \qquad d_{64} = -d_{44}, \qquad d_{44} = -\frac{4}{\omega^2 - 4\gamma^2}.$$

Introduce the corresponding notations also into the expression for C:

$$C = -\tfrac{1}{2}(C_1\alpha_1 + C_2\alpha_2), \qquad C_1 = -\frac{4(4\Omega^2 - 1)}{\omega^2} + \frac{16\Omega^2}{\omega^2 - 1},$$

$$C_2 = -\frac{8}{\Omega\omega(\omega^2 - 1)}. \tag{7}$$

In (6) and (7) it is assumed that $\beta = 1$ and $a = 1$. The perturbed density ensues by formula (12), §2, Chapter IV:

$$\varepsilon\rho_1 = -\tfrac{1}{2}\varepsilon\rho_0(b^2\mu^2 D_5 + \gamma^2 c^2 D_6) + \varepsilon\rho_0 C. \tag{8}$$

The equation for the perturbed boundary of the ellipsoid is

$$\frac{x^2}{a^2} + \frac{y^2}{b^2} + \frac{z^2}{c^2} = 1 - \varepsilon(D_1 x^2 + 2iD_2 xy + D_3 y^2 + D_4 z^2). \tag{9}$$

In this case, the potential perturbation Φ_1 can be calculated by writing an exact potential of a homogeneous ellipsoid with the density $\rho_0 + \rho_1$ in the new boundaries of (9) and by subtracting from it the unperturbed potential Φ_0.

We turn the plane (x, y) by an angle δ, so that the new coordinate axes (x', y') are the main axes of the ellipsoid (9):

$$x = x' + \delta y', \qquad y = y' - \delta x', \qquad \delta = i\varepsilon D_2 \frac{a^2 b^2}{a^2 - b^2}. \tag{10}$$

The equation of perturbed ellipsoid in the coordinates x', y', $z' = z$ takes on the form

$$\frac{(x')^2}{(a')^2} + \frac{(y')^2}{(b')^2} + \frac{(z')^2}{(c')^2} = 1,$$

$$(a')^2 = a^2(1 - \varepsilon a^2 D_1), \qquad (b')^2 = b^2(1 - \varepsilon b^2 D_3), \tag{11}$$

$$(c')^2 = c^2(1 - \varepsilon c^2 D_4).$$

The potential of the rotated ellipsoid is

$$\Phi(x', y', z') = \pi G(\rho_0 + \rho_1)[\alpha'_0(x')^2 + \beta'_0(y')^2 + \gamma'_0 z^2] + \text{const}, \tag{12}$$

$$\alpha'_0 = F\left(\frac{(a')^2}{(b')^2}, \frac{(a')^2}{(c')^2}\right), \qquad \beta'_0 = F\left(\frac{(b')^2}{(a')^2}, \frac{(b')^2}{(c')^2}\right), \qquad \gamma'_0 = F\left(\frac{(c')^2}{(a')^2}, \frac{(c')^2}{(b')^2}\right). \tag{13}$$

Expansions α'_0, β'_0, γ'_0 over ε have the form

$$\alpha'_0 = \alpha_0 + F_{a_1}(a_1, a_3)\varepsilon a_2 + F_{a_3}(a_1, a_3)\varepsilon a_4 \equiv \alpha_0 + \varepsilon\alpha_{(1)},$$

$$\beta'_0 = \beta_0 + F_{b_1}(b_1, b_3)\varepsilon b_2 + F_{b_3}(b_1, b_3)\varepsilon b_4 \equiv \beta_0 + \varepsilon\beta_{(1)}, \tag{14}$$

$$\gamma'_0 = \gamma_0 + F_{c_1}(c_1, c_3)\varepsilon c_2 + F_{c_3}(c_1, c_3)\varepsilon c_4 \equiv \gamma_0 + \varepsilon\gamma_{(1)}.$$

Here the following notations are introduced:

$$a_1 = \frac{a^2}{b^2}, \quad a_2 = -\frac{a^2}{b^2}(a^2 D_1 - b^2 D_3), \quad a_3 = \frac{a^2}{c^2}, \quad a_4 = -\frac{a^2}{c^2}(a^2 D_1 - c^2 D_4),$$

$$b_1 = \frac{b^2}{a^2}, \quad b_2 = \frac{b^2}{a^2}(a^2 D_1 - c^2 D_4), \quad b_3 = \frac{b^2}{c^2}, \quad b_4 = -\frac{b^2}{c^2}(b^2 D_3 - c^2 D_4),$$

$$c_1 = \frac{c^2}{a^2}, \quad c_2 = \frac{c^2}{a^2}(a^2 D_1 - c^2 D_4), \quad c_3 = \frac{c^2}{b^2}, \quad c_4 = \frac{c^2}{b^2}(b^2 D_3 - c^2 D_4),$$

$$F_u(u, v) \equiv \frac{\partial F}{\partial u}, \quad F_v(u, v) \equiv \frac{\partial F}{\partial v}. \tag{15}$$

Proceeding in (12) again to the coordinates (x, y): $x' = x - \delta y, y' = y + \delta x$, we obtain Φ_1:

$$\varepsilon \Phi_1 = \varepsilon \pi G (\alpha_0 x^2 + \beta_0 y^2 + \gamma_0 z^2) \rho_1 + \varepsilon \pi G \rho_0$$
$$\times [\alpha_{(1)} x^2 + \beta_{(1)} y^2 + \gamma_{(1)} z^2 + 2\delta x y (\beta_0 - \alpha_0)]. \tag{16}$$

A comparison with the initial expression in (1) for Φ_1 gives the following uniform system of algebraic equations relative to $\alpha_1, \alpha_2, \alpha_3$, and α_4:

$$\alpha_1 = \pi G \rho_1 + \pi G \rho_0 \alpha_{(1)}, \quad -i\alpha_2 = \pi G \rho_0 \delta (\beta_0 - \alpha_0),$$
$$\alpha_3 = \pi G \beta_0 \rho_1 + \pi G \rho_0 \beta_{(1)}, \quad \alpha_4 = \pi G \gamma_0 \beta_1 + \pi G \rho_0 \gamma_{(1)}. \tag{17}$$

The required dispersion equation ensues from (17) by equating the determinant of the system to zero:

$$f(\omega, b, c) \equiv \det \|r_{ik}\| = 0, \tag{18}$$

where the elements of the determinant r_{ik} are

$$r_{11} = \frac{2}{\rho} + \alpha_0 (d_{51} + c_1) - 2A_1, \quad r_{12} = \alpha_0 (d_{52} + c_2) - 2A_2,$$

$$r_{13} = \alpha_0 d_{53} - 2A_3, \quad r_{14} = \alpha_0 d_{64} - 2A_4,$$

$$r_{21} = d_{21}, \quad r_{22} = \frac{2(1 - b^2)}{\rho(\beta_0 - \alpha_0)} - d_{22}, \quad r_{23} = d_{23}, \quad r_{24} = 0,$$

$$r_{31} = \beta_0 (d_{51} + c_1) - 2B_1, \quad r_{32} = \beta_0 (d_{52} + c_2) - 2B_2, \tag{19}$$

$$r_{33} = \frac{2}{\rho} + \beta_0 d_{53} - 2B_3, \quad r_{34} = \beta_0 d_{64} - 2B_4,$$

$$r_{41} = \gamma_0 (d_{51} + c_1) - 2C_1, \quad r_{42} = \gamma_0 (d_{52} + c_2) - 2C_2,$$

$$r_{43} = \gamma_0 d_{53} - 2C_3, \quad r_{44} = \frac{2}{\rho} + \gamma_0 d_{64} - 2C_4.$$

In (19), the following notations are used

$$A_1 = F_{a_1}(a_1, a_3)\frac{1}{b^2}(d_{31} - d_{11}) - F_{a_3}(a_1, a_3)\frac{1}{c^2}d_{11},$$

$$A_2 = F_{a_1}(a_1, a_3)\frac{1}{b^2}(d_{32} - d_{12}) - F_{a_3}(a_1, a_3)\frac{1}{c^2}d_{12},$$

$$A_3 = F_{a_1}(a_1, a_3)\frac{1}{b^2}(d_{33} - d_{13}) - F_{a_3}(a_1, a_3)\frac{1}{c^2}d_{13},$$

$$A_4 = F_{a_3}(a_1, a_3)\frac{1}{c^2}d_{44},$$

$$B_1 = F_{b_1}(b_1, b_3)b^2(d_{11} - d_{31}) - F_{b_3}(b_1, b_3)\frac{b^2}{c^2}d_{31},$$

$$B_2 = F_{b_1}(b_1, b_3)b^2(d_{12} - d_{32}) - F_{b_3}(b_1, b_3)\frac{b^2}{c^2}d_{32},$$

$$B_3 = F_{b_1}(b_1, b_3)b^2(d_{13} - d_{33}) - F_{b_3}(b_1, b_3)\frac{b^2}{c^2}d_{33},$$

$$B_4 = F_{b_3}(b_1, b_3)\frac{b^2}{c^2}d_{44},$$

$$C_1 = F_{c_1}(c_1, c_3)c^2 d_{11} + F_{c_3}(c_1, c_3)\frac{c^2}{b^2}d_{31},$$

$$C_2 = F_{c_1}(c_1, c_3)c^2 d_{12} + F_{c_3}(c_1, c_3)\frac{c^2}{b^2}d_{32},$$

$$C_3 = F_{c_1}(c_1, c_3)c^2 d_{13} + F_{c_3}(c_1, c_3)\frac{c^2}{b^2}d_{33},$$

$$C_4 = -F_{c_1}(c_1, c_3)c^2 d_{44} - F_{c_3}(c_1, c_3)\frac{c^2}{b^2}d_{44},$$

$$\rho \equiv (6\alpha_0 + 2\beta_0)^{-1}, \qquad \Omega^2 = 2\rho\alpha_0, \qquad \gamma^2 = 2\rho\gamma_0,$$

$$\mu^2 = 1 - 8\rho b^2 \alpha_0.$$

Dispersion equation (18) has been investigated numerically. First of all, the values of the determinant (18) were calculated on the real axis of the complex plane ω for different values of the semiaxes b, c. In particular, in this way, we determined the real zeros of this determinant. As their number, at some values of b and c, decreased by the value multiple of 2, this meant the appearance of complex pairs of the roots. In this case, the integrals over the contour γ lying wholly in the upper semiplane were calculated (since we were interested only in unstable solutions):

$$\frac{1}{2\pi i} \oint_\gamma \frac{f'(\omega)}{f(\omega)} d\omega = N, \qquad \frac{1}{2\pi i} \oint_\gamma \frac{f'(\omega)}{f(\omega)} \omega \, d\omega = \sum_l n_l \omega_l, \qquad (20)$$

where N is the number of zeros inside γ, taking into account their multiplicity n_l.[5] If only one root ($N = 1$) appeared in the contour γ, then the second formula (20) gave its value.

The results are given in Fig. 27.

§7 WKB Solutions of the Poisson Equation Taking into Account the Preexponential Terms and Solution of the Kinetic Equation in the Postepicyclic Approximation

In this section, we deal with, following Shu [325], the solutions of the system of Vlasov's equations for gravitating disks within an accuracy of two orders of magnitude with respect to the parameters $1/kr$ and $\varepsilon = c_r/r\Omega$, which are assumed to be small.

7.1 The Relation Between the Potential and the Surface Density

Let us obtain the asymptotic solution of the Poisson equation for the disk with an accuracy of the first two orders of magnitude with respect to $(kr)^{-1}$. Let the surface density be $\sigma_1 = \sigma(r) \exp[-i(\omega t - m\varphi)]$, let the potential similarly be $\Phi_1 = \Phi(r, z) \exp[-i(\omega t - m\varphi)]$, and in the plane of the disk let $\Phi(r, z = 0) = A(r) \exp[i\psi(r)]$. By integrating the Poisson equation

$$\left[\frac{\partial^2}{\partial r^2} + \frac{1}{r} \frac{\partial}{\partial r} + \frac{\partial^2}{\partial z^2} - \frac{m^2}{r^2} \right] \Phi(r, z) = 4\pi G\sigma(r)\, \delta(z) \qquad (1)$$

over z between $(0 - \varepsilon)$ and $(0 + \varepsilon)$ and then presetting $\varepsilon \to 0$, we obtain the familiar boundary condition

$$\sigma(r) = \frac{1}{4\pi G} \left[\frac{\partial \Phi}{\partial z} (r, z) \right] \Bigg|_{z=-0}^{z=+0}. \qquad (2)$$

The function $\Phi(r, z)$ may be dependent only on $|z|$ as a consequence of the symmetry of the problem with respect to the reflection $z \to -z$. It is natural to introduce the transformation $\Phi \to W$ customary for the cylindrical geometry:

$$W(r, z) = r^{1/2}\Phi(r, z). \qquad (3)$$

---

[5] From (6), it is evident that all poles of $f(\omega, b, c)$ lie on the real ω-axis.

Then Eqs. (1) and (2) are rewritten in the form

$$\frac{\partial^2 W}{\partial r^2} + \frac{\partial^2 W}{\partial |z|^2} - \frac{m^2 - \frac{1}{4}}{r^2} W = 0, \qquad \text{for } |z| > 0, \tag{4}$$

$$\sigma(r) = \frac{r^{-1/2}}{2\pi G} \left[\frac{\partial W}{\partial |z|} (r, |z|) \right] \Bigg|_{|z|=0}. \tag{5}$$

These equations must be solved with the boundary conditions

$$W(r, 0) = r^{1/2} A(r) e^{i\psi(r)}, \tag{6}$$

$$W(r, z) \to 0 \quad \text{as } |z| \to \infty. \tag{7}$$

We shall study here only the part of the disk system, in which $|kr| \gg 1$. In it, the "term of curvature" $\sim [(m^2 - \frac{1}{4})/r^2]W$ has a smallness of the order of $(kr)^{-2}$ in comparison with the first two terms, and therefore it may be omitted. Thus, within an accuracy of the second order of magnitude of the WKB approximation, the exact Poisson equation is coincident with the following:

$$\frac{\partial^2 W}{\partial r^2} + \frac{\partial^2 W}{\partial |z|^2} = 0, \qquad |z| \geq 0, \tag{8}$$

[the boundary conditions are, as before, (6) and (7)]. Continuation of the solution to (6) from the plane $z = 0$ in $|z| > 0$ is trivial, if one makes use of the theory of the functions of a complex variable. To continue the solution, it is necessary simply to make a substitution: $r \to r \pm i|z|$ in $W(r, 0)$. The correct selection of the sign is dictated by the boundary conditions at infinity $(z \to \pm \infty)$. In accordance with the approximation adopted, we can change the condition in (7) for the restraint that, for small $|z|$, the value $W(r, |z|)$ decreases from its value $W(r, 0)$ at $|z| = 0$. If $r + ip|z|$ with $p = \pm 1$ provides a correct selection of the sign and if $\psi(r)$ is a rapidly varying function, this condition, taking into account (6), may be presented in the form

$$\text{Im}[\psi(r + ip|z|)] > 0, \qquad |z| > 0. \tag{9}$$

Expansion for small $|z|$ yields

$$p = \pm 1 = \text{sgn } \psi'(r) \equiv \text{sgn } k. \tag{10}$$

Thus, one may obtain

$$W(r, |z|) = (r + ip|z|)^{1/2} A(r + ip|z|) e^{i\psi(r + ip|z|)}, \tag{11}$$

$$\sigma(r) = -\frac{|k| \Phi(r)}{2\pi G} \left\{ 1 - \frac{i}{kr} \frac{d \ln[r^{1/2} A(r)]}{d \ln r} \right\}. \tag{12}$$

7.2 Calculations of the Response of a Stellar Disk to an Imposed Perturbation of the Potential

We start from the general expression derived in Section 4.2, Chapter V, for the perturbation of the distribution function of the stars,

$$f_1 = \left(\frac{\partial f_0}{\partial E}\Phi_1 + \chi\right)e^{-i(\omega t - m\varphi)}, \tag{1}$$

where

$$\chi(r, E, L) = -\frac{\omega\,\partial f_0/\partial E + m\,\partial f_0/\partial L}{2\sin(\omega\tau_{12} - m\varphi_{12})}\int_{-\tau_{12}}^{\tau_{12}}\Phi_1(r'(\tau))e^{-i[\omega\tau - m\varphi'(\tau)]}\,d\tau. \tag{2}$$

In these formulae, $f_0(E, L)$ is the unperturbed distribution function, $2\tau_{12}(E, L)$ is the period of radial oscillations of the star with the energy E and the angular momentum L, and $2\varphi_{12}(E, L)$ is the azimuthal angular displacement of the star for the time $2\tau_{12}$. The functions $r'(\tau)$ and $\varphi'(\tau)$ describe the unperturbed orbit of the star, such that its position at the times $\tau = \pm\tau_{12}$ is $(r, \pm\varphi_{12})$:

$$\frac{dr'}{dt} = \Pi_0(r', E, L), \qquad r' = r \qquad \text{at } \tau = \pm\tau_{12}, \tag{3}$$

$$\frac{d\varphi'}{dt} = \frac{L}{r'^2}, \qquad \varphi' = \pm\varphi_{12} \quad \text{at } \tau = \pm\tau_{12}, \tag{4}$$

where the function $\Pi_0(r, E, L)$ in Eq. (3) is the radial velocity of the star: $\Pi_0 = \{2[E - \Phi_0(r)] - L^2/r^2\}^{1/2}$.

Assume that, in the unperturbed state, the stars satisfy the "modified Schwarzschild distribution" [325]:

$$f_0(E, L) = \begin{cases} P_0(r_0)\exp[-\mathscr{E}/c_0^2(r_0)], & \mathscr{E} < -E(r_0), \\ 0 & \mathscr{E} > -E(r_0), \end{cases} \tag{5}$$

where the functions $P_0(r_0)$ and $c_0(r_0)$ may be expressed through the surface density and the dispersion of radial velocities, while the "epicyclic" integrals r_0 and $\mathscr{E}$ are defined as the functions of E and L from the equations

$$r_0^2\Omega(r_0) = L, \qquad \mathscr{E} = E - E_c(r_0),$$

$$E_c = \tfrac{1}{2}r_0^2\Omega^2(r_0) + \Phi_0(r_0), \qquad r_0\Omega^2(r_0) = \frac{\partial\Phi_0}{\partial r_0}. \tag{6}$$

Equation (5) may be considered as the parametric representation of the unperturbed distribution function through (r, c_r, c_φ), if the peculiar velocities are determined by the natural relations:

$$c_r = \{2[\mathscr{E} - \varepsilon_c(r_0, r)]\}^{1/2},$$

$$c_\varphi = \frac{r_0^2}{r}\Omega(r_0) - r\Omega(r), \tag{7}$$

and the $\varepsilon_c(r_0, r)$ function is given by the equation

$$\varepsilon_c(r_0, r) = \Phi_0(r) - \Phi_0(r_0) + \frac{r_0^2 \Omega^2(r_0)}{2} \left(\frac{r_0^2}{r^2} - 1 \right),\qquad (8)$$

so that, for example, $\varepsilon_c(r_0, r_0) = 0$. Under the transformation $(E, L) \to (\mathscr{E}, r_0)$ the partial derivatives are transformed thus:

$$\frac{\partial}{\partial E} = \frac{\partial}{\partial \mathscr{E}}, \qquad \frac{\partial}{\partial L} = \frac{2\Omega(r_0)}{r_0 \varkappa^2(r_0)} \frac{\partial}{\partial r_0} - \Omega(r_0) \frac{\partial}{\partial \mathscr{E}}. \qquad (9)$$

Assume now that the second dimensionless parameter $\varepsilon = c_0(r_0)/r_0 \varkappa(r_0)$ is also small: $\varepsilon \ll 1$. This assumption seems to be well satisfied in the disk parts of most of the spiral galaxies. With an accuracy of two orders with respect to ε, one of the terms of (2) has the form, with due regard for (5) and (9)

$$-\left(\omega \frac{\partial f_0}{\partial E} + m \frac{\partial f_0}{\partial L} \right) = v(r_0) \frac{\varkappa(r_0)}{c^2(r_0)} f_0(\varepsilon, r_0). \qquad (10)$$

In the same approximation, the velocity dispersion $c_r(r_0)$ is $c_0(r_0)$, while the unperturbed distribution (5) is reduced to the following:

$$f_0 = \frac{2\Omega(r_0)}{\varkappa(r_0)} \cdot \frac{\sigma_0(r_0)}{2\pi c_r^2(r_0)} \exp\left[-\frac{\tilde{\varepsilon}^2}{2\varepsilon^2(r_0)} \right], \qquad (11)$$

where the "eccentricity" $\tilde{\varepsilon} = \sqrt{2\mathscr{E}}/r_0 \varkappa(r_0)$ is introduced. Assume further that the parameters ε and $(kr)^{-1}$ have the same order of smallness: $\varepsilon \sim (kr)^{-1}$.

To calculate the response of the surface density up to the second order of magnitude in this asymptotical approximation, we require that all formulae be true within an accuracy of the second order with respect to ε or $|kr|^{-1}$. An exception is the relation providing the radial orbit $r' = r'(\tau)$; for it there is required an accuracy of the third order of magnitude with respect to ε (since the two orders with respect to ε represent the usual epicyclic approximation). This allows one to calculate $\Phi_1(r'(\tau))$ (the potential on the unperturbed orbit) within an accuracy of the second order. For small ε, the most part of the stars is contained in the part of the phase space, in which $\tilde{\varepsilon} \leq \varepsilon(r_0)$. Therefore, in the determination of the stellar orbit from Eqs. (3) and (4), it is necessary to derive expressions true only up to a corresponding order with respect to $\tilde{\varepsilon}$.

Symmetrical expansion yielding stellar orbits of this kind, can be obtained by using the parametric representation [325]. In this representation, one radial period $2\tau_{12}$ corresponds to the change by 2π of the radial phase coordinate θ, defined by the relation

$$\Pi_0(r', E, L) = r_0 \varkappa(r_0)\tilde{\varepsilon} \sin \theta. \qquad (12)$$

If another dimensionless time $s = \varkappa(r_0)\tau$ is introduced, then the parametric representation of the orbits required may be obtained by expanding

Eqs. (3) and (4), which leads to the relations

$$s - s_0 = \theta - 2B_2(r_0)\tilde{\varepsilon} \sin \theta, \tag{13}$$

$$\frac{r_0}{r'} = 1 + \tilde{\varepsilon} \cos \theta + A_2(r_0)\tilde{\varepsilon}^2 \cos^2 \theta, \tag{14}$$

$$\varphi' - \varphi_0 = \frac{\Omega(r_0)}{\varkappa(r_0)} [\theta + 2A_2(r_0)\tilde{\varepsilon} \sin \theta], \tag{15}$$

where $A_2(r_0) = \frac{1}{2}[1 + \frac{2}{3} d \ln \varkappa/d \ln r_0]$, and $B_2(r_0) = 1 - A_2(r_0)$, while the phase constants s_0 and φ_0 will be chosen so that $s = -\pi$ and $\varphi' = -\pi\Omega(r_0)/\varkappa(r_0)$, when $r' = r$ [see (3) and (4)]. Within an accuracy of the second order with respect to $\tilde{\varepsilon}$, from Eqs. (13)–(15) it follows that s and $\varphi'\varkappa(r_0)/\Omega(r_0)$ change by 2π when θ changes by 2π; therefore, the radial period of oscillations and the change in azimuth $2\varphi_{12}$ for that period are

$$2\tau_{12} = \frac{2\pi}{\varkappa(r_0)}, \qquad 2\varphi_{12} = \frac{2\pi\Omega(r_0)}{\varkappa(r_0)}. \tag{16}$$

It is convenient to express all the values through (r, ξ, η, s), where

$$\xi = \tilde{\varepsilon} \sin s_0, \qquad \eta = \tilde{\varepsilon} \cos s_0.$$

In particular, eliminating θ from Eqs. (13)–(15), one may represent the required orbits in the form

$$r' = r(1 - R_1 - R_2), \qquad \varphi' = \frac{\Omega(r_0)}{\varkappa(r_0)} \left[s - 2\left(\frac{\partial R_1}{\partial s} + \xi\right) \right], \tag{17}$$

where

$$R_1 = \eta(1 + \cos s) + \xi \sin s, \qquad R_2 = B_2(r)R_1^2 - [1 + 2B_2(r)]\eta R_1. \tag{18}$$

We shall use the variables (ξ, η) instead of (v_r, v_φ). Accordingly, we shall need the Jacobian of transformation equal in the approximation used to

$$\left| \frac{\partial(v_r, v_\varphi)}{\partial(\xi, \eta)} \right| = \frac{r_0^4 \varkappa^3(r_0)}{2r\Omega(r_0)}, \tag{19}$$

$$r_0 = r(1 - \eta). \tag{20}$$

The mass element in the same approximation is given by the formula

$$f_0 dv_r\, dv_\varphi = \frac{r_0^2}{r} \sigma_0(r_0) \exp\left[-\frac{\xi^2 + \eta^2}{2\varepsilon^2(r_0)} \right] \frac{d\xi\, d\eta}{2\pi\varepsilon^2(r_0)}. \tag{21}$$

The integration range over ξ and η may be extended from $-\infty$ to $+\infty$ and the errors arising will be exponentially small. Integrating (21) over the velocities v_r and v_φ, we obtain the density amplitude

$$\sigma(r) = \int\!\!\!\int_{-\infty}^{\infty} \frac{d\xi\, d\eta}{2\pi\varepsilon^2(r_0)} \cdot \frac{\sigma_0(r_0)}{r^2\varepsilon^2(r_0)\varkappa^2(r_0)} \exp\left[-\frac{\xi^2 + \eta^2}{2\varepsilon^2(r_0)} \right]$$

$$\times \left\{ -\Phi(r) + \frac{\nu(r_0)\pi}{\sin \nu(r_0)\pi} \cdot \frac{1}{2\pi} \int_{-\pi}^{\pi} \Phi(r')e^{i\nu(r_0)s} \left[1 + im\frac{2\Omega}{\varkappa}\left(\frac{\partial R_1}{\partial s} + \xi\right) \right] ds \right\}. \tag{22}$$

We shall further restrict ourselves to the analysis of the regions of the disk sufficiently far from the resonances. We expand the potential $\Phi(r')$ by assuming that $\varepsilon^{-1} \sim |kr|$:

$$\Phi(r') = \Phi(r)e^{-ikrR_1}\left[1 - R_1\frac{d\ln A}{d\ln r} + ikr\left(\frac{R_1^2}{2}\frac{d\ln k}{d\ln r} - R_2\right)\right]. \quad (23)$$

Substituting (23) into (22) and expanding near $r = r_0$, we obtain [by using also (20)]

$$\sigma(r) = -\frac{\sigma_0(r)\Phi(r)}{\varepsilon^2(r)r^2\varkappa^2(r)}\left\{1 - \frac{\pi v(r)}{\sin \pi v(r)}[\langle 1|g_v\rangle + \langle \tilde{G}|g_v\rangle]\right\}, \quad (24)$$

where the Dirac notations

$$\langle h|g_v\rangle = \frac{1}{2\pi}\int_{-\pi}^{\pi} ds \iint_{-\infty}^{\infty} hg_v \, d\xi \, d\eta \quad (25)$$

are employed, and the g_v and $\tilde{G}$ functions are

$$g_v = \frac{\exp[iv(r)s - ikrR_1]}{2\pi\varepsilon^2(r)}\exp\left[-\frac{\xi^2 + \eta^2}{2\varepsilon^2(r)}\right], \quad (26)$$

$$\tilde{G} = -\eta\frac{d}{d\ln r}\ln\left(\frac{\sigma_0}{\varepsilon^2\varkappa^2}\frac{v\pi}{\sin v\pi}g_v\right) - R_1\frac{d\ln A}{d\ln r}$$

$$+ im\frac{2\Omega}{\varkappa}\left(\frac{\partial R_1}{\partial s} + \xi\right) + ikr\left[-\eta R_1\frac{d\ln kr}{d\ln r} + \frac{R_1^2}{2}\frac{d\ln k}{d\ln r} - R_2\right]. \quad (27)$$

Such differentiations with respect to $\ln r$ are commutative with the operations "bra" and "ket"; therefore,

$$\langle \tilde{G}|g_v\rangle = -\langle \eta|g_v\rangle\frac{d}{d\ln r}\ln\left(\frac{\sigma_0}{\varkappa^2\varepsilon^2}\frac{v\pi\langle \eta|g_v\rangle}{\sin v\pi}\right) - \langle R_1|g_v\rangle\frac{d\ln A}{d\ln r}$$

$$+ im\frac{2\Omega}{\varkappa}\left[\left\langle\frac{\partial R_1}{\partial s}\Big|g_v\right\rangle + \langle\xi|g_v\rangle\right] + ikr\left[-\langle\eta R_1|g_v\rangle\frac{d\ln kr}{d\ln r}\right.$$

$$\left.+ \tfrac{1}{2}\langle R_1^2|g_v\rangle\frac{d\ln k}{d\ln r} - \langle R_2|g_v\rangle\right]. \quad (28)$$

Direct calculation shows that (the prime means differentiation with respect to x at $v = \text{const}$)

$$\langle 1|g_v\rangle = G_v(x),$$

$$\langle \eta|g_v\rangle = \tfrac{1}{2}\langle R_1|g_v\rangle = \frac{i}{kr}\, xG_v'(x), \quad (29)$$

$$\langle \eta R_1|g_v\rangle = -\langle R_2|g_v\rangle = \tfrac{1}{2}\langle R_1^2|g_v\rangle \quad (30)$$

$$= -\frac{1}{(kr)^2}[xG_v'(x) + 2x^2G_v''(x)], \quad (31)$$

$$\langle\xi|g_v\rangle = -\left\langle\frac{\partial R_1}{\partial s}\Big|g_v\right\rangle = \frac{1}{kr}\left[\frac{\sin v\pi}{v\pi} - vG_v(x)\right], \quad (32)$$

where

$$G_v(x) = \frac{1}{2\pi} \int_{-\pi}^{\pi} \cos vs \, e^{-x(1 + \cos s)} \, ds, \tag{33}$$

and $x = \varepsilon^2 k^2 r^2 = k^2 c_r^2/\varkappa^2$.

Using (29)–(32), one may make sure that the sums in the square brackets in (28) are zero, while the first two terms may be combined into the following:

$$\langle \tilde{G} | g_v \rangle = -\frac{i}{kr} x G_v'(x) \frac{d}{d \ln r} \ln \left[\frac{\sigma_0}{\varkappa^2} kr A^2 \frac{v\pi}{\sin v\pi} G_v'(x) \right]. \tag{34}$$

Then Eq. (24) may be written thus:

$$\sigma(r) = -\frac{k^2 \Phi \sigma_0}{\varkappa^2(1 - v^2)} \mathscr{F}_v(x) \left\{ 1 - \frac{i}{kr} D_v(x) \frac{d}{d \ln r} \ln \left(\frac{\sigma_0}{\varkappa^2} \frac{k\mathscr{F}_v}{1 - v^2} D_v r A^2 \right) \right\}, \tag{35}$$

where $\mathscr{F}_v(x)$ is the reduction factor in (22) Section 4.1, Chapter V, and

$$D_v(x) = -(1 - v^2) \frac{v\pi}{\sin v\pi} \frac{G_v'(x)}{\mathscr{F}_v(x)} \frac{\partial}{\partial \ln x} \ln[x\mathscr{F}_v(x)]. \tag{36}$$

§8 On the Derivation of the Nonlinear Dispersion Equation for a Collisionless Disk

In this section, we give some details and comments on the derivation of the nonlinear dispersion equation (14), Section 1.3, Chapter VII.

For the equilibrium distribution function

$$f = \frac{1}{\pi} e^{-v^2} \tag{1}$$

(which is made dimensionless, as described in the main text) the linear correction for the harmonic with the wave number k is (cf. §4, Chapter V) [271]

$$f_k^{(1)} = -2\Phi_k f \left\{ 1 - \frac{v\pi}{\sin v\pi} \right.$$

$$\left. \times \frac{1}{2\pi} \int_{-\pi}^{\pi} \exp[-i(vx + kv_x \sin x + kv_y(1 + \cos x))] \, dx \right\}, \tag{2}$$

where Φ_k is the corresponding harmonic of the perturbed potential; v_x and v_y are the velocities in the locally Cartesian coordinate system and v is the dimensionless frequency.

The perturbation of the distribution function in the second order of perturbation theory consists of three terms, which will be denoted f_{2k},

$\tilde{f}_{2k}$, and $f_0^{(2)}$. The first two terms are proportional to Φ_k^2 and Φ_{2k}, respectively; they must be found from the equations

$$- 2i(v - kv \cos \varphi) f_{2k} + \frac{\partial f_{2k}}{\partial \varphi} = ik\Phi_k \frac{\partial f_k^{(1)}}{\partial v_x}, \tag{3}$$

$$- 2i(v - kv \cos \varphi) \tilde{f}_{2k} + \frac{\partial \tilde{f}_{2k}}{\partial \varphi} = 2ik\Phi_{2k} \frac{\partial f}{\partial v_x} = -4ik\Phi_{2k} fv \cos \varphi. \tag{4}$$

The equation for $f_0^{(2)}$ (zero harmonic) is

$$2\varepsilon f_0^{(2)} + \frac{\partial f_0^{(2)}}{\partial \varphi} = ik\left(\Phi_k \frac{\partial f_k^{(1)*}}{\partial v_x} - \Phi_k^* \frac{\partial f_k^{(1)}}{\partial v_x}\right), \tag{5}$$

where ε is the positive imaginary additive to the frequency v, corresponding to the adiabatic inclusion of perturbation as $t \to -\infty$: $v = v + i\varepsilon$, $\varepsilon \to +0$.

A more detailed form of Eq. (3) is as follows:

$$- 2i(v - kv \cos \varphi) f_{2k} + \frac{\partial f_{2k}}{\partial \varphi} = -2ik\Phi_k^2 \left\{ -2v \cos \varphi \right.$$

$$+ \frac{v\pi}{\sin v\pi} \frac{1}{2\pi} \int_{-\pi}^{\pi} dx (2v \cos \varphi + ik \sin x) \exp[-i(vx$$

$$\left. + kv \cos \varphi \sin x + kv \sin \varphi (1 + \cos x)] \right\}, \tag{6}$$

where we have used (2).

Establish the rule for the solution of the equation of a somewhat more general form

$$- iN(v - kv \cos \varphi) F + \frac{\partial F}{\partial \varphi} = A(\varphi). \tag{7}$$

We solve Eq. (7) by the method of variation of the constant. We have

$$F = e^{iN(v\varphi - kv \sin \varphi)} \left[\int_0^{\varphi} e^{-iN(vy - kv \sin y)} A(y) \, dy + \lambda \right], \tag{8}$$

where $\lambda = \lambda(v)$ is the "constant" (independent of the φ function). We impose the periodicity condition $F(\varphi) = F(\varphi + 2\pi)$

$$e^{iN \cdot 2\pi v} \left\{ \int_0^{\varphi} + \int_{\varphi}^{\varphi + 2\pi} + \lambda \right\} = \int_0^{\varphi} + \lambda. \tag{9}$$

The integral $\int_{\varphi}^{\varphi + 2\pi}$, by substituting for $y = x + \varphi$, is reduced to

$$e^{-iNv\varphi} \int_0^{2\pi} e^{-iN[vx - kv \sin(x + \varphi)]} A(x + \varphi) \, dx. \tag{10}$$

This allows one to express the expression $(\int_0^\varphi + \lambda)$, which stands in brackets in (8), through the integral (10), and we finally obtain

$$F = \frac{e^{2iNv\pi}}{1 - e^{2iNv\pi}} e^{-iNkv\sin\varphi} \int_0^{2\pi} e^{-iN[vx-kv\sin(x+\varphi)]} A(x + \varphi) \, dx. \quad (11)$$

In case if $A(\varphi) = \cos\varphi$, we write the solution in another way, similarly to (2):

$$F = \frac{1}{iNkv} \left[1 - \frac{Nv\pi}{\sin Nv\pi} \cdot \frac{1}{2\pi} \int_{-\pi}^{\pi} e^{-iN[vx+kv_x\sin x + kv_y(1+\cos x)]} \, dx \right]. \quad (12)$$

Thus, for f_{2k} we find

$$f_{2k} = -2ik\Phi_k^2 f \left[-\frac{1}{ik} \left(1 - \frac{2v\pi}{\sin 2v\pi} \cdot \frac{1}{2\pi} \int_{-\pi}^{\pi} \exp\{-2i(vx \right. \right.$$

$$\left. + kv_x \sin x + kv_y[1 + \cos x])\} \, dx \right) + \frac{v\pi}{\sin v\pi} \cdot \frac{1}{2\pi} \frac{e^{4iv\pi}}{1 - e^{2iv\pi}}$$

$$\times \int_0^{2\pi} dy \int_{-\pi}^{\pi} dx [2v\cos(\varphi + y) + ik\sin x] \cdot \exp\{ikv[-2v\sin\varphi$$

$$+ 2\sin(y + \varphi) - \sin x \cos(\varphi + y) - (1 + \cos x)\sin(\varphi + y)]$$

$$\left. - iv(x + 2y)\} \right]. \quad (13)$$

In this formula, we can proceed to the limit $v \to 0$ by opening the uncertainty according to l'Hôpital rule, then we obtain

$$f_{2k} = -2ik\Phi_k^2 \left\{ -\frac{1}{ik} [1 - J_0(2kv)e^{-2ikv\sin\varphi}] \right.$$

$$+ \frac{1}{(2\pi)^2} \int_0^{2\pi} dy \int_{-\pi}^{\pi} dx (v_x \cos y - v_y \sin y + \tfrac{1}{2}ik \sin x)(x + 2y)$$

$$\left. \times \exp[ik(\alpha v_x + \beta v_y)] \right\}, \quad (14)$$

where we denoted

$$\begin{aligned} \alpha &= 2\sin y - \sin x \cos y - (1 + \cos x)\sin y, \\ \beta &= -2 + 2\cos y + \sin x \sin y - (1 + \cos x)\cos y. \end{aligned} \quad (15)$$

In a similar way, Eq. (4) for $\tilde{f}_{2k}$

$$\tilde{f}_{2k} = -2\Phi_{2k} f \left(1 - \frac{2v\pi}{\sin 2v\pi} \frac{1}{2\pi} \int_{-\pi}^{\pi} e^{-2i[vx+kv_x\sin x + kv_y(1+\cos x)]} \, dx \right) \quad (16)$$

is solved.

Equation (5) has the form

$$2\varepsilon f_0^{(2)} + \frac{\partial f_0^{(2)}}{\partial\varphi} = A(\varphi), \quad (17)$$

where

$$A(\varphi) = -2ik|\Phi_k|^2 \frac{1}{2\pi \sin v\pi} \frac{v\pi}{\sin v\pi} \int_{-\pi}^{\pi} dx[(2v \cos \varphi - ik \sin x)$$

$$\times e^{i\{vx + kv[\sin \varphi + \sin(x + \varphi)]\}}$$

$$- (2v \cos \varphi + ik \sin x)e^{-i\{vx + kv[\sin \varphi + \sin(x + \varphi)]\}}]. \tag{18}$$

In a conventional way, the periodical solution of Eq. (17):

$$f_0^{(2)} = \frac{e^{-4\pi\varepsilon}}{1 - e^{-4\pi\varepsilon}} \int_0^{2\pi} A(y + \varphi)e^{2\varepsilon y} \, dy \tag{19}$$

was found. In (19), one has further to perform the limiting transition $\varepsilon \to +0$. By opening the uncertainty arising here [one has then to bear in mind that the frequency v has a small imaginary part: $v \to v + i\varepsilon$, and the bracket in (18) is multiplied by $\exp(\varepsilon x)$], we obtain

$$f_0^{(2)} = -\frac{1}{2\pi} 2ik|\Phi_k|^2 \frac{1}{2\pi \sin v\pi} \frac{v\pi}{\sin v\pi} f \int_0^{2\pi} dy \int_{-\pi}^{\pi} dx(y + \tfrac{1}{2}x)$$

$$\times \{[2v \cos(y + \varphi) - ik \sin x]e^{i\{vx + kv[\sin(\varphi + y) + \sin(x + y + \varphi)]\}}$$

$$- [2v \cos(y + \varphi) - ik \sin x]e^{-i\{vx + kv[\sin(\varphi + y) + \sin(x + y + \varphi)]\}}\}. \tag{20}$$

To calculate the surface density, we need the integrals of the form

$$\iint (av_x + bv_y)e^{ik(\alpha v_x + \beta v_y)}f \, dv_x \, dv_y = \tfrac{1}{2}ik(a\alpha + b\beta)e^{-(k^2/4)(\alpha^2 + \beta^2)}. \tag{21}$$

In particular, for $\sigma_{2k}^{(2)} = \int f_{2k} \, dv_x \, dv_y$, we obtain by using (14) and (21):

$$\sigma_{2k}^{(2)} = B_k\Phi_k^2, \qquad B_k = 2\left\{(1 - I_0(2k^2)e^{-2k^2})\right.$$

$$- \frac{ik}{(2\pi)^2} \int_0^{2\pi} dy \int_{-\pi}^{\pi} dx \, (x + 2y)\tfrac{1}{2}ik(\alpha \cos y - \beta \sin y + \sin x)$$

$$\left. \times e^{-(k^2/4)(\alpha^2 + \beta^2)}\right\}. \tag{22}$$

It is easy to check that for α and β from (15)

$$\alpha \cos y - \beta \sin y + \sin x = 2 \sin y,$$
$$\alpha^2 + \beta^2 = 6 - 2 \cos x - 4 \cos y + 4 \cos(x + y),$$

so that

$$B_k = 2\left(1 - I_0(2k^2)e^{-2k^2} + \frac{k^2}{(2\pi)^2} \int_0^{2\pi} dy \int_{-\pi}^{\pi} dx \, (x + 2y) \sin y\right.$$

$$\left. \times \exp\{-\tfrac{1}{2}k^2[3 - \cos x + 2 \cos(x + y) - 2 \cos y]\}\right). \tag{23}$$

In the third order of perturbation theory, the calculations are similar, but more cumbersome. For example, to find the contribution of the third order to the harmonic of the perturbed distribution function (with the wave number k), due to f_{2k}, first there is the equation

$$-i(v - kv \cos \varphi) f_k^{(3)} + \frac{\partial f_k^{(3)}}{\partial \varphi} = -ik\Phi_{-k} \frac{\partial f_{2k}}{\partial v_x}. \tag{24}$$

For the arbitrary frequency v we obtain

$$f_k^{(3)} = -2ik|\Phi_k|^2 \Phi_k \cdot f \cdot \left(-\frac{2}{ik} \left(1 - \frac{v\pi}{\sin v\pi} \frac{1}{2\pi} \right) \right.$$

$$\times \int_{-\pi}^{\pi} e^{-i(vx + kv_x \sin x + kv_y(1 + \cos x))} \right) + \frac{2v\pi}{\sin 2v\pi} \cdot \frac{1}{2\pi} \frac{e^{2iv\pi}}{1 - e^{2iv\pi}}$$

$$\times \int_0^{2\pi} dy \int_{-\pi}^{\pi} dx \, [2v \cos (y + \varphi) + 2ik \sin x]$$

$$\times \exp\{-2ivx + ikv[\mu \cos(y + \varphi) + \lambda \sin(y + \varphi)]$$

$$- ikv \sin \varphi - i[vy - kv \sin(\varphi + y)]\}$$

$$- \frac{2ik}{2\pi} \frac{v\pi}{\sin v\pi} \frac{e^{4iv\pi}}{1 - e^{4iv\pi}} \frac{e^{2iv\pi}}{1 - e^{2iv\pi}} \int_0^{2\pi} dz \int_0^{2\pi} dy \int_{-\pi}^{\pi} dx \, \{[ik\alpha$$

$$- 2v \cos(z + \varphi)][v \cos y \cos(z + \varphi) - v \sin y \sin(z + \varphi) + \tfrac{1}{2}ik \sin x]$$

$$+ \cos y\} \times \exp\{-iv(z + 2y + x) + ikv[\alpha \cos(z + \varphi)$$

$$+ \beta \sin(z + \varphi) - \sin \varphi - \sin(z + \varphi)]\} \Big),$$

$$\mu \equiv -2 \sin x,$$

$$\lambda \equiv -2(1 + \cos x). \tag{25}$$

As $v \to 0$, we therefore find

$$f_k^{(3)} = -2ik|\Phi_k|^2 \Phi_k f \left\{ -\frac{2}{ik} \left[1 - \frac{1}{2\pi} e^{-ikv \sin \varphi} J_0(kv) \right] \right.$$

$$+ \frac{1}{(2\pi)^2} \int_0^{2\pi} dy \int_{-\pi}^{\pi} dx \, (y + 2x)[2 \cos y \, v_x - 2 \sin y \, v_y + 2ik \sin x]$$

$$\times \exp[ik(A_0 v_x + B_0 v_y)] - \frac{1}{2} \frac{ik}{(2\pi)^3} \int_0^{2\pi} dz \int_0^{2\pi} dy \int_{-\pi}^{\pi} dx (z + 2y + x)^2$$

$$\times \{\cos y + [ik\alpha - 2 \cos z \, v_x + 2 \sin z \, v_y] \cdot [\tfrac{1}{2}ik \sin x \, v_x \cos (z + y)$$

$$- v_y \sin(z + y)] \exp[ik(Av_x + Bv_y)]\}, \tag{26}$$

where the notations are introduced

$$A_0 = \mu \cos y + \lambda \sin y + \sin y, \qquad B_0 = -\mu \sin y + \lambda \cos y - 1 + \cos y,$$

$$A = \alpha \cos z + \beta \sin z + \sin z, \qquad B = -\alpha \sin z + \beta \cos z - 1 + \cos z.$$

By integrating (25) or (26) over the velocity according to the general formula (21), we obtain the corresponding contribution to the perturbed surface density $\sigma_k^{(3)}$, proportional to $|\Phi_k|^2\Phi_k$. The proportionality coefficient connecting the total perturbed surface density $\sigma_k^{(3)}$ with $|\Phi_k|^2\Phi_k$ (calculated in the third order of perturbation theory), we denote C_k:

$$\sigma_k^{(3)} = C_k|\Phi_k|^2\Phi_k. \tag{27}$$

The contribution to C_k from (26) proves to equal

$$C_{k_1} = I_1 + I_2 + I_3, \tag{28}$$

$$I_1 = 4\left[1 - I_0\left(\frac{k^2}{2}\right)e^{-k^2/2}\right], \tag{29}$$

$$I_2 = \frac{2ik}{(2\pi)^2}\int_0^{2\pi} dy \int_{-\pi}^{\pi} dx\,(y + 2x)(-\cos y \cdot ikA_0 + \sin y$$

$$\cdot ikB_0 - 2ik \sin x)e^{-k^2(A_0^2 + B_0^2)/4}$$

$$= \frac{2k^2}{(2\pi)^2}\int_0^{2\pi} dy \int_{-\pi}^{\pi} dx\,(y + 2x)\sin y\, e^{-k^2[3 + 2\cos x + \cos y + 2\cos(x + y)]/2}, \tag{30}$$

$$I_3 = -\frac{1}{2}\frac{k^4}{(2\pi)^3}\int_0^{2\pi} dz \int_0^{2\pi} dy \int_{-\pi}^{\pi} dx(z + 2y + x)^2 \sin z[\sin y + \sin(y + z)]$$

$$\times \exp\{-\tfrac{1}{2}k^2[2 - \cos x - \cos y + \cos(x + y) - \cos(y + z)$$

$$+ \cos(x + y + z) + \cos z]\}. \tag{31}$$

The contribution from $\tilde{f}_{2k}$ must be determined from the equation

$$-i(v - kv)\tilde{f}_k^{(3)} + \frac{\partial \tilde{f}_k^{(3)}}{\partial \varphi} = -ik\Phi_{-k}\frac{\partial \tilde{f}_{2k}}{\partial v_x} + 2ik\Phi_{2k}\frac{\partial f_{-k}^{(1)}}{\partial v_x}. \tag{32}$$

Finding from (32) $\tilde{f}_k^{(3)}$ and integrating over velocities, we obtain the corresponding perturbation of the surface density

$$\tilde{\sigma}_k^{(3)} = D_k\Phi_{2k}\Phi_{-k}, \tag{33}$$

where

$$D_k = 4\left[1 - I_0\left(\frac{k^2}{2}\right)e^{-k^2/2}\right] - \frac{2k^2}{(2\pi)^2}\int_0^{2\pi} dy \int_{-\pi}^{\pi} dx\,(2x + y)\sin y$$

$$\times e^{-k^2(3 + 2\cos x + \cos y + 2\cos(x + y)]/2}$$

$$+ \frac{4k^2}{(2\pi)^2}\int_0^{2\pi} dy \int_{-\pi}^{\pi} dx(y - x)\sin y\, e^{-k^2[3 + 2\cos x - 2\cos y - \cos(x + y)]/2} \tag{34}$$

Finally, the last contribution to $f_k^{(3)}$, from $f_0^{(2)}$, is found from the equation

$$-i(v - kv\cos\varphi)f_{k_0}^{(3)} + \frac{\partial f_{k_0}^{(3)}}{\partial \varphi} = ik\Phi_{k_0}\frac{\partial f_0^{(2)}}{\partial v_x}. \tag{35}$$

The perturbation of surface density corresponding to $f_{k_0}^{(3)}$ is obtained as a result of simple but cumbersome calculations in the form

$$\sigma_{k_0}^{(3)} = C_{k_2} \Phi_k |\Phi_k|^2,$$

$$C_{k_2} = \frac{2k^4}{4\pi} P_2(v) Q_1(v) \int_0^{2\pi} dz \int_0^{2\pi} dy \int_{-\pi}^{\pi} dx\, e^{-ivz} \Big[(x + 2y + z)$$

$$\times (E_5 e^{ivx} - E_6 e^{-ivx}) - i\frac{\partial Q_1/\partial v}{Q_1}(E_5 e^{ivx} + E_6 e^{-ivx}) \Big], \qquad (36)$$

where

$$Q_1 = \frac{1}{2\pi} \frac{v\pi}{\sin v\pi}, \qquad P_2 = \frac{e^{2iv\pi}}{1 - e^{2iv\pi}},$$

$$E_5 = \exp\{-\tfrac{1}{2}k^2[2 + \cos y + \cos(x + y) + \cos x - \cos z - \cos(y + z)$$
$$- \cos(x + y + z)\} \times \sin z \cdot [\sin(y + z) - \sin y], \qquad (37)$$

$$E_6 = \exp\{-\tfrac{1}{2}k^2[2 - \cos y - \cos(x + y) + \cos x - \cos z + \cos(y + z)$$
$$+ \cos(x + y + z)\} \times \sin z[\sin(y + z) - \sin y].$$

As $v \to 0$

$$C_{k_2} = \frac{k^4}{(2\pi)^3} \int_0^{2\pi} dz \int_0^{2\pi} dy \int_{-\pi}^{\pi} dx\, (x + 2y + z)[E_5(z - x) - E_6(z + x)].$$

We now show how the nonlinear equation is constructed for the potential harmonic Φ_k.

To begin with, calculate the frequency v for $v^2 \ll x^2$. Expanding the linear dispersion equation (23), Section 4.1, Chapter V,

$$\frac{k_T}{|k|} x = 1 - \frac{v\pi}{\sin v\pi} \cdot \frac{1}{2\pi} \int_{-\pi}^{\pi} e^{-x(1 + \cos s)} \cos vs\, ds \qquad (38)$$

($k_T = x^2/2\pi G\sigma_0 = 1/2\pi G\sigma_0$, $x = k^2 c_r^2/x^2 = k^2/2$ in the units adopted here) for $v^2 \ll 1$, we obtain:

$$v_k^2 = \frac{1 - I_0(k^2/2)e^{-k^2/2} - k/4\pi G\sigma_0}{\tfrac{1}{6}\pi^2 e^{-k^2/2} I_0(k^2/2) - (1/2\pi) \int_0^{\pi} e^{-k^2(1 + \cos s)/2} s^2\, ds}. \qquad (39)$$

On the other hand, our nonlinear dispersion equation [(14) in the main text] may be reduced to the form

$$-\Big\{ \frac{k}{2} \cdot \frac{1}{2\pi G\sigma_0} - \Big[1 - I_0\Big(\frac{k^2}{2}\Big)e^{-k^2/2}\Big]\Big\}\Phi_k - \frac{1}{2} R_k |\Phi_k|^2 \Phi_k = 0, \qquad (40)$$

or, if divided by the denominator in (39),

$$v_k^2 \Phi_k = \frac{1}{2} \frac{R_k |\Phi_k|^2 \Phi_k}{\tfrac{1}{6}\pi^2 I_0(k^2/2)e^{-k^2/2} - (1/2\pi) \int_0^{\pi} e^{-k^2(1 + \cos s)/2} s^2\, ds}. \qquad (41)$$

Hence it is obvious that the required nonlinear equation for $\Phi_k(t)$ is

$$\frac{\partial^2 \Phi_k}{\partial t^2} + v_k^2 \Phi_k = \alpha_k |\Phi_k|^2 \Phi_k, \tag{42}$$

where

$$\alpha_k = \frac{1}{2}\left[C_k - \frac{(2\pi G \sigma_0/|2k|)B_k D_k}{1 + (2\pi G_0/|2k|)A_{2k}} \right]$$

$$\times \left[\frac{\pi^2}{6} I_0\left(\frac{k^2}{2}\right) e^{-k^2/2} - \frac{1}{2\pi} e^{-k^2(1+\cos s)/2} s^2 \, ds \right]^{-1}. \tag{43}$$

Recall the meaning of the coefficients B, C, D involved in (42): B_k is the coefficient in $\sigma_{2k}^{(2)}$ before Φ_k^2 (calculated in the second order of perturbation theory), D_k is the proportionality coefficient between σ_k and $\Phi_{2k}\Phi_{-k}$ (calculated in the third order), C_k is the coefficient in σ_k before $\Phi_{-k}\Phi_k^2$ (in the third order). In conclusion, let us give the final expressions for all values used in (43):

$$A_{2k_0} = -2\left(1 - Q_2 \int_{-\pi}^{\pi} e^{-2ivx - 2k_0^2(1+\cos x)} \, dx\right), \tag{44}$$

$$B_{k_0} = 2\left(1 - Q_2 \int_{-\pi}^{\pi} e^{-2ivx - 2k_0^2(1+\cos x)} \, dx\right.$$

$$\left. + 2k_0^2 Q_1 P_4 \int_0^{2\pi} dy \int_{-\pi}^{\pi} dx \, E_1 e^{-iv(x+2y)}\right), \tag{45}$$

$$D_{k_0} = 4\left(1 - Q_1 \int_{-\pi}^{\pi} e^{-ivx - k_0^2(1+\cos x)/2} \, dx\right)$$

$$- Q_2 P_2 \cdot 2k_0^2 \int_0^{2\pi} dy \int_{-\pi}^{\pi} dx \, e^{-iv(2x+y)} E_2$$

$$+ 4Q_1 P_2 k_0^2 \int_0^{2\pi} dy \int_{-\pi}^{\pi} dx \, e^{-iv(y-x)} E_3 = 2B k_0, \tag{46}$$

$$C_{k_0} = 4\left(1 - Q_1 \int_{-\pi}^{\pi} e^{-ivx - k_0^2(1+\cos x)/2} \, dx\right)$$

$$+ 2Q_2 P_2 k_0^2 \int_0^{2\pi} dy \int_{-\pi}^{\pi} dx \, e^{-iv(y+2x)} E_2$$

$$- 2Q_1 P_2 P_4 k_0^4 \int_0^{2\pi} dz \int_0^{2\pi} dy \int_{-\pi}^{\pi} dx \, e^{-iv(x+2y+z)} E_4$$

$$+ \frac{k_0^4}{2\pi} P_2 Q_1 \int_0^{2\pi} dz \int_0^{2\pi} dy \int_{-\pi}^{\pi} dx \left[(x + 2y)(E_5 e^{ivx} - E_6 e^{-ivx})\right.$$

$$\left. - \frac{i(\partial Q_1/\partial v)}{Q_1} \cdot (E_5 e^{ivx} + E_6 e^{-ivx}) e^{-ivz}\right]. \tag{47}$$

Here the following notations are introduced:

$$Q_n = \frac{1}{2\pi} \frac{nv\pi}{\sin nv\pi}, \qquad P_n = \frac{e^{inv\pi}}{1 - e^{inv\pi}}, \tag{48}$$

$$E_1(x, y) = \sin y \cdot \exp\{-\tfrac{1}{2}k_0^2[3 - \cos x - 2\cos y + 2\cos(x + y)]\}, \tag{49}$$

$$E_2(x, y) = \sin y \cdot \exp\{-\tfrac{1}{2}k_0^2[3 + 2\cos x + \cos y + 2\cos(x + y)]\}, \tag{50}$$

$$E_3(x, y) = \sin y \cdot \exp\{-\tfrac{1}{2}k_0^2[3 + 2\cos x - 2\cos y - \cos(x + y)]\}, \tag{51}$$

$$E_4(x, y, z) = \sin z \, [\sin y + \sin(y + z)] \exp[-\tfrac{1}{2}k_0^2(2 - \cos x - \cos y$$
$$+ \cos(x + y) - \cos(y + z) + \cos(x + y + z) + \cos z)], \tag{52}$$

$$E_5(x, y, z) = \sin z \, [\sin(y + z) - \sin y] \exp\{-\tfrac{1}{2}k_0^2[2 + \cos y + \cos(x + y)$$
$$+ \cos x - \cos z - \cos(y + z) - \cos(x + y + z)]\}, \tag{53}$$

$$E_6(x, y, z) = -E_5(x, y + \pi, z). \tag{54}$$

In the case $k_0 = k_0^*$, $v \to 0$ instead of (44)–(47) we have simpler expressions:

$$A_{2k_0} = -2[1 - I_0(2k_0^2)e^{-2k_0^2}], \tag{55}$$

$$B_{k_0} = 2\{1 - I_0(2k_0^2)e^{-2k_0^2}\} + \frac{k_0^2}{(2\pi)^2} \int_0^{2\pi} dy \int_{-\pi}^{\pi} dx(x + 2y)E_1, \tag{56}$$

$$D_{k_0} = 2B_{k_0}, \tag{57}$$

$$C_{k_0} = 4\left[1 - I_0\left(\frac{k_0^2}{2}\right)e^{-k_0^2/2}\right] + \frac{2k_0^2}{(2\pi)^2} \int_0^{2\pi} dy \int_{-\pi}^{\pi} dx(y + 2x)E_2$$
$$- \frac{k_0^4}{2(2\pi)^3} \int_0^{2\pi} dz \int_0^{2\pi} dy \int_{-\pi}^{\pi} dx(x + 2y + z)^2 E_4$$
$$+ \frac{k_0^4}{(2\pi)^3} \int_0^{2\pi} dz \int_0^{2\pi} dy \int_{-\pi}^{\pi} dx \, (x + 2y)[E_5(z - x) - E_6(z + x)], \tag{58}$$

where $I_0(x)$ is the Bessel function of imaginary argument. The computations give the following values ($k_0 = 1.377$):

$$A_{2k_0} = -1.574, \qquad B_{k_0} = 0.230, \qquad D_{k_0} = 2B_{k_0}, \qquad C_{k_0} = 0.206 \tag{59}$$

§9 Calculation of the Matrix Elements for the Three-Waves Interaction

Let us use formulae (20), Section 4.1, Chapter VII (for notations see in the same place):

$$\mathcal{H}^{(3)} = \tfrac{1}{2}\sigma_0 \int [\tau V_1^2 + 2(V_1 V_2) + \tfrac{1}{3}c^2(\gamma - 2)\tau^3] \, dr,$$

$$V_1 = \nabla\varphi + \sqrt{\frac{\varkappa}{\sigma_0}}\, d, \qquad V_2 = \frac{\lambda\nabla\mu - \mu\nabla\lambda}{2\sigma_0} - \sqrt{\frac{\varkappa}{4\sigma_0}}\, \tau d, \tag{1}$$

where

$$d = -(\lambda l_x + \mu l_y).$$

It may be easily checked that

$$\lambda \nabla \mu - \mu \nabla \lambda = \frac{1}{\varkappa} \{[d \operatorname{rot}[d\varkappa]] + (d\nabla)[d\varkappa]\};$$

consequently,

$$V_2 = \frac{1}{2\sigma_0 \varkappa} \{[d \operatorname{rot}[d\varkappa]] + (d\nabla)[d\varkappa]\} - \sqrt{\frac{\varkappa}{4\sigma_0}} \tau d.$$

Let us express the integral $\mathscr{H}^{(3)}$ through the Fourier components:

$$\mathscr{H}^{(3)} = \frac{\sigma_0}{4\pi} \int dk\, dk_1\, dk_2 \left[\tau_k(V_{k_1} V_{k_2}) + i\left(V_k, \frac{[d_{k_1}[k_2[d_{k_2}\varkappa]]]}{\varkappa\sigma_0} \right) \right.$$

$$+ i\left(V_k, \frac{(d_{k_1} k_2)[d_{k_2}\varkappa]}{\varkappa\sigma_0} \right) - \sqrt{\frac{\varkappa}{\sigma_0}} \tau_k(V_{k_1} d_{k_2})$$

$$\left. + \frac{\gamma - 2}{3} c^2 \tau_k \tau_{k_1} \tau_{k_2} \right] \delta(k + k_1 + k_2).$$

Reducing the vector products and performing the symmetrization over k_1 and k_2, we obtain

$$\mathscr{H}^{(3)} = \frac{\sigma_0}{4\pi} \int dk\, dk_1\, dk_2 \left\{ \tau_k(V_{k_1} V_{k_2}) + i \frac{(V_k, k_2 - k_1)(\varkappa[d_{k_1} d_{k_2}])}{2\varkappa\sigma_0} \right.$$

$$\left. - \sqrt{\frac{\varkappa}{4\sigma_0}} \tau_k[(V_{k_1} d_{k_2}) + (V_{k_2} d_{k_1})] + \frac{\gamma - 2}{3} c^2 \tau_k \tau_{k_1} \tau_{k_2} \right\}$$

$$\times \delta(k + k_1 + k_2). \tag{2}$$

With the aid of Eqs. (27) (subsection 4.1, Chapter VII) we first express the Fourier components through the normal variables a_k, and after this we shall be occupied with the term-by-term calculation of the right-hand side of (2):

$$d_k = \frac{1}{|k|\sqrt{2\varkappa\omega_k^2}} \{(\varkappa^2 k - i\omega_k[\varkappa k])a_k - (\varkappa^2 k + i\omega_k[\varkappa k])a^*_{-k}\},$$

$$N = 4\sqrt{2\sigma_0^3 \omega_k^3 \omega_{k_1}^3 \omega_{k_2}^3} |k||k_1||k_2|.$$

(1) $$\tau_k(V_{k_1} V_{k_2}) = \frac{2}{N} k^2 \omega_k \omega_{k_1} \omega_{k_2}(a_k + a^*_{-k})\{[(\omega_{k_1}\omega_{k_2} - \varkappa^2)(k_1 k_2)$$

$$+ i(\omega_{k_1} - \omega_{k_2})(\varkappa[k_1 k_2])]a_{k_1} a_{k_2} + [(\omega_{k_1}\omega_{k_2} - \varkappa^2)(k_1 k_2)$$

$$- i(\omega_{k_1} - \omega_{k_2})(\varkappa[k_1 k_2])]a^*_{-k_1} a^*_{-k_2} - [(\omega_{k_1}\omega_{k_2} + \varkappa^2)(k_1 k_2)$$

$$+ i(\omega_{k_1} + \omega_{k_2})(\varkappa[k_1 k_2])]a^*_{-k_1} a_{k_2} - [(\omega_{k_1}\omega_{k_2} + \varkappa^2)(k_1 k_2)$$

$$- i(\omega_{k_1} + \omega_{k_2})(\varkappa[k_1 k_2])]a_{k_1} a^*_{-k_2}\};$$

(2) $\quad \dfrac{i(V_k, k_2 - k_1)(\varkappa[d_{k_1}d_{k_2}])}{2\varkappa\sigma_0} = \dfrac{\omega_k}{N}\{[((k_1 - k_2)[\varkappa k])$

$+ i\omega_k(k, k_1 - k_2)]a_k + [((k_1 - k_2)[\varkappa k]) - i\omega_k(k, k_1 - k_2)]a^*_{-k}\}$
$\times \{[(\omega_{k_1}\omega_{k_2} - \varkappa^2)(\varkappa[k_1k_2]) - i\varkappa^2(\omega_{k_1} - \omega_{k_2})(k_1k_2)]a_{k_1}a_{k_2}$
$+ [(\omega_{k_1}\omega_{k_2} - \varkappa^2)(\varkappa[k_1k_2]) + i\varkappa^2(\omega_{k_1} - \omega_{k_2})(k_1k_2)]a^*_{-k_1}a^*_{-k_2}$
$+ [(\omega_{k_1}\omega_{k_2} + \varkappa^2)(\varkappa[k_1k_2]) - i\varkappa^2(\omega_{k_1} + \omega_{k_2})(k_1k_2)]a^*_{-k_1}a_{k_2}$
$+ [(\omega_{k_1}\omega_{k_2} + \varkappa^2)(\varkappa[k_1k_2]) + i\varkappa^2(\omega_{k_1} + \omega_{k_2})(k_1k_2)]a_{k_1}a^*_{-k_2}\};$

(3) $\quad -\sqrt{\dfrac{\varkappa}{4\sigma_0}}\,\tau_k[(V_{k_1}d_{k_2}) + (V_{k_2}d_{k_1})] = -\dfrac{\omega_k}{N}k^2(a_k + a^*_{-k})$

$\times \{[\varkappa^2(\omega_{k_1} - \omega_{k_2})^2(k_1k_2) + i(\omega_{k_1} - \omega_{k_2})(\omega_{k_1}\omega_{k_2} - \varkappa^2)(\varkappa[k_1k_2])]$
$\times a_{k_1}a_{k_2} + [\varkappa^2(\omega_{k_1} - \omega_{k_2})^2(k_1k_2) - i(\omega_{k_1} - \omega_{k_2})(\omega_{k_1}\omega_{k_2} - \varkappa^2)$
$\times (\varkappa[k_1k_2])]a^*_{-k_1}a^*_{-k_2} - [\varkappa^2(\omega_{k_1} + \omega_{k_2})^2(k_1k_2) + i(\omega_{k_1} + \omega_{k_2})$
$\times (\omega_{k_1}\omega_{k_2} + \varkappa^2)(\varkappa[k_1k_2])]a^*_{-k_1\varkappa k_2} - [\varkappa^2(\omega_{k_1} + \omega_{k_2})^2(k_1k_2)$
$- i(\omega_{k_1} + \omega_{k_2})(\omega_{k_1}\omega_{k_2} + \varkappa^2)(\varkappa[k_1k_2])]a_{k_1}a^*_{-k_2}\};$

(4) $\quad \dfrac{\gamma - 2}{3}c^2\tau_k\tau_{k_1}\tau_{k_2} = \dfrac{2(\gamma - 2)}{3N}c^2k^2k_1^2k_2^2\omega_k\omega_{k_1}\omega_{k_2}$

$\times (a_k + a^*_{-k})(a_{k_1} + a^*_{-k_1})(a_{k_2} + a^*_{-k_2}).$

It is easily seen that the expressions obtained have two important properties. First, under the products of the normal variables, differing from each other only by substitution of a_{k_i} for $a^*_{-k_i}$ (for example, under $a^*_{-k}a_{k_1}a_{k_2}$ and $a_k a^*_{-k_1}a^*_{-k_2}$), we have the coefficients complex conjugate each to other. Second, all such coefficients are invariant with respect to simultaneous substitution $k \to -k$, $k_1 \to -k_1$, $k_2 \to -k_2$, since they involve the products of only even numbers of vectors k_i, and $\omega_{-k} = \omega_k$. Therefore, later on we may restrict ourselves to the calculation of the terms containing the products

$$a_k a_{k_1}a_{k_2}, \qquad a^*_{-k}a_{k_1}a_{k_2}, \qquad a_k a^*_{-k_1}a_{k_2}, \qquad a_k a_{k_1}a^*_{-k_2}.$$

The remaining terms are the complex conjugate ones.

Taking into account this circumstance, we summarize the expressions obtained above, performing the symmetrization over k, k_1, k_2 and using the fact that, due to the δ function contained in (2), $k + k_1 + k_2 = 0$, so that

$$[k_1k_2] = [kk_1] = -[kk_2].$$

We obtain

$$\mathcal{H}^{(3)} = \tfrac{1}{3}\int dk\, dk_1\, dk_2(V_{kk_1k_2}a_k a_{k_1}a_{k_2} + V^{(1)}_{kk_1k_2}a^*_{-k}a_{k_1}a_{k_2}$$

$$+ V^{(2)}_{kk_1k_2}a_k a^*_{-k_1}a_{k_2} + V^{(3)}_{kk_1k_2}a_k a_{k_1}a^*_{-k_2} + \text{c.c.})\delta(k + k_1 + k_2), \quad (3)$$

where c.c. denotes the terms complex conjugate to those written,

$$V_{kk_1k_2} = U_{kk_1k_2} + U_{k_1kk_2} + U_{k_2k_1k},$$

$$U_{kk_1k_2} = \frac{\sigma_0}{4\pi N} \{2\omega_k\omega_{k_1}\omega_{k_2}\varkappa^2[k_1k_2]^2 + 2\varkappa^4\omega_k([kk_1][kk_2])$$

$$+ 2\omega_k\omega_{k_1}^2\omega_{k_2}^2 k^2(k_1k_2) + \tfrac{2}{3}(\gamma - 2)c^2k^2k_1^2k_2^2\omega_k\omega_{k_1}\omega_{k_2}$$
$$+ \varkappa^2[\omega_k^2\omega_{k_1}((kk_1)(k_1k_2) + \tfrac{1}{2}(kk_2)(k_2^2 - k^2 - k_1^2))$$
$$+ \omega_k^2\omega_{k_2}((kk_2)(k_1k_2) + \tfrac{1}{2}(kk_1)(k_1^2 - k^2 - k_2^2))]$$
$$- 2i\omega_k^2\omega_{k_1}\omega_{k_2} \times (k_1^2 - k_2^2)(\varkappa[k_1k_2])$$
$$- i\varkappa^2[(\omega_{k_1}^2 - \omega_{k_2}^2)(\varkappa[k_1k_2])(k_1k_2)$$
$$+ \omega_{k_1}\omega_{k_2}(k_1^2 - k_2^2)(\varkappa[k_1k_2])\}, \tag{4}$$

and

$$V^{(1)}_{kk_1k_2} = V^*_{-kk_1k_2}, \qquad V^{(2)}_{kk_1k_2} = V^*_{k-k_1k_2}, \qquad V^{(3)}_{kk_1k_2} = V^*_{kk_1-k_2}. \tag{5}$$

Substituting in (3) $k_i \rightarrow -k_i$ in the necessary cases and taking into account (5) and the symmetry of $V_{kk_1k_2}$ with respect to all the indexes, we obtain

$$\mathscr{H}^{(3)} = \int dk dk_1 dk_2 \{\tfrac{1}{3}V_{kk_1k_2} a_k a_{k_1} a_{k_2} \delta(k + k_1 + k_2)$$

$$+ V^*_{kk_1k_2} a_k^* a_{k_1} a_{k_2} \delta(k - k_1 - k_2) + \text{c.c.}\}.$$

§ 10 Derivation of the Formulas for the Boundaries of Wave Numbers Range Which May Take Part in a Decay

Let us write down the dispersion equation for the gas disk in the dimensionless form:

$$v_q^2 = (q - 1)^2 + Q^2 - 1, \qquad q = \frac{k}{k_0}, \qquad v_q = \frac{\omega_k}{ck_0},$$

$$Q = \frac{\varkappa}{ck_0}, \qquad k_0 = \frac{\pi G\sigma_0}{c^2}.$$

Decay conditions are $\mathbf{q}_1 = \mathbf{q} + \mathbf{q}_3$ and $v_1 = v_2 + v_3$. We denote

$$v_0^2 = Q^2 - 1, \; \tilde{q}^+ = \tfrac{3}{2}Q^2, \qquad q^\pm = 1 \pm \sqrt{3(Q^2 - 1)};$$

$$q_0^\pm = 1 \pm \tfrac{1}{2}[1 + \sqrt{(3Q^2 - 2)(Q^2 - 1)/(2 - Q^2)}].$$

It may be shown that $q_0^- < q^-$ and $q^+ \leq q_0^+$ as $Q^2 \leq \tfrac{4}{3}$ and $\tilde{q}^+ \leq q_0^+$ for $Q^2 \geq \tfrac{4}{3}$ (the equalities correspond to $Q^2 = \tfrac{4}{3}$).

1. $Q^2 \leq \tfrac{4}{3}$: For $0 \leq q_1 \leq q_0^-$ boundaries are determined from the conditions $v_1 = v_2 + v_3$ and $q_1 = |q_2 - q_3|$ and described by the formulae

$$|q_2 - 1| = \frac{1}{2}\left[\sqrt{q_1^2 + \frac{4(2 - Q^2)q_1(q_1 - 1) + Q^2(4 - 3Q^2)}{Q^2 - 2q_1}} \pm q_1\right]. \tag{1}$$

As $q_1 \to 0$ the outer $(+)$ and inner $(-)$ boundaries merge and $|q_2 - 1| \to \sqrt{1 - 3Q^2/4}$, and for $q_1 = q_0^-$ the inner boundaries merge $(q_2 = 1)$, see Fig. 101. Continuation of the outer boundaries for $q_0^- \leq q_1 \leq q^-$ are determined by the equation $v_1 = v_2 + v_0$:

$$(q_2 - 1)^2 = q_1^2 - 2q_1 + Q^2 - 2\sqrt{(Q^2 - 1)(q_1^2 - 2q_1 + Q^2)} = v_1^2 - 2v_0 v_1. \tag{2}$$

For $q_1 \geq q^+$, the upper boundary is described by (2) (the larger root), and the lower, by the smaller root in (2) for $q^+ \leq q_1 \leq q_0^+$ and by the formula $(v_1 = v_2 + v_3; q_1 = q_2 + q_3)$

$$q_2 = \frac{1}{2}\left[q_1 - \sqrt{\frac{(2q_1 - 3Q^2)(q_1^2 - 2q_1 + Q^2)}{2q_1 - 4 + Q^2}}\right] \tag{3}$$

for $q_1 \geq q_0^+$.

2. *For $Q^2 \geq \frac{4}{3} q_1 \geq q^+$*: The lower boundary is described by (3). If $Q^2 \geq 2$, then the upper boundary is determined by the equations $v_1 = v_2 + v_3$ and $q_1 = q_2 + q_3$:

$$q_2 = \frac{1}{2}\left[q_1 + \sqrt{\frac{(2q_1 - 3Q^2)(q_1^2 - 2q_1 + Q^2)}{2q_1 - 4 + Q^2}}\right]. \tag{4}$$

If $\frac{4}{3} \leq Q^2 < 2$, then for $\tilde{q}^+ \leq q_1 \leq q_0^+$, the upper boundary is described by (4), and for $q_1 \geq q_0^+$ by (2).

Asymptotics (2), (3), (4) are given in the main text $(q_1 \to \infty)$.

§ 11 Derivation of the Kinetic Equation for Waves

We shall start from Eq. (2), Section 4.3, Chapter VII:

$$\frac{\partial a_k}{\partial t} + i\omega_k a_k = -i \int dk_1 dk_2 [V_{kk_1k_2}^* a_{k_1} a_{k_2} \delta(k - k_1 - k_2)$$

$$+ 2V_{kk_1k_2} a_{k_1}^* a_{k_2} \delta(k + k_1 - k_2) + V_{kk_1k_2}^* a_{k_1}^* a_{k_2}^* \delta(k + k_1 + k_2)]$$

$$- 6i \int dk_1 dk_2 dk_3 (W_{kk_1k_2k_3} + W_{kk_1k_2k_3}^*)$$

$$\times a_{k_1}^* a_{k_2} a_{k_3} \delta(k + k_1 - k_2 - k_3),$$

in which only the terms necessary later on are left. Let us introduce the amplitudes A_k

$$a_k = A_k e^{-i\omega_k t}.$$

They obviously obey the equation

$$\frac{\partial A_k}{\partial t} = -i \int dk_1 dk_2 \{V^*_{kk_1k_2} A_{k_1} A_{k_2} \exp[i(\omega_k - \omega_{k_1} - \omega_{k_2})t]\delta(k - k_1 - k_2)$$

$$+ 2V_{kk_1k_2} A^*_{k_1} A_{k_2} \exp[i(\omega_k + \omega_{k_1} - \omega_{k_2})t]\delta(k + k_1 - k_2)$$
$$+ V^*_{kk_1k_2} A^*_{k_1} A^*_{k_2} \exp[i(\omega_k + \omega_{k_1} + \omega_{k_2})t]\delta(k + k_1 + k_2)\}$$

$$- 6i \int dk_1 dk_2 dk_3 (W_{kk_1k_2k_3} + W^*_{kk_1k_2k_3})A^*_{k_1} A_{k_2} A_{k_3}$$

$$\times \exp[i(\omega_k + \omega_{k_1} - \omega_{k_2} - \omega_{k_3})t]\delta(k + k_1 - k_2 - k_3). \tag{1}$$

By neglecting the wave interaction, the solution of the Eq. (1) is

$$A_k^{(0)}(t) = A_k^{(0)}(0) = \text{const.}$$

We shall seek the solution in the form of the expansion in powers $A_k^{(0)}$, assuming that $A_k^{(0)}$ are the exact values of the amplitudes for $t = 0$:

$$A_k = A_k^{(0)} + A_k^{(1)} + A_k^{(2)} + \cdots, \qquad A_k^{(0)} = A_k(0). \tag{2}$$

In the first approximation

$$A_k^{(1)} = - \int dk_1 dk_2 \left(\frac{V^*_{kk_1k_2} A_{k_1}^{(0)} A_{k_2}^{(0)}}{\omega_k - \omega_{k_1} - \omega_{k_2}} \{\exp[i(\omega_k - \omega_{k_1} - \omega_{k_2})t] - 1\} \right.$$

$$\times \delta(k - k_1 - k_2) + \frac{2V_{kk_1k_2} A_{k_1}^{(0)*} A_{k_2}^{(0)}}{\omega_k + \omega_{k_1} - \omega_{k_2}}$$

$$\times \{\exp[i(\omega_k + \omega_{k_1} - \omega_{k_2})t] - 1\}$$

$$\times \delta(k + k_1 - k_2) + \frac{V^*_{kk_1k_2} A_{k_1}^{(0)*} A_{k_2}^{(0)*}}{\omega_k + \omega_{k_1} + \omega_{k_2}}$$

$$\left. \times \{\exp[i(\omega_k + \omega_{k_1} + \omega_{k_2})t] - 1\} \times \delta(k + k_1 + k_2) \right). \tag{3}$$

In the second approximation

$$A_k^{(2)} = -i \int_0^t dt' \int dk_1 dk_2 \left(2V^*_{kk_1k_2} A_{k_1}^{(1)} A_{k_2}^{(0)} \exp[i(\omega_k - \omega_{k_1} - \omega_{k_2})t'] \right.$$

$$\times \delta(k - k_1 - k_2) + 2V_{kk_1k_2}(A_{k_1}^{(1)*} A_{k_2}^{(0)} + A_{k_1}^{(1)} A_{k_2}^{(0)})$$

$$\times \exp[i(\omega_k + \omega_{k_1} - \omega_{k_2})t']$$

$$\times \delta(k + k_1 - k_2) + 2V^*_{kk_1k_2} A_{k_1}^{(1)*} A_{k_2}^{(0)*} \exp[i(\omega_k + \omega_{k_1} + \omega_{k_2})t']$$

$$\times \delta(k + k_1 + k_2) - 6' \int dk_1 dk_2 dk_3 (W_{kk_1k_2k_3} + W^*_{kk_1k_2k_3})$$

$$\times A_{k_1}^{(0)*} A_{k_2}^{(0)} A_{k_3}^{(0)} \frac{\exp[i(\omega_k + \omega_{k_1} - \omega_{k_2} - \omega_{k_3})t'] - 1}{\omega_k + \omega_{k_1} - \omega_{k_2} - \omega_{k_3}}$$

$$\left. \times \delta(k + k_1 - k_2 - k_3) \right). \tag{4}$$

Substituting (3) into (4), we obtain (for brevity, we write only those terms which will contribute to the kinetic equation, i.e., the terms containing the products like $A_{k_1}^{(0)*} A_{k_2}^{(0)} A_{k_3}^{(0)}$):

$$
A_k^{(2)} = 2 \int dk_1 dk_2 dq_1 dq_2 \Bigg\{ \frac{2 V_{kk_1k_2}^* V_{k_1 q_1 q_2} A_{q_1}^{(0)*} A_{q_2}^{(0)} A_{k_2}^{(0)}}{\omega_{k_1} + \omega_{q_1} - \omega_{q_2}}
$$

$$
\times \left(\frac{\exp[i(\omega_k - \omega_{k_2} + \omega_{q_1} - \omega_{q_2})t] - 1}{\omega_k - \omega_{k_2} + \omega_{q_1} - \omega_{q_2}} \right.
$$

$$
\left. - \frac{\exp[i(\omega_k - \omega_{k_1} - \omega_{k_2})t] - 1}{\omega_k - \omega_{k_1} - \omega_{k_2}} \right)
$$

$$
\times \delta(k - k_1 - k_2)\delta(k_1 + q_1 - q_2)
$$

$$
+ \frac{2 V_{kk_1k_2} V_{k_1 q_1 q_2}^* A_{q_1}^{(0)} A_{q_2}^{(0)*} A_{k_2}^{(0)}}{\omega_{k_1} + \omega_{q_1} - \omega_{q_2}}
$$

$$
\times \left(\frac{\exp[i(\omega_k - \omega_{k_2} - \omega_{q_1} + \omega_{q_2})t] - 1}{\omega_k - \omega_{k_2} - \omega_{q_1} + \omega_{q_2}} \right.
$$

$$
\left. - \frac{\exp[i(\omega_k + \omega_{k_1} - \omega_{k_2})t] - 1}{\omega_k + \omega_{k_1} - \omega_{k_2}} \right)
$$

$$
\times \delta(k + k_1 - k_2)\delta(k_1 + q_1 - q_2)
$$

$$
+ \frac{V_{kk_1k_2} V_{k_2 q_1 q_2}^* A_{q_1}^{(0)} A_{q_2}^{(0)} A_{k_1}^{(0)*}}{\omega_{k_2} - \omega_{q_1} - \omega_{q_2}}
$$

$$
\times \left(\frac{\exp[i(\omega_k + \omega_{k_1} - \omega_{q_2} - \omega_{q_1})t] - 1}{\omega_k + \omega_{k_1} - \omega_{q_1} - \omega_{q_2}} \right.
$$

$$
\left. - \frac{\exp[i(\omega_k + \omega_{k_1} - \omega_{k_2})t] - 1}{\omega_k + \omega_{k_1} - \omega_{k_2}} \right)
$$

$$
\times \delta(k + k_1 - k_2)\delta(k_2 - q_1 + q_2)
$$

$$
+ \frac{V_{kk_1k_2}^* V_{k_1 q_1 q_2} A_{q_1}^{(0)} A_{q_2}^{(0)} A_{k_1}^{(0)*}}{\omega_{k_1} + \omega_{q_1} + \omega_{q_2}}
$$

$$
\times \left(\frac{\exp[i(\omega_k + \omega_{k_2} - \omega_{q_1} - \omega_{q_2})t] - 1}{\omega_k + \omega_{k_2} - \omega_{q_1} - \omega_{q_2}} \right.
$$

$$
\left. - \frac{\exp[i(\omega_k + \omega_{k_1} + \omega_{k_2})t] - 1}{\omega_k + \omega_{k_1} + \omega_{k_2}} \right)
$$

$$
\times \delta(k + k_1 + k_2)\delta(k_1 + q_1 + q_2) - 6 \int dk_1 dk_2 dk_3 (W_{kk_1k_2k_3}
$$

$$
+ W_{kk_1k_2k_3}^*) A_{k_1}^{(0)*} A_{k_2}^{(0)} A_{k_3}^{(0)} \frac{\exp[i(\omega_k + \omega_{k_1} - \omega_{k_2} - \omega_{k_3})t] - 1}{\omega_k + \omega_{k_1} - \omega_{k_2} - \omega_{k_3}}
$$

$$
\times \delta(k + k_1 - k_2 - k_3) \Bigg\} + R, \tag{5}
$$

where R denotes the terms which are nonessential for the future.

We calculate now the time change of $|A_k|^2$, with an accuracy to terms of second order:

$$|A_k|^2 \equiv A_k A_k^* = |A_k^{(0)}|^2 + (A_k^{(0)} A_k^{(1)*} + A_k^{(0)*} A_k^{(1)})$$
$$+ |A_k^{(1)}|^2 + (A_k^{(0)*} A_k^{(2)} + A_k^{(0)} A_k^{(2)*}). \quad (6)$$

Let us perform then the averaging (6) over the oscillation phases. Here, obviously, it is necessary to consider the intervals of time t such that $|\omega_k t| \gg 1$. We denote the averaging by angle brackets. Assuming the hypothesis on the randomness of the oscillation phases, we shall consider that all the correlations of the odd orders are equal to zero; from the pair correlations only the following,

$$\langle A_k A_{k_1}^* \rangle = n_k \delta(k - k_1), \quad (7)$$

are nonzero, and fourth correlations split into the products of the pair correlations, for example,

$$\langle A_k^* A_{k_1}^* A_{k_2} A_{k_3} \rangle = \langle A_k^* A_{k_2} \rangle \langle A_{k_1}^* A_{k_3} \rangle + \langle A_k^* A_{k_3} \rangle \langle A_{k_1}^* A_{k_2} \rangle. \quad (7')$$

Then we obtain from (6) $[A_k(0) = A_k^{(0)}]$:

$$\langle |A_k(t)|^2 \rangle - \langle |A_k^{(0)}|^2 \rangle = \langle |A_k^{(1)}|^2 \rangle + \langle A_k^{(0)*} A_k^{(2)} + A_k^{(0)} A_k^{(2)*} \rangle. \quad (8)$$

The calculations of the terms on the right-hand side of (8), with the help of (3) and (5) and taking into account (7) and (7'), give

$$\langle A_k^{(1)} A_{k_0}^{(1)*} \rangle = \delta(k - k_0) \int dk_1 dk_2 \bigg(8 |V_{kk_1 k_2}|^2 n_{k_1} n_{k_2}$$

$$\times \left\{ \frac{\sin^2[\frac{1}{2}(\omega_k - \omega_{k_1} - \omega_{k_2})t]}{(\omega_k - \omega_{k_1} - \omega_{k_2})^2} \delta(k - k_1 - k_2) \right.$$

$$+ \frac{2 \sin^2[\frac{1}{2}(\omega_k + \omega_{k_1} - \omega_{k_2})t]}{(\omega_k + \omega_{k_1} - \omega_{k_2})^2} \delta(k + k_1 - k_2)$$

$$+ \left. \frac{\sin^2[\frac{1}{2}(\omega_k + \omega_{k_1} + \omega_{k_2})t]}{(\omega_k + \omega_{k_1} + \omega_{k_2})^2} \delta(k + k_1 + k_2) \right\}$$

$$+ 16 V_{kk_1 k_2}^* V_{kk_2 k_2} \frac{\sin^2(\frac{1}{2}\omega_k t)}{\omega_k^2} \delta(k) \bigg). \quad (9)$$

Since, as already noted above, $|\omega_k t| \gg 1$, we may use the formula

$$\pi \delta(x) = \lim_{\alpha \to \infty} \frac{\sin^2 \alpha x}{\alpha x^2}, \quad (10)$$

which gives one of the representations of the δ function. The last term in (9) is proportional to $\delta(K) \cdot \delta(\omega_k)$. But since, for $k = 0$, $\omega_k \neq 0$ this term vanishes. Therefore, later on we shall omit such terms

$$\langle A_{k_0}^{(0)*}A_k^{(2)} + A_{k_0}^{(0)}A_k^{(2)*}\rangle = 8\delta(k - k_0)\int dk_1 dk_2 |V_{kk_1k_2}|^2$$

$$\times \left\{ \frac{2\sin^2[\frac{1}{2}(\omega_k + \omega_{k_1} - \omega_{k_2})t]}{(\omega_k + \omega_{k_1} - \omega_{k_2})^2} n_k n_{k_2}\delta(k + k_1 - k_2) \right.$$

$$+ \frac{\sin^2[\frac{1}{2}(\omega_k + \omega_{k_1} + \omega_{k_2})t]}{(\omega_k + \omega_{k_1} + \omega_{k_2})} \times (n_k n_{k_1} + n_k n_{k_2})\delta(k + k_1 + k_2)$$

$$- \frac{\sin^2[\frac{1}{2}(\omega_k - \omega_{k_1} - \omega_{k_2})t]}{(\omega_k - \omega_{k_1} - \omega_{k_2})^2} (n_k n_{k_1} + n_k n_{k_2})\delta(k - k_1 - k_2)$$

$$\left. - \frac{2\sin^2[\frac{1}{2}(\omega_k + \omega_{k_1} - \omega_{k_2})t]}{(\omega_k + \omega_{k_1} - \omega_{k_2})^2} n_k n_{k_1}\delta(k + k_1 - k_2) \right\}$$

$$+ 48\delta(k - k_0)\int dk_1 dk_2 dk_3 (W_{kk_1k_2k_3} + W_{kk_1k_2k_3}^*)n_k n_{k_1}$$

$$\times \frac{\sin^2[\frac{1}{2}(\omega_k + \omega_{k_1} - \omega_{k_2} - \omega_{k_3})t]}{\omega_k + \omega_{k_1} - \omega_{k_2} - \omega_{k_3}}$$

$$\times \delta(k_1 - k_3)\delta(k + k_1 - k_2 - k_3). \tag{11}$$

Substitution of (9) and (11) into (8) gives [taking into account (10)]

$$\langle A_k(t)A_{k_0}^*(t)\rangle - \langle A_k^{(0)}A_{k_0}^{(0)*}\rangle$$

$$= 4\pi t\delta(k - k_0)\int dk_1 dk_2 |V_{kk_1k_2}|^2\{(n_{k_1}n_{k_2} - n_k n_{k_1} - n_k n_{k_2})$$

$$\times \delta(\omega_k - \omega_{k_1} - \omega_{k_2})\delta(k - k_1 - k_2) + 2(n_{k_1}n_{k_2} + n_k n_{k_2} - n_k n_{k_1})$$

$$\times \delta(\omega_k + \omega_{k_1} - \omega_{k_2})\delta(k + k_1 - k_2) + (n_{k_1}n_{k_2} + n_k n_{k_1} + n_k n_{k_2})$$

$$\times \delta(\omega_k + \omega_{k_1} + \omega_{k_2})\delta(k + k_1 + k_2)\}$$

$$+ 6\int dk_1 dk_2 dk_3 (W_{kk_1k_2k_3} + W_{kk_1k_2k_3}^*)n_k n_{k_1}$$

$$\times (\omega_k + \omega_{k_1} - \omega_{k_2} - \omega_{k_3})\delta(\omega_k + \omega_{k_1} - \omega_{k_2} - \omega_{k_3})$$

$$\times \delta(k_1 - k_3)\delta(k + k_1 - k_2 - k_3). \tag{12}$$

The last term vanishes since it contains the product of a type $x \cdot \delta(x)$; therefore, the derivation of the kinetic equation may be produced, at the very beginning, starting from the equation which doesn't involve the terms, describing the four wave interaction. Finally, since in the problem considered $\omega_k > 0$ for any k, the term containing $\delta(\omega_k + \omega_{k_1} + \omega_{k_2})$ doesn't contribute, which corresponds physically to impossibility of the spontaneous production of "quanta" from the vacuum. Taking into account (7), we have

$$\langle A_k(t)A_{k_0}^*(t)\rangle - \langle A_k^{(0)}A_{k_0}^{(0)*}\rangle = \delta(k - k_0)[n_k(t) - n_k(0)].$$

As was indicated in Section 4.2, Chapter VII, under derivation of the limits of appreciability for the kinetic equation for waves, the characteristic time τ of the spectrum evolution is much more than the inverse maximum increment of the decay instability

$$\tau \gg \frac{1}{\tilde{\gamma}_{\max}}.$$

On the other hand, the wave amplitudes are assumed to be sufficiently small so that $\tilde{\gamma}_{\max} \ll \omega_k$. Consequently, we can choose the time t so that the inequalities

$$\frac{t}{\tau} \ll 1 \quad \text{and} \quad \omega_k t \gg 1$$

perform simultaneously. In this case

$$\frac{n_k(t) - n_k(0)}{t} = \frac{dn_k}{dt},$$

and, with only nonzero terms remaining in (12), we obtain the kinetic equation for waves:

$$\frac{dn_k}{dt} = 4\pi \int dk_1 dk_2 |V_{kk_1k_2}|^2 [(n_{k_1}n_{k_2} - n_k n_{k_1} - n_k n_{k_2})$$

$$\times \delta(\omega_k - \omega_{k_1} - \omega_{k_2})\delta(k - k_1 - k_2) + 2(n_{k_1}n_{k_2} + n_k n_{k_2} - n_k n_{k_1})$$

$$\times \delta(\omega_k + \omega_{k_1} - \omega_{k_2})\delta(k + k_1 - k_2)].$$

§12 Table of Non-Jeans Instabilities (with a Short Summary)

In Table X we use the following notations: ω, k = the frequency and the wave number of oscillations, where $\omega = \text{Re } \omega + i\gamma$, $k^2 = k_r^2 + m^2/r^2$ (in the case of cylindrical symmetry), k_r = the radial component of the wave vector, m = the azimuthal wave number; Jeans frequency $\omega_0 = \sqrt{4\pi G\rho}$, where ρ is the medium density, G is the gravity constant; $A = (\rho_1 - \rho_2)/(\rho_1 + \rho_2)$, where ρ_1 and ρ_2 are densities above and below the boundary of density discontinuity (the value a wide); $v_\perp$ = the thermal velocity of medium (the sound velocity); $v_0 = \Omega R$, where Ω = the angular velocity of the cylinder and R = the cylinder radius; v = the beam velocity (along the z-axis); $T_\perp$ = the transversal (relative to the axis of rotation) temperature, $T_\parallel$ = the longitudinal temperature; the parameters of the beam are marked by indexes "1". The remaining notations are given in the table.

   Non-Jeans instabilities (NJI) of the gravitating medium are, by definition, instabilities which may excite in the gravitating systems provided that Jeans instability (JI) is completely absent or develops much slower than NJI.

Table X Table of non-Jeans instabilities.

Type of instability	Instability conditions	Oscillation branch	Maximum increment	Reference of the work in which the instability was first described
		I. Hydrodynamic instabilities		
(1) Kelvin–Helmholtz instability				1. A. G. Morozov and A. M. Fridman, Preprint of Sib. Izmir. SO Akad. Nauk SSSR, No. 5, Irkutsk (1974)
(a) plane jump of a velocity (along the z-axis)	$ka \ll 1$, $v_0 \quad v_{0x}$, $k = k(k_x, 0, k_z)$	Re $\omega \sim kv_0$	$\gamma \simeq kc_s \gg \omega_0$	
(b) cylindrical jump of a velocity (along $r = r_0$)	$k_r r_0 \gg m$, $k_r a \ll 1$	Re $\omega \simeq -\Omega$	$\gamma \simeq m\Omega \gg \omega_0$ for $m \gg 1$	2. A. G. Morozov, V. G. Fainstein, and A. M. Fridman, in *Dynamics and Evolution of Stellar Systems*, M.-L., Vago, Ed., 1975, p. 238
(2) Flute-like instability				
(a) plane jump of a density (along z-axis)	$ka \ll 1$, $k = k(k_x, 0, k_z)$, $gA > 0$, $g = d\Phi_0/dz$	Re $\omega = 0$	$\gamma \sim \sqrt{kgA}$	3. A. G. Morozov, V. G. Fainstein, and A. M. Fridman, Dokl. Akad. Nauk SSSR, **228**, 1072 (1976)
(b) cylindrical jump of a density (along $r = r_0$)	(1) $ka \quad 1$, $m^2 \gg k_r^2 r_0^2$, $gA > 0$, $g = d\Phi_0/dr - \Omega^2 r$, $A = (\rho_1 - \rho_2)/(\rho_1 + \rho_2)$	Re $\omega \sim m\Omega$; more exactly (for Ω = const): Re $\omega = (m - A)\Omega$	$\gamma \sim \sqrt{k_r gA}$	
	(2) $ka \gg 1$, $gd \ln \rho_0/dr > 0$	Re $\omega \sim m\Omega$	$\gamma = \dfrac{m}{Kr}\sqrt{g}\sqrt{\dfrac{d \ln \rho_0}{dr}} \gg \omega_0$ for $m \gg kr$	4. A. G. Morozov, V. G. Fainstein, and A. M. Fridman, Zh. Eksp. Teor. Fiz **71**, 1249 (1976).

	Conditions	$\mathrm{Re}\,\omega$	γ	References
(3) Beam instability	$\alpha = n_1/n \ll 1$ $v > v_T + v_{T_1}$, $v_T^2 \gg v_0^2,\ v_{T_1}^2 \gg \alpha v_0^2$, $\lvert\omega + 2\Omega_0 - k_z v\rvert \gg \lvert k_z\rvert v_{T_1}$	$\mathrm{Re}\,\omega = \omega_0$	$\gamma = (\alpha^{1/2}/2^{5/4})\omega_0$	5. A. G. Morozov, V. L. Polyachenko, V. G. Fainstein, and A. M. Fridman, Astron Zh. **53**, 735 (1976). A. B. Mikhailovsky and A. M. Fridman, Zh. Eksp. Teor. Fiz. **61** 457 (1971)
(4) Beam-gradient instability	$\dfrac{1}{\Omega}\dfrac{d\Omega}{dr} > 0,\quad m = 2$	$\mathrm{Re}\,\omega = 0$	$\gamma \sim r\Omega/\Omega$	G. S. Bisnovatyi-Kogan, Ya. B. Zel'dovich, R. Z. Sagdeev, and A. M. Fridman. Zh Prikl. Mekh. Tekh. Fiz. 3, 3 (1969)
(5) Temperature-gradient instability	$T_\perp = 0;\ T_z,\ \dfrac{dT_z}{dr} \neq 0$ (a) $\rho_0 = $ const. $v_T > v_0,\ K_z \to 0$ $\lvert\mathrm{Re}\,\omega - m\Omega\rvert \ll mv_T/2r\Omega \cdot \lvert dv_T/dr\rvert$	$\mathrm{Re}\,\omega \sim m\Omega$	$\gamma \sim (m\Omega/rv_T\,dv_T/dr)^{1/3}$	V. L. Polyachenko and I. G. Shukhman, Astron, Zh. **50**, 649 (1973)
	(b) $d\rho_0/dr \neq 0$ $d\ln T_z/d\ln\rho < 0$, $d\ln T_z/d\ln\rho > 1$	$\mathrm{Re}\,\omega \sim m\Omega$	$\gamma \sim (m\Omega/rv_T\,dv_T/dr)^{1/3}$	G. S. Bisnovatyi-Kogan and A. B. Mikhailovsky, Astron. Zh. **50**, 312 (1973)

Table X (*contd.*)

(6) Fire-hose instability				
(a) flat nonrotating layer	$k_\perp h > 0.265,$ $\alpha = \sqrt{\frac{3}{2}} v_{T_x} / \omega_0 h > 2.9$ for $\lambda \gg h(h\alpha \lesssim \lambda \lesssim 3h\alpha^2)$	$\mathrm{Re}\,\omega = 0$	$\gamma \sim \omega_0$	1. R. M. Kulsrud and J. W.-K. Mark, Astrophys. J. **160**, 471 (1970) 2. V. L. Polyachenko and I. G. Shukhman, Pis'ma Astron. Zh. **3** (6) (1977)
(b) cylinder	$\alpha = \frac{1}{2}\sqrt{3}\, v_{T_z}/\Omega R > 1.1;$ $k_z R < 0.55$	$\mathrm{Re}\,\omega = 0$	$\gamma \sim \omega_0$	V. L. Polyachenko, Astron. Zh. **57**, (1980)
(c) ellipsoid at rest	c = minor semiaxis, a = major semiaxis	$\mathrm{Re}\,\omega = 0$	$\gamma \sim \omega_0$	V. L. Polyachenko and I. G. Shukhman, Astron. Zh. **56**, 724 (1979)

II. Kinetic instability

(7) Beam instability	$\alpha = n_1/n \ll 1,$ $v_{T_1}/v \gg \alpha^{1/2},$ $v_T > v_0,$ $v_{T_1}/v_0 > \alpha^{1/2}$	$\mathrm{Re}\,\omega = k_z v - 2\Omega$	$\gamma \sim \alpha\omega_0 v/v_{T_1}$	A. B. Mikhailovsky and A. M. Fridman, Zh. Eksp. Teor. Fiz **61**, 457 (1971) V. L. Polyachenko and I. G. Shukhman, Astron. Zh. **50**, 649 (1973)
(8) Temperature-gradient instability	$\rho = \mathrm{const},$ $\alpha = \rho_1/\rho \ll 1,$ $k_z v_{T_1} \ll \alpha\Omega \ll k_z v_{T_2}$ $m^2 > 4\alpha^2 k_z^2 r_0^2$	$\mathrm{Re}\,\omega = m\Omega$	$\gamma \sim \alpha^2\Omega^2/|k_z| v_{T_2}$	V. L. Polyachenko, A. M. Fridman, I. G. Shukhman, Monthly Notices Roy. Astron. Soc. (1983), in press.
(9) Cone instability	$\omega - l_1\Omega_1 - l_2\Omega_2 - l_3\Omega_3 = 0$			

In the table we enumerate all the hydrodynamic and kinetic NJI known at present. The kinetic instabilities distinguish themselves in that only a rather small group of so-called *resonance* particles takes part in their generation. For this reason the energy capacity of the kinetic instabilities, as a rule, is considerably less than the energy capacity of the hydrodynamic instabilities: The latter are generated by all the particles of the phase space.

The necessary and sufficient condition of the *Kelvin–Helmholtz instability* (KHI) is the occurrence of a rather sharp drop in the medium velocity. The classical example of KHI yields the instability arising at the boundary between two media, one of which is moved respective to another medium (see Fig. 136). The most intensive KHI corresponds to the incompressible medium which is the limiting case of the compressible medium as the sound velocity $c_s \to \infty$. It follows from this fact that KHI builds up more strongly in a hotter medium than in a colder one. This is the main difference between KHI and JI: The latter is stabilized in a hot medium. The growth rate of KHI may considerably exceed the increment of JI in the region of just those short waves which are not important for the growth of JI.

The physical meaning of KHI is as follows. The disturbance amplitude decreases exponentially in both sides from the boundary. Therefore, "the

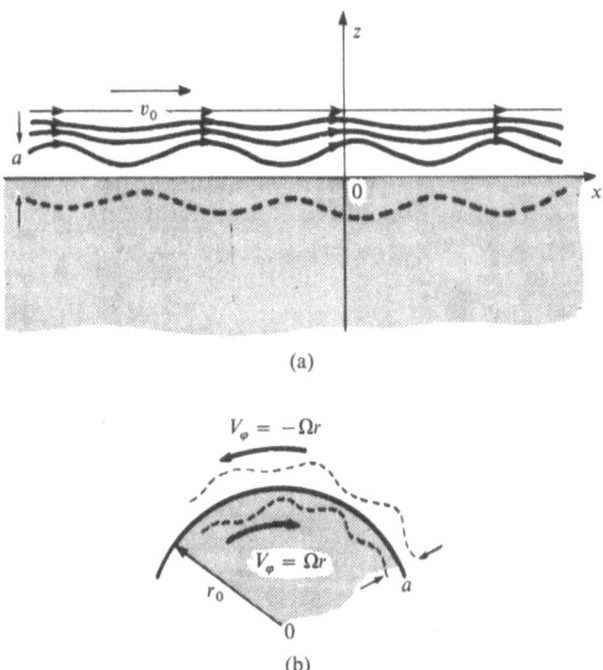

(a)

(b)

Figure 136. (a) Plane jump of a velocity. The medium in the semispace $z < 0$ is at rest, the medium in the semispace $z > 0$ is moving with the velocity v_0, parallel to x-axis; a is the thickness of the transition layer which is perturbed. KHI leads to growth of perturbation amplitude. (b) Cylindrical velocity jump.

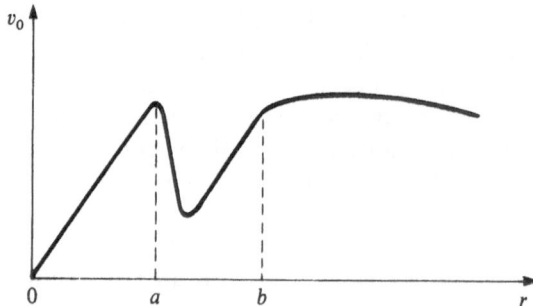

Figure 137. Two-humped rotation curve of the flat subsystem of a spiral galaxy.

scene" plays completely either directly at the disturbed boundary or in the narrow vicinity of the boundary. Outside this region the flow may be assumed as practically undisturbed. This means that in the moving component of the fluid, due to conservation of the total flux through the variable "cross section," the velocity must increase immediately above "humps" produced by the boundary displacements. Thus the reason for increasing the velocity above "the humps" is the same as in a narrowed section of a water-pipe. From Bernoulli's law (i.e., from the conservation of the total pressure: statical + dynamical) it follows that the statical pressure decreases above the humps of the boundary disturbances in the moving component, due to an increase of the dynamical pressure in this region. So a gradient of the statical pressure arises that compels the medium to be moved in such a manner that the early created humps some more increase.

The energy reservoir of KHJ is the kinetic energy of the relative movement of the medium in the vicinity of the disturbed boundary. Evidently, the concept of "the division surface" itself for the continuous function $v_0(z)$ [Fig. 136(a)] (when the tangential discontinuity is absent), is conditional. The best known example of the region, where the condition of KHI is probably satisfied, is the interval ab (Fig. 137) of a sharp drop of the azimuthal velocity of rotation for a number of spiral galaxies.

The necessary and sufficient conditions for the *flute-like* instability (FI) are the following: (1) the presence of a density gradient A; (2) the presence of an "effective" force of the weight g (per unit mass); and (3) the fact that direction of density gradient must be opposite to the vector of the "effective" force of the weight. The effective force of the weight may be produced by various means: by proper or external gravitational fields, by centrifugal force, and so on, in case the nature of this force is not important. The classical example of FI is the instability of the situation when the "heavy" fluid is placed above the "light" fluid (see Fig. 138). Similar to KHI, the increment of FI may greatly exceed the increment of JI. The present name of this instability—"flute-like"—came from plasma physics (where the stability of the boundary between two plasma components—heavy and light—is defined by the form of magnetic surfaces) and characterizes exactly the dynamics of the development of FI. The appearance of an initial perturbation

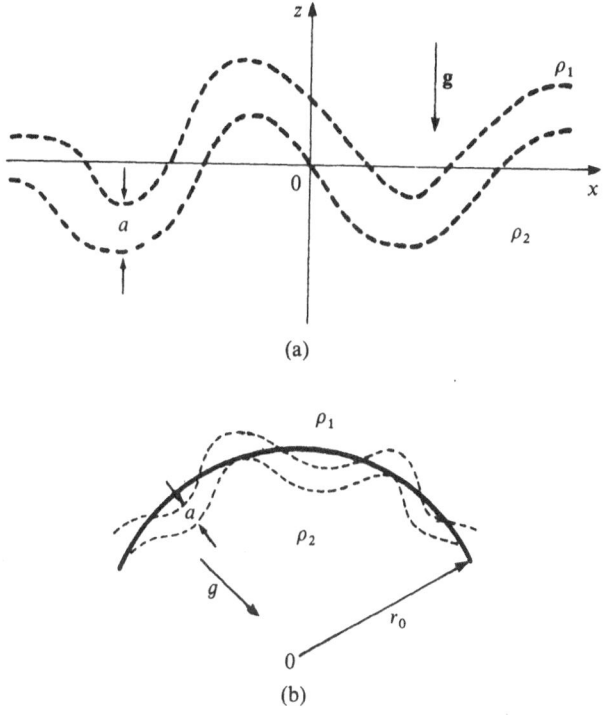

(a)

(b)

Figure 138. (a) Plane jump of a density, $\rho_1 < \rho_2$. (b) Cylindrical jump of a density, $\rho_1 > \rho_2$.

at the boundary of separation between "heavy" and "light" fluids in a form of a flute leads to increasing its dimension in the region of the light component and to forcing the latter out into the region of the heavy component. In the medium without the magnetic field, FI has another name—Reley–Teylor instability (RTI). The necessary and sufficient conditions of FI are fulfilled, for example, in a some region of Galaxy disk (see Fig. 139).

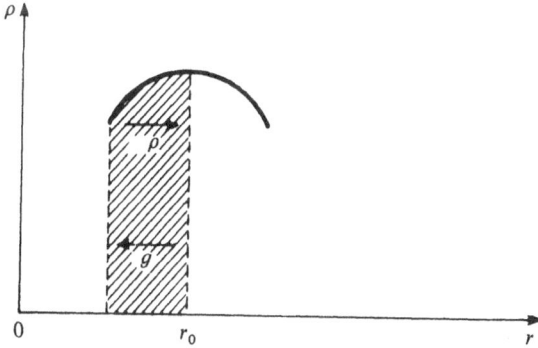

Figure 139. The presence of a density maximum permits the development of FI on the left side from the maximum point r_0. The possible instability region is dashed, here the necessary and sufficient condition of FT is satisfied.

The building-up mechanism of the *hydrodynamic beam instability* (HBI) is essentially different from the *kinetic beam instability* (KBI—see below) and is as follows. Let us have the beam, moving in the medium, undergo perturbations. Then the beam (modulated in a density) when it is passing near an arbitrary point of the medium, serves as the source of the disturbed gravitational field connected with the disturbed beam density by the Poisson equation. The field perturbation causes in turn the perturbation of the medium density.

If the beam velocity v is less than its thermal velocity v_{T_1}, then the perturbation of the beam density will spread with the thermal velocity of the beam. If the beam velocity v is less than the thermal velocity of the medium, then the density "response," arising in the medium, spreads with the thermal velocity of the medium. In the case when the beam velocity exceeds the summary sound velocity in the beam and medium, $v > v_T + v_{T_1}$, the density perturbations of the beam and medium don't go out of each point of its appearance; then they accumulate and that just leads to HBI. Hence we obtain the necessary condition of HBI:

$$v > v_T + v_{T_1} \tag{1}$$

(which is exactly coincident with the analogous criterion for HBI in the case of plasma). Thus, the mechanism of the perturbation growth for HBI is analogous to the mechanism of the growth of a density jump at the front of a shock wave, arising in a supersound movement of a body in the medium. Apart from condition (1) it is necessary to satisfy the condition that the increment of JI was much less than the increment of HBI. This is so if

$$v_0^2 > v_{T_1}^2, v_T^2, \tag{2}$$

where $v_0 = R\Omega$, R is the radius of the cylindrical system.

Contrary to the oscillative (Re $\omega \neq 0$) HBI, the *beam-gradient instability* (BGI) is aperiodical. It was investigated for the case of two oppositely rotating equal-density systems with cylindrical symmetry, which were enclosed each into the other. The instability arose only for the azimuthal mode $m = 2$ in the region where the value of the rotational velocity Ω increases outwards, $d\Omega/dr > 0$. The fact that similar increasing occurs in the region ab in Fig. 137 is not excluded. The relative rotational velocity of the flat and nonflat constituents of the spiral galaxies, which exceeds their thermal velocities, create the necessary conditions for the appearance of HBI and BGI.

The *temperature-gradient instability* (TGI) was investigated in the rotating cylindrical systems provided the transversal temperature is much less than the rotational energy: $T_\perp \ll W_r$, and $dT_z/dr \neq 0$ and finally JI is absent: $v_0^2 \gg v_T^2$. The density gradient has the stabilizing influence for TGI. For $d\rho_0/dr = 0$, TGI occurs for the practically arbitrary temperature gradient. For $d\rho_0/dr \neq 0$, TGI is absent in the region $0 < \eta < 1$, where $\eta = \partial \ln T/\partial \ln \rho$. The necessary condition of TGI is as follows:

$$(m/2r)v_T|dv_T/dr| \gg |\text{Re } \omega - m\Omega|.$$

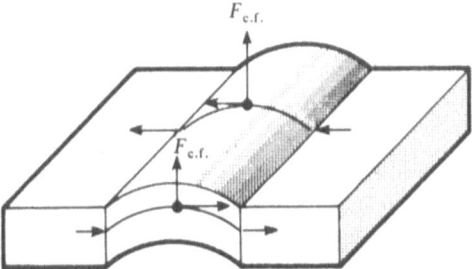

Figure 140. The flat layer with a perturbation. $F_{\text{c.f.}} = mv_\perp^2/R$, where R is the curvature radius, and m is the particle mass.

Since, however, $\text{Re}\,\omega \sim m\Omega$, then the necessary condition of TGI may be simply satisfied: The resonance frequencies must occur for the arbitrarily small (but finite) temperature gradient. For the excitation of waves with frequencies far from the resonance ones, correspondingly more temperature gradients are necessary. The most unstable are the long waves along the rotation axis with a large azimuthal number m.

The nature of TGI is analogous to the universal instability (UI) of a plasma. The stability region for the long waves in a plasma ($0 < \eta < 2$) exceeds by a factor of 2 the analogous stability region ($0 < \eta < 1$) in a gravitating medium, since particles of both negative and positive charge (electrons and ions), contribute to the real part of the dielectric permeability while the gravitating medium consists of the particles with the one sign of "a charge."

The *fire-hose instability* (FHI) may be easily observed if we bend the flexible hose with the flowing water: the hose would then bend already without any help. The centrifugal force, which is the reason of FHI, does not depend on the direction of flux, i.e., if one bends a flat layer as is shown in Fig. 140, then the centrifugal forces for all the molecules are directed outside along the radius of curvature. The less the radius of curvature the more the centrifugal force F_{cf}, consequently, with a decrease in the perturbation wavelength λ, F_{cf} increases, and for the wavelengths $\lambda < \lambda_1$, one may exceed the gravitational returning force F_g, having the opposite direction. Thus the range of the development FHI, $\lambda < \lambda_1$, lies in the region of the wavelengths stable, according to Jeans, short waves.

The considerations described above determine the lower boundary of FHI, which develops for

$$\lambda < \lambda_1 \simeq 3h\alpha^2,$$

$$\alpha^2 = \frac{\langle v_x^2 \rangle}{\langle v_\perp^2 \rangle} = \frac{3v_{Tx}^2}{2\omega_0^2 h^2} = \text{the anisotropy of the velocity distribution.}[6]$$

---

[6] Here and in the table the exact value of numerical coefficients is given for the homogeneous (in density) systems. Since the boundaries in λ ($h\alpha \lesssim \lambda \lesssim 3h\alpha^2$) are given at the long wavelengths limit, $\lambda \gg h$ (in the general case, see Fig. 141) the concrete form of the function $\rho(r)$ leaves the numerical coefficients practically unchanged.

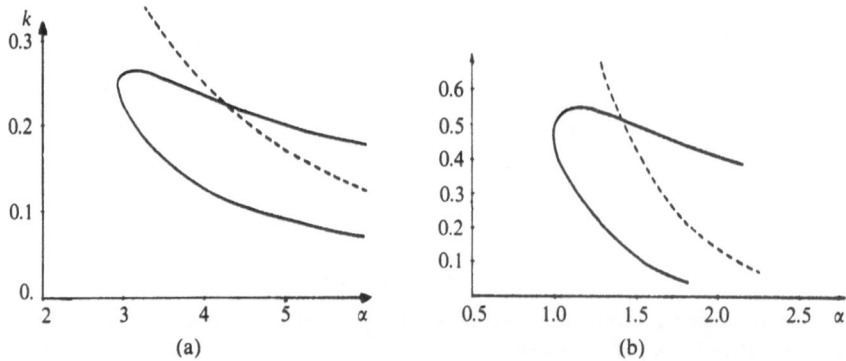

Figure 141. (a) The range of FHI for the gravitating flat layer. Dotted line is the boundary for J.I. in this case. (b) The range of FHI for the gravitating cylinder. Dotted line is the boundary for J.I. in this case.

The range of unstable wavelengths is also restricted from above: $\lambda > \lambda_2 \sim h\alpha$. Indeed, if $\lambda/v_{Tx} \ll 1/\omega_0$, a particle, during the time of one oscillation $1/\omega_0$, has time to pass many wavelengths, i.e., such perturbations are smoothed out by the thermal movement of the particles. The range of FHI for the flat layer is represented in Fig. 141 (a). For the gravitating cylinder an analogous picture is given in Fig. 141.

The *kinetic beam instability* (KBI) may become manifest only when stronger HBI are absent, i.e., provided $\alpha^{1/2} \ll v_{T_1}/v$ $(\alpha = n_1/n)$.

The distribution function of a particle ensemble in the presence of a beam is represented in Fig. 142. Consider the interaction between the wave and the particles. Let the wave have the frequency ω and the wave number k. The most effective interaction occurs with those particles whose velocities are close to the phase velocity of the wave (ω/k). Such particles are called resonance particles; they occupy the narrow interval in the velocity space $(\omega/k - \Delta v_z, \ \omega/k + \Delta v_z)$ $\Delta v_z \ll \omega/k$. If ω/k lies within the range, where $\partial f_0/\partial v < 0$, then the wave "collides" more frequently with those particles which have velocities smaller than the wave velocity, in comparison with the particles exceeding the wave in velocity (since each volume unit contains,

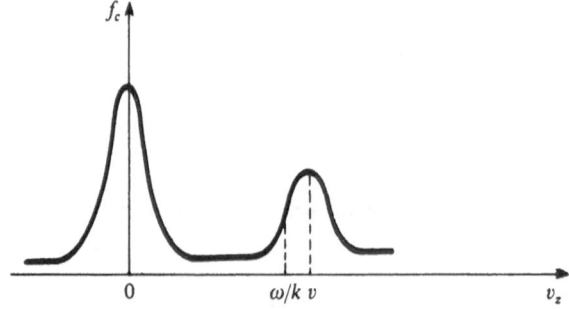

Figure 142. Beam-like distribution function.

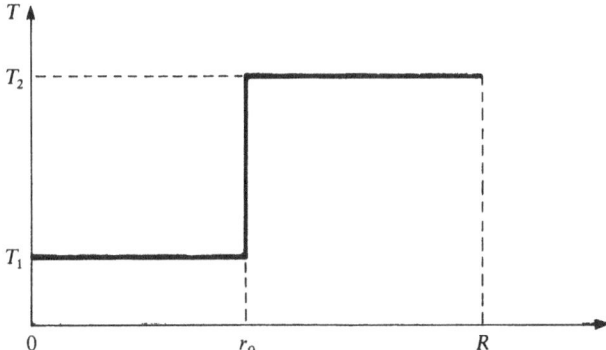

Figure 143. Temperature jump in the homogeneous (in a density) cylinder of the radius R.

for $\partial f_0/\partial v < 0$, slower particles more than faster ones). As a result, the energy of the wave decreases, i.e., the wave is decaying. On the other hand, if the phase velocity of the wave is situated in the range where $\partial f_0/\partial v > 0$, as shown in Fig. 142, then such a wave collides more frequently with those particles whose velocities are higher than the wave phase velocity. Therefore, in this case the wave will on the average collect the energy, i.e., the wave amplitude will increase. The physics of KBI consists of just this. The resonance condition which must be satisfied here is the following: $\mathrm{Re}\,\omega - k_z v + m\Omega = 0$. The strongest KBI develops for $m = 2$.

The *kinetic temperature-gradient instability* (KT–GI) may arise in the simplest case of a homogeneous rotating cylindrical system when a small portion of particles with the density $\alpha\rho$ ($\alpha \ll 1$, $\rho =$ the medium density) has the temperature jump (see Fig. 143). The remaining mass of the medium has the fixed temperature T. The conditions, when the instability occurs with the maximum growth rate, are reflected in Table X.

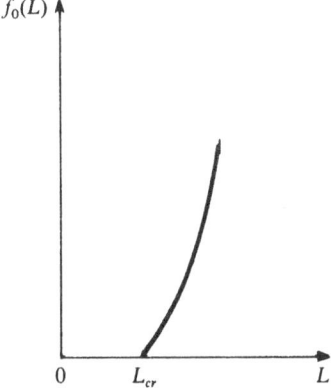

Figure 144. The graph of the distribution function $f_0(L) = \int f_0(E, L)\, dE$ in the vicinity of the critical value of the angular momentum L_{cr}.

The *kinetic cone instability* (KCI) may arise, for example, in the spherically symmetrical stellar system in the presence of the massive compact central body with mass M_h. Assume that r_R is the Roche limit for stars: inside r_R tidal forces from the central body exceed the forces of self-gravitation of the star. Thus, all the stars whose trajectories lie inside the radius r_R, are destroyed. The distribution function of stars in the spherically symmetrical system depends on two integrals of motion: energy E and angular momentum L. In the presence of the central body, stars with small momenta are absent (with momenta $L < L_{cr}$). This means that in the region $L_{cr} + \varepsilon L_{cr}$ (where $\varepsilon \ll 1$) $\partial f_0 / \partial L > 0$ (Fig. 144). Repeating then all the reasonings for KBI, we are led to the conclusion that in this region of momenta the system must be unstable for the suitable sign of the wave energy. KCI leads to anomalously fast (compared to Newtonian collisions) filling in the "empty" region of small values of momenta due to the diffusion flow in the phase space E, L.

References

[1] T. A. Agekyan, *Stars, Galaxies, Metagalaxy*, Nauka, Moscow, 1966 (in Russian).

[2] T. A. Agekyan, Vestn. Leningr. Univ. **1**, 152 (1962) (in Russian).

[3] C. W. Allen, *Astrophysical Quantities*, 3d ed., Athlone Press, London.

[4] V. A. Antonov, Astron. Zh. **37**, 918 (1960) [Sov. Astron. **4**, 859 (1961)].

[5] V. A. Antonov, Vestn. Leningr. Univ. **13**, 157 (1961) (in Russian).

[6] V. A. Antonov, Vestn. Leningr. Univ. **19**, 96 (1962) (in Russian).

[7] V. A. Antonov, in *Itogi Nauki, Ser. Astron.: Kinematika i Dinamika Zvezdnykh Sistem* (*Scientific Findings, Astron. Ser.: Kinematics and Dynamics of Stellar Systems*), VINITI, Moscow, 1968 (in Russian).

[8] V. A. Antonov, Uch. Zap. Leningr. Univ. No. 359, 64 (1971) (in Russian).

[9] V. A. Antonov, Dokl. Akad. Nauk SSSR **209**(3), 584 (1973) [Sov. Phys.— Dokl. **18** (3), 159 (1973)].

[10] V. A. Antonov and E. M. Nezhinskii, Uch. Zap. Leningr. Univ. **363**, 122 (1973) (in Russian).

[11] V. A. Antonov, in *Dinamika Galaktik i Zvezdnykh Skoplenii* (*Dynamics of Galaxies and Star Clusters*), Nauka, Alma-Ata, 1973 (in Russian).

[12] V. A. Antonov and S. N. Nuritdinov, Vestn. Leningr. Univ. **7**, 133 (1975) (in Russian).

[12a] V. A. Antonov and S. N. Nuritdinov, Astron. Zh. **54**, 745 (1977).

[13] V. A. Antonov, in *Itogi Nauki, Ser. Astron., T. 10: Ravnovesije i Ustoichivost' Gravitirujushchikh Sistem* (*Scientific Findings, Astron. Ser.): Equilibrium and Stability of Gravitating Systems*), Nauka, Moscow, 1975 (in Russian).

[13a] V. A. Antonov, in *Dinamika i Evolutsija Zvezdnykh Sistem* (*Dynamics and Evolution of Stellar Systems*), VAGO, Moscow and Leningrad, 1975, p. 269 (in Russian).

[14] V. A. Antonov, Uch. Zap. Leningr. Univ., Tr. Astron. Obs. **24**, 98 (1968) (in Russian).

[15] P. Appell, *Figures d'equilibre d'une masse liquide homogene en rotation*, Russian translation, ONTI, Leningrad and Moscow, 1936.

[16] H. Arp, in *Astrofizika* (*Astrophysics*), Nauka, Moscow, 1961 (in Russian).

[17] W. Baade, *Evolution of Stars and Galaxies*, Cecilia Payne-Gaposhkin, Ed., Harvard University Press, Cambridge, Mass. (Russian translation: Mir, Moscow, 1966.)

[18] R. Bellman, *Stability Theory of Differential Equations*, McGraw-Hill, New York, 1953 (Russian translation: IL, Moscow, 1954).

[19] G. S. Bisnovatyi-Kogan, Ya. B. Zel'dovich, and A. M. Fridman, Dokl. Akad. Nauk SSSR **182**, 794 (1968) [Sov. Phys.—Dokl. **13**, 960 (1969)].

[20] G. S. Bisnovatyi-Kogan, Ya. B. Zel'dovich, R. Z. Sagdeev, and A. M. Fridman, Zh. Prikl. Mekh. Tekh. Fiz. **3**, 3 (1969) (in Russian).

[21] G. S. Bisnovatyi-Kogan and Ya. B. Zel'dovich, Astrofizika **5** (3), 425 (1969) [Astrophysics **5**(3), 198–200 (1969)].

[22] G. S. Bisnovatyi-Kogan and Ya. B. Zel'dovich, Astrofizika **5** (2), 223 (1969) [Astrophysics **5**(3), 105–109 (1969).

[23] G. S. Bisnovatyi-Kogan and Ya. B. Zel'dovich, Astrofizika **6** (3), 387 (1970) [Astrophysics **6**(3), 207–212 (1973)].

[24] G. S. Bisnovatyi-Kogan and Ya. B. Zel'dovich, Astron, Zh. **47**, 942 (1970) [Sov. Astron. **14**, 758 (1971)].

[25] G. S. Bisnovatyi-Kogan, Astrofizika **7**, 121 (1971) [Astrophysics **7**, 70 (1971)].

[26] G. S. Bisnovatyi-Kogan, Astron. Zh. **49**, 1238 (1972) [Sov. Astron. **16**, 997 (1973)].

[27] G. S. Bisnovatyi-Kogan and S. I. Blinnikov, preprint, Inst. Prikl. Mat. Akad. Nauk SSSR **34**, Moscow, 1972 (in Russian).

[28] G. S. Bisnovatyi-Kogan and A. B. Mikhailovskii, Astron. Zh. **50**, 312 (1973) [Sov. Astron. **17**, 205 (1973)].

[29] G. S. Bisnovatyi-Kogan, Pis'ma Astron. Zh. **1**(9), 3 (1975) [Sov. Astron. Lett. **1**(5), 177 (1975)].

[30] M. S. Bobrov, *Kol'tsa Saturna* (*Saturn's Rings*), Nauka, Moscow, 1970 (in Russian).

[31] A. A. Vedenov, E. P. Velikhov, and R. Z. Sagdeev, Usp. Fiz. Nauk **73**, 701 (1961) [Sov. Phys.—Usp. **4**, 332 (1961)].

[32] Yu.-I. K. Veltmann, in *Itogi Nauki, Ser. Astron.: Kinematika i Dinamika Zvezdnykh Sistem* (*Scientific Findings, Astron. Ser.: Kinematics and Dynamics of Stellar Systems*), VINITI, Moscow, 1968 (in Russian).

[33] N. Ya. Vilenkin, *Spetsial'nyje Funktsii i Teorija Group* (*Special Functions and Group Theory*), Nauka, Moscow, 1965 (in Russian).

[34] M. A. Vlasov, *Pis'ma Zh. Eksp. Teor. Fiz.* **2**, 274 (1965) [JETP Lett. **2**, 174 (1965)].

[35] G. de Vaucouleurs, in *Strojenije Zvezdnykh Sistem* (*The Structure of Stellar Systems*), IL, Moscow, 1962 (in Russian).

[36] B. A. Vorontsov-Velyaminov, *Atlas i Katalog Vzaimodeistvuyushchikh Galaktik* (*Atlas and Catalogue of Interconnected Galaxies*), Gos. Astron. Inst. imeni P. K. Shternberg, Moscow, 1959 (in Russian).

[37] B. A. Vorontsov-Velyaminov, *Vnegalakticheskaya Astronomia* (*Extragalactic Astronomy*), Nauka, Moscow, 1972 (in Russian).

[38] S. K. Vsekhsvjatskii, in *Problemy Sovremennoi Kosmogonii* (*Problems of Modern Cosmogony*), V. A. Ambartsumjan, Ed., Nauka, Moscow, 1969 (in Russian).

[39] A. A. Galeev and R. Z. Sagdeev, Vopr. Teor. Plazmy (Plasma Theory Problems) **7**, 3 (1973) (in Russian).

[40] I. F. Ginzburg, V. L. Polyachenko, and A. M. Fridman, Astron. Zh. **48**, 815 (1971) [Sov. Astron. **15**, 643 (1972)].

[41] I. M. Glazman, *Direct Methods of Qualitative Spectral Analysis*, IPST, Jerusalem, 1965 (Engl. translation).

[42] I. S. Gradshtein and I. M. Ryznik, *Table of Integrals, Series, and Products*, Academic Press, New York, 1965 (Engl. translation).

[43] L. E. Gurevich, Vopr. Kosmog. (The Problems of Cosmogony), **2**, 150 (1954) (in Russian).

[44] L. E. Gurevich, Astron. Zh. **46**, 304 (1969) [Sov. Astron, **13**, 241 (1969)].

[45] A. G. Doroshkevich and Ya. B. Zel'dovich, Astron. Zh. **40**, 807 (1963) [Sov. Astron. **7**, 615 (1964)].

[46] B. M. Dzyuba and V. B. Yakubov, Astron. Zh. **47**, 3 (1970) [Sov. Astron. **14**, 1 (1970)].

[47] Ya. B. Zel'dovich and M. A. Podurets, Astron. Zh. **42**, 963 (1965) [Sov. Astron. **9**, 742 (1966)].

[48] Ya. B. Zel'dovich and I. D. Novikov, *Relativistic Astrophysics*, translated by David Arnett, University of Chicago Press, Chicago, 1971 (Engl. translation).

[48a] Ya. B. Zel'dovich and I. D. Novikov, *Strojenije i Evolutsija Vselennoi* (*The Structure and Evolution of the Universe*), Nauka, Moscow, 1975 (in Russian).

[49] Ya. B. Zel'dovich and I. D. Novikov, Preprint, Inst. Prikl. Mat. Akad. Nauk SSSR **23**, Moscow, 1970 (in Russian).

[50] Ya. B. Zel'dovich, V. L. Polyachenko, A. M. Fridman, and I. G. Shukhman, preprint, Inst. Zemn. Magn. Ionsf. Rasprostr. Radiovoln Sibir. Otd. Akad. Nauk SSSR, No. 7–72, Irkutsk, 1972 (in Russian).

[51] G. M. Idlis, Astron. Zh. **29**, 694 (1952) (in Russian).

[52] G. M. Idlis, in *Itogi Nauki, Ser. Astron., Kinematika i Dinamika Zvezdnykh Sistem* (*Scientific Findings, Astron. Ser.: Kinematics and Dynamics of Stellar Systems*), VINITI, Moscow, 1968 (in Russian).

[53] B. B. Kadomtsev, Vopr. Teor. Plazmy (The Problems of Plasma Theory) **2**, 132 (1963) (in Russian).

[54] B. B. Kadomtsev, A. B. Mikhailovskii, and A. V. Timofeev, Zh. Eksp. Teor. Fiz. **47**, 2266 (1964) Sov. Phys.—[JETP **20**, 1517 (1965)].

[55]   E. Kamke, *Differentialgleichungen Reeller Funktionen*, Chelsea Publishing Co., New York, 1947 (Russian translation: Nauka, Moscow, 1965).

[55a] V. I. Karpman, *Nelinejnyje Volny v Dispergirujushchikh Sredakh* (*Nonlinear Waves in Dispersing Media*), Nauka, Moscow, 1973 (in Russian).

[56]   S. V. Kovalevskaya, in *S. V. Kovalevskaya, Nauchnyje Raboty* (*Scientific Reports*), Izd. Akad. Nauk SSSR, Moscow, 1948 (in Russian).

[57]   V. I. Korchagin and L. S. Marochnik, Astron. Zh. **52**, 15 (1975) [Sov. Astron. **19**, 8 (1975)].

[58]   G. G. Kuz'min, Publ. Tartus. Astron. Obs. **32**, 211 (1952) (in Russian).

[59]   G. G. Kuz'min, Publ. Tartus. Astron. Obs. **35**, 285 (1956) (in Russian).

[60]   G. G. Kuz'min, Astron. Zh. **33**, 27 (1956) (in Russian).

[61]   G. G. Kuz'min, Izv. Akad. Nauk Eston. SSR **5**, 91 (1956) (in Russian).

[62]   G. G. Kuz'min and Yu.-I. K. Veltmann, Publ. Tartus. Astron. Obs. **36**, 5 (1967) (in Russian).

[63]   M. A. Lavrentjev and B. V. Shabat, *Metody Teorii Funktsii Kompleksnogo Peremennogo* (*Methods of the Theory of Functions of a Complex Variable*), Fizmatgiz, Moscow, 1958 (in Russian).

[64]   H. Lamb, *Hydrodynamics*, Cambridge University Press, Cambridge, 1932 (Russian translation: Gostekhizdat, Moscow, 1947).

[65]   L. D. Landau, Zh. Eksp. Teor. Fiz. **16**, 574 (1946) (in Russian).

[66]   L. D. Landau, Izv. Akad. Nauk SSSR, Ser. Fiz. **17**, 51 (1953) (in Russian).

[67]   L. D. Landau and E. M. Lifshitz, *Fluid Mechanics*, Pergamon Press, London, and Addison-Wesley Publishing Co., Reading, Mass., 1959 (Engl. translation).

[68]   L. D. Landau and E. M. Lifshitz, *Quantum Mechanics*, Pergamon Press, Oxford and New York, and Addison-Wesley Publishing Co., Reading, Mass., 1965 (Engl. translation).

[69]   L. D. Landau and E. M. Lifshitz, *Mechanics*, Pergamon Press, Oxford and New York, 1976 (Engl. translation).

[70]   L. D. Landau and E. M. Lifshitz, *The Classical Theory of Fields*, Pergamon Press, Oxford and New York, 1971 (Engl. translation).

[71]   N. N. Lebedev, *Problems in Mathematical Physics*, Pergamon Press, Oxford and New York, 1966 (Engl. translation).

[72]   V. I. Lebedev, M. N. Maksumov, and L. S. Marochnik, Astron. Zh. **42**, 709 (1965) [Sov. Astron. **9**, 549 (1966)].

[73]   B. Lindblad, in *Strojenije Zvezdnykh Sistem* (*The Structure of Stellar Systems*), IL, Moscow, 1962 (Russian translation).

[74]   C. C. Lin, *The Theory of Hydrodynamic Stability*, Cambridge University Press, Cambridge, 1966 (Russian translation: IL, Moscow, 1958).

[75]   E. M. Lifshitz, Zh. Eksp. Teor. Fiz. **16**, 587 (1946) (in Russian).

[76]   E. M. Lifshitz and I. M. Khalatnikov, Usp. Fiz. Nauk **30**, 391 (1963) [Sov. Phys.—Usp. **6**, 495 (1964)].

[77]   L. Lichtenshtein, *Figury Ravnovesija Vrashchayushchejsja Zhidkosti* (*Equilibrium Configurations of Rotating Fluid*), Nauka, Moscow, 1965 (in Russian).

[78] A. M. Lyapunov, *Selected Works*, Izd. Akad. Nauk SSSR, Moscow, 1954–1965 (in Russian).

[78a] R. K. Mazitov, Prikl. Mat. Tekh. Fiz. **1**, 27 (1965) (in Russian).

[79] V. A. Mazur, A. B. Mikhailovskii, A. L. Frenkel, and I. G. Shukhman, preprint, Instituta Atomnoi Energii, No. 2693, 1976 (in Russian).

[80] M. N. Maksumov, Dokl. Akad. Nauk Tadzh. SSR **13**, 15 (1970) (in Russian).

[81] M. N. Maksumov, Bul. Inst. Astrofiz. Akad. Nauk Tadzh. SSR **64**, 3 (1974) (in Russian).

[82] M. N. Maksumov, Bul. Inst. Astrofiz. Akad. Nauk Tadzh. SSR **64**, 22 (1974) (in Russian).

[83] M. N. Maksumov and Yu. I. Mishurov, Bul. Inst. Astrofiz. Akad. Nauk Tadzh. SSR **64**, 16 (1974) (in Russian).

[84] L. S. Marochnik and A. A. Suchkov, Usp. Fiz. Nauk **112**(2), 275 (1974) [Sov. Phys.—Usp. **17**(1), 85 (1974)].

[85] A. B. Mikhailovskii, A. L. Frenkel', and A. M. Fridman, Zh. Eksp. Teor. Fiz. **73**, 20 (1977) [Sov. Phys.—JETP **46**, 9 (1977)].

[86] A. B. Mikhailovskii, *Theory of Plasma Instabilities*, Consultants Bureau, New York, 1974, Vol. I (Engl. translation).

[87] A. B. Mikhailovskii, A. M. Fridman, and Ya. G. Epel'baum, Zh. Eksp. Teor. Fiz. **59**, 1608 (1970) [Sov. Phys.—JETP **32**, 878 (1971)].

[88] A. B. Mikhailovskii and A. M. Fridman, Zh. Eksp. Teor. Fiz. **61**, 457 (1971) [Sov. Phys.—JETP **34**, 243 (1972)].

[89] A. B. Mikhailovskii, *Theory of Plasma Instabilities, Vol. 2, Instabilities of an Inhomogeneous Plasma*, Consultants Bureau, New York, 1974 (Engl. translation).

[89a] A. B. Mikhailovskii, V. I. Petviashvili, and A. M. Fridman, Astron. Zh. **56**, 279 (1979).

[90] A. B. Mikhailovskii and A. M. Fridman, Astron. Zh. **50**, 88 (1973) [Sov. Astron. **17**, 57 (1973)].

[90a] A. B. Mikhailovskii, V. I. Petviashvili, and A. M. Fridman, Pis'ma Zh. Eksp. Teor. Fiz. **26**, 129 (1977) [JETP Lett. **24**(2), 43 (1976)].

[91] A. G. Morozov, V. L. Polyachenko, and I. G. Shukhman, preprint, Inst. Zemn. Magn. Ionosf. Rasprostr. Radiovoln Sibir. Otd. Akad. Nauk SSSR, No. 3–72, Irkutsk, 1972, (in Russian).

[92] A. G. Morozov, V. L. Polyachenko, and I. G. Shukhman, preprint, Inst. Zemn. Magn. Ionosf. Rasprostr. Radiovoln Sibir. Otd. Akad. Nauk SSSR, No. 6–72, Irkutsk, 1972, (in Russian).

[93] A. G. Morozov, V. L. Polyachenko, and I. G. Shukhman, preprint, Inst. Zemn. Magn. Ionosf. Rasprostr. Radiovoln Sibir. Otd. Akad. Nauk SSSR, No. 1–73, Irkutsk, 1973, (in Russian).

[94] A. G. Morozov and A. M. Fridman, Astron. Zh. **50**, 1028 (1973) [Sov. Astron. **17**, 651 (1974)].

[95] A. G. Morozov, V. L. Polyachenko, and I. G. Shukhman, preprint, Inst. Zemn. Magn. Ionosf. Rasprostr. Radiovoln Sibir. Otd. Akad. Nauk SSSR, No. 5–74, Irkutsk, 1974, (in Russian).

[96]   A. G. Morozov, V. L. Polyachenko, and I. G. Shukhman, Astron. Zh. **51**, 75 (1974) [Sov. Astron. **18**, 44 (1974)].

[97]   A. G. Morozov, V. L. Polyachenko, A. M. Fridman, and I. G. Shukhman, in *Dinamika i Evolutsija Zvezdnykh Sistem* (*Dynamics and Evolution of Stellar Systems*), VAGO, Akad. Nauk SSSR, Moscow, 1975 (in Russian).

[98]   A. G. Morozov, V. L. Polyachenko, and I. G. Shukhman, preprint, Inst. Zemn. Magn. Ionosf. Rasprostr. Radiovoln Sibir. Otd. Akad. Nauk SSSR, No. 3–75, Irkutsk, 1975, (in Russian).

[98a]   A. G. Morozov and A. M. Fridman, in *Dinamika i Evolutsija Zvezdnykh Sistem* (*Dynamics and Evolution of Stellar Systems*), VAGO, Akad. Nauk SSSR, Moscow, 1975, p. 238 (in Russian).

[99]   A. G. Morozov, V. L. Polyachenko, V. G. Fainshtein, and A. M. Fridman, Astron. Zh. **53**, 946 (1976) [Sov. Astron. **20**, 535 (1976)].

[99a]   A. G. Morozov, V. L. Fainshtein, and A. M. Fridman, Dokl. Akad. Nauk SSSR **231**, 588 (1976) [Sov. Phys.—Dokl. **21**(11), 661 (1976)].

[100]   A. G. Morozov, V. G. Fainshtein, and A. M. Fridman, Zh. Eksp. Teor. Fiz. **71**, 1249 (1976) [Sov. Phys.—JETP **44**, 653 (1976)].

[101]   K. F. Ogorodnikov, *Dynamics of Stellar Systems*, Pergamon, Oxford, 1965 (Engl. translation).

[102]   L. M. Ozernoi and A. D. Chernin, Astron. Zh. **44**, 321 (1967) [Sov. Astron. **11**, 907 (1968)].

[103]   L. M. Ozernoi and A. D. Chernin, Astron. Zh. **45**, 1137 (1968) [Sov. Astron. **12**, 901 (1969)].

[104]   J. H. Oort, In *Strojenije Zvezdnykh Sistem* (*The Structure of Stellar Systems*), IL, Moscow, 1962 (Russian translation).

[105]   M. Ya. Pal'chik, A. Z. Patashinskii, V. K. Pienus, and Ya. G. Epel'baum, preprint, Instituta Yadernoi Fiziki Sibir. Otd. Akad. Nauk SSSR, 99–100, Novosibirsk, 1970 (in Russian).

[106]   M. Ya. Pal'chik, A. Z. Patashinskii, and V. K. Pienus, preprint, Instituta Yadernoi Fiziki Sibir. Otd. Akad. Nauk SSSR, 100, Novosibirsk, 1970 (in Russian).

[107]   A. G. Pakhol'chik, Astron. Zh. **39**, 953 (1962) [Sov. Astron. **6**, 741 (1963)].

[108]   S. B. Pikel'ner, *Osnovy Kosmicheskoi Elektrodinamiki* (*Principles of Cosmical Electrodynamics*), Fizmatgiz, Moscow, 1961 (in Russian).

[108a] V. L. Polyachenko, V. S. Synakh, and A. M. Fridman, Astron. Zh. **48**, 1174 (1971) [Sov. Astron. **15**, 934 (1972)].

[109]   V. L. Polyachenko and A. M. Fridman, Astron. Zh. **48**, 505 (1971) [Sov. Astron. **15**, 396 (1971)].

[110]   V. L. Polyachenko and A. M. Fridman, Astron. Zh. **49**, 157 (1972) [Sov. Astron. **16**, 123 (1972)].

[111]   V. L. Polyachenko and I. G. Shukhman, preprint, Inst. Zemn. Magn. Ionosf. Rasprostr. Radiovoln Sibir. Otd. Akad. Nauk SSSR, 1–72, Irkutsk, 1972 (in Russian).

[112]   V. L. Polyachenko and I. G. Shukhman, preprint, Inst. Zemn. Magn. Ionosf. Rasprostr. Radiovoln Sibir. Otd. Akad. Nauk SSSR, 2–72, Irkutsk, 1972, (in Russian).

[113] V. L. Polyachenko and I. G. Shukhman, Astron. Zh. **50**, 97 (1973) [Sov. Astron. **17**, 62 (1973)].

[114] V. L. Polyachenko and I. G. Shukhman, Astron. Zh. **50**, 649 (1973) [Sov. Astron. **17**, 413 (1973).

[115] V. L. Polyachenko and I. G. Shukhman, Astron. Zh. **50**, 721 (1973) [Sov. Astron. **17**, 460 (1974)].

[116] V. L. Polyachenko, kandidatskaja dissertatsija (doctoral dissertation), Leningrad, 1973 (in Russian).

[117] V. L. Polyachenko, Dokl. Akad. Nauk SSSR **229**, 1335 (1976) [Sov. Phys.—Dokl. **21**(8), 417 (1976)].

[118] V. L. Polyachenko and I. G. Shukhman, Pis'ma Astron. Zh. **3**, 199 (1977) [Sov. Astron. Lett. **3**, 105 (1977)].

[119] V. S. Safronov, *Evolutsija Doplanetnogo Oblaka i Obrazovanije Zemli i Planet* (*The Evolution of a Protoplanetary Cloud and Formation of the Earth and Planets*), Nauka, Moscow, 1969 (in Russian).

[120] V. P. Silin and A. A. Rukhadze, *Elektromagnitnyje Svoistva Plazmy i Plazmopodobnykh Sred* (*The Electromagnetic Properties of Plasma and Plasmalike Media*), Gosatomoizdat, Moscow, 1961 (in Russian).

[121] L. J. Slater, *Confluent Hypergeometric Functions*, Cambridge University Press, Cambridge, 1960 (Russian translation: Comp. Centr. Akad. Nauk SSSR, Moscow, 1966).

[122] H. B. Sawyer Hogg, in *Strojenije Zvezdnykh Sistem* (*The Structure of Stellar Systems*), IL, Moscow, 1962 (Russian translation).

[123] Th. H. Stix, *The Theory of Plasma Waves*, McGraw-Hill, New York, 1962 (Russian translation: Atomoizdat, Moscow, 1966).

[124] V. S. Synakh, A. M. Fridman, and I. G. Shukhman, Dokl. Akad. Nauk SSSR **201**(4), 827 (1971) [Sov. Phys.—Dokl. **16**(12), 1062 (1972)].

[125] V. S. Synakh, A. M. Fridman, and I. G. Shukhman, Astrofizika **8**(4), 577 (1972) [Astrophysics **8**(4), 338 (1972)].

[126] S. I. Syrovatskii, Tr. Fiz. Inst. Akad. Nauk **8**, 13 (1956) (in Russian).

[127] M. F. Subbotin, *Kours Nebesnoi Mekhaniki* (*Celestial Mechanics*), GTTI, Moscow–Leningrad, 1949, Vol. 3 (in Russian).

[128] A. V. Timofeev, preprint, Inst. Atomnoi Energii im. Kurchatova, Moscow, 1968 (in Russian).

[129] A. V. Timofeev, Usp. Fiz. Nauk **102**, 185 (1970) [Sov. Phys.—Usp. **13**(5), 632 (1971)].

[130] A. N. Tikhonov and A. A. Samarskii, *Equations in Mathematical Physics*, Macmillan, New York, 1963 (Engl. translation).

[131] B. A. Trubnikov, Vopr. Teor. Plazmy (The Problems of Plasma Theory) **1**, 98 (1963) (in Russian).

[132] E. T. Whittaker and G. N. Watson, *A Course of Modern Analysis*, The University Press, New York, 1947, Vol. 1 (Russian translation: GIFML, Moscow, 1959).

[133] E. T. Whittaker and G. N. Watson, *A Course of Modern Analysis*, The University Press, New York, 1947, Vol. 2 (Russian translation: GIFML, Moscow, 1963).

[134]   V. N. Fadeeva and N. M. Terentjev, *Tablitsy Znachenii Integrala Verojatnosti ot Kompleksnogo Argumenta* (*Tables of Integral Values of Probability of a Complex Argument*), GITTL, Moscow, 1954 (in Russian).

[135]   Ya. B. Fainberg, At. Energ. (At. Energy) 11, 391 (1961) (in Russian).

[136]   E. L. Feinberg, Usp. Fiz. Nauk 104, 539 (1971) [Sov. Phys.—Usp. 14, 455 (1972)].

[137]   V. G. Fesenkov, Astron. Zh. 28, 492 (1951) (in Russian).

[138]   D. A. Frank-Kamenetskii, *Lektsii po Fizike Plazmy* (*Lectures on Plasma Physics*), Atomizdat, Moscow, 1968 (in Russian).

[139]   A. M. Fridman, in *Itogi Nauki, Ser. Astron., T. 10: Ravnovesije i Ustoichivost' Gravitirujushchikh Sistem* (*Scientific Findings, Astron. Ser.: Equilibrium and Stability of Gravitating Systems*), Moscow, 1975 (in Russian).

[140]   A. M. Fridman, Astron. Zh. 43, 327 (1966) [Sov. Astron. 10, 261 (1966)].

[141]   A. M. Fridman, Astron. Zh. 48, 910 (1971) [Sov. Astron. 15, 720 (1972)].

[142]   A. M. Fridman, Astron. Zh. 48, 320 (1971) [Sov. Astron. 15, 250 (1971)].

[143]   A. M. Fridman and I. G. Shukhman, Dokl. Akad. Nauk SSSR, 202, 67 (1972) [Sov. Phys.—Dokl. 17, 44 (1972)].

[144]   A. M. Fridman, doktorskaja dissertatsija (doctoral thesis), Moscow, 1972 (in Russian).

[145]   L. G. Khazin and E. E. Shnol', Dokl. Akad. Nauk SSSR, 185, 1018 (1969) [Sov. Phys.—Dokl. 14, 332 (1969)].

[146]   F. Zwicky, in *Strojenije Zvezdnykh Sistem* (*The Structure of Stellar Systems*), IL, Moscow, 1962 (Russian translation).

[147]   S. Chandrasekhar, *Principles of Stellar Dynamics*, Chicago University Press, Chicago, 1942 (Reprinted: Dover, New York, 1960) (Russian translation: IL, Moscow, 1948).

[148]   S. Chandrasekhar, *Ellipsoidal Figures of Equilibrium*, Yale University Press, New Haven, 1969 (Russian translation: Mir, Moscow, 1973).

[149]   V. D. Shafranov, Vopr. Teor. Plazmy (The Problems of Plasma Theory) 3, 3 (1963) (in Russian).

[150]   M. Schwarzschild, *Structure and Evolution of the Stars*, Princeton University Press, Princeton, N.J., 1958 (Reprinted: Dover, New York, 1965) (Russian translation: IL, Moscow, 1961).

[151]   O. Yu. Schmidt, *Chetyre Lektsii o Teorii Proiskhozhdenija Zemli* (Four Lectures on the Theory of Origin of the Earth), Izd. Akad. Nauk SSSR, Moscow, 1950 (in Russian).

[152]   E. E. Shnol', Astron. Zh. 46, 970 (1969) [Sov. Astron. 13, 762 (1970)].

[153]   I. G. Shukhman, kanadidatskaja dissertatsija (doctoral dissertation), Leningrad, 1973 (in Russian).

[154]   I. G. Shukhman, Astron. Zh. 50, 651 (1973) [Sov. Astron. 17, 415 (1973)].

[155]   A. Einstein, *Sobr. Sochin., T. 2* (*Selected Works, Vol. 2*), Nauka, Moscow, 1967 (in Russian).

[156]   L. E. El'sgoltz, *Differential Equations and the Calculus of Variations*, Mir, Moscow, 1970 (in English).

[157] E. Yanke, F. Emde, and F. Loesh, *Spetsialnyje Funktsii. Formuly, Grafiki, Tablitsy* (*Special Functions. Formulas, Graphs, and Tables*), Nauka, Moscow, 1964 (in Russian).

[158] M. Aggarswal and S. P. Talwar, Monthly Notices Roy. Astron. Soc. **146**, 187 (1969).

[159] E. S. Avner and I. R. King, Astron. J. **72**, 650 (1967).

[160] B. Barbanis and K. H. Prendergast, Astron. J., **72**(2), 215 (1967).

[161] J. M. Bardeen and R. V. Wagoner, Astrophys. J. **158**(2), 65 (1969).

[162] J. M. Bardeen and R. V. Wagoner, Astrophys. J., **167**(3), 359 (1971).

[163] L. Bel, Astrophys. J. **155**, 83 (1969).

[164] H. P. Berlage, Proc. K. Ned. Akad. Wet. Amsterdam **51**, 965 (1948).

[165] H. P. Berlage, Proc. K. Ned. Akad. Wet. Amsterdam **53**, 796 (1948).

[166] A. B. Bernstein, F. A. Frieman, H. D. Kruskal, and R. M. Kulsrud, Proc. Roy. Soc. London **17**, 244 (1958).

[167] P. Bodenheimer and J. P. Ostriker, Astrophys. J. **180**, 159 (1973).

[168] W. B. Bonnor, Appl. Math. **8**, 263 (1967).

[169] W. H. Bostick, Rev. Mod. Phys. **30**, 1090 (1958).

[170] J. C. Brandt, Astrophys. J. **131**, 293 (1960).

[171] J. C. Brandt, Monthly Notices Roy. Astron. Soc. **129**, 309 (1965).

[172] G. H. Bryan, Philos. Trans. **180**, 187 (1888).

[173] E. M. Burbidge, G. R. Burbidge, and K. H. Prendergast, Astrophys. J. **130**, 739 (1959).

[174] E. M. Burbidge, G. R. Burbidge, and K. H. Prendergast, Astrophys. J. **137**, 376 (1963).

[175] E. M. Burbidge, G. R. Burbidge, and K. H. Prendergast, Astrophys. J. **140**, 80, 1620 (1964).

[176] E. M. Burbidge and G. R. Burbidge, Astrophys. J. **140**, 1445 (1964).

[177] B. F. Burke, Astron. J. **62**, 90 (1957).

[178] W. B. Burton, Bull. Astron. Netherl. **18**, 247 (1966).

[179] G. L. Camm, Monthly Notices Roy. Astron. Soc. **101**, 195 (1941).

[180] G. L. Camm, Monthly Notices Roy. Astron. Soc. **112**(2), 155 (1952).

[181] G. Carranza, G. Courtes, Y. Georgellin, and G. Monnet, C. R. Acad. Sci. Paris **264**, 191 (1967).

[182] G. Carranza, G. Courtes, Y. Georgellin, G. Monnet, and A. Pourcelot, Ann. Astrophys. **31**, 63 (1968).

[183] G. Carranza, R. Crillon, and G. Monnet, Astron. Astrophys. **1**, 479 (1969).

[184] K. M. Case, Phys. Fluids **3**, 149 (1960).

[185] S. Chandrasekhar and E. Fermi, Astrophys. J. **118**, 113 (1953).

[186] S. Chandrasekhar, *Hydrodynamics and Hydromagnetic Stability*, Clarendon Press, Oxford, 1961.

[187] G. Contopoulos, Astrophys. J. **163**, 181 (1971).

[188] G. Contopoulos, Astrophys. Space Sci. **13**(2), 377 (1971).

[189]   G. Contopoulos, Astrophys. J. **160**, 113 (1970).

[190]   G. Courtes and R. Dubout-Crillon, Astron. Astrophys. **11**(3), 468 (1971).

[191]   M. Crezé and M. O. Mennessier, Astron. Astrophys. **27**(2), 281 (1973).

[192]   J. M. A. Danby, Astron. J. **70**, 501 (1965).

[193]   G. Danver, Ann. Obs. Lund. **10**, 134 (1942).

[194]   M. E. Dixon, Astrophys. J. **164**, 411 (1971).

[194a] J. P. Doremus and M. R. Feix, Astron. Astrophys. **29**(3), 401 (1973).

[195]   O. J. Eggen, D. Lynden-Bell, and A. R. Sandage, Astrophys. J. **136**, 748 (1962).

[196]   A. S. Eddington, Monthly Notices Roy. Astron. Soc. **75**(5), 366 (1915).

[197]   A. S. Eddington, Monthly Notices Roy. Astron. Soc. **76**(7), 572 (1916).

[198]   G. Elwert and D. Z. Hablick, Astrophys. J., **61**, 273 (1965).

[199]   S. I. Feldman, and C. C. Lin. Stud. Appl. Math. **52**, 1 (1973).

[200]   E. Fermi, Progr. Theor. Phys. **5**, 570 (1950).

[201]   K. C. Freeman, Monthly Notices Roy. Astron. Soc. **130**, 183 (1965).

[202]   K. C. Freeman, Monthly Notices Roy. Astron. Soc. **133**(1), 47 (1966).

[203]   K. C. Freeman, Monthly Notices Roy. Astron. Soc. **134**(1), 1 (1966).

[204]   K. C. Freeman, Monthly Notices Roy. Astron. Soc. **134**(1), 15 (1966).

[205]   K. C. Freeman, Astrophys. J. **160**(3), 811 (1970).

[206]   K. C. Freeman and G. de Vaucouleurs, Astron. J. **71**(9), 855 (1966).

[207]   M. Fujimoto, Publ. Astron. Soc. Jpn. **15**, 107 (1963).

[208]   M. Fujimoto, *IAU Symposium No. 29*, D. Reidel, Dordrecht, 1966.

[209]   P. Goldreich and D. Lynden-Bell, Monthly Notices Roy. Astron. Soc. **130**, (2–3), 97 (1965).

[210]   P. Goldreich and D. Lynden-Bell, Monthly Notices Roy. Astron. Soc. **130**, (2–3), 125 (1965).

[211]   J. Guibert, Astron. Astrophys. **30**(3), 353 (1974).

[212]   D. ter Haar, Rev. Mod. Phys. **22**, 119 (1950).

[213]   A. P. Henderson, Ph.D. thesis, University of Maryland, 1967.

[214]   M. Henon, Ann. Astrophys. **29**(2), 126 (1959).

[215]   F. Hohl, Astron. J. **73**(5), 98, 611 (1968).

[216]   M. Henon, Bull Astron, **3**, 241 (1968).

[217]   M. Henon, Astron. Astrophys. **24**(2), 229 (1973).

[218]   R. W. Hockney and D. R. K. Brownrigg, Monthly Notices Roy. Astron. Soc. **167**(2), 351 (1974).

[219]   F. Hohl, J. Comput. Phys. **9**, 10 (1972).

[220]   F. Hohl, Astrophys. J. **168**, 343 (1971).

[221]   R. J. Hosking, Austr. J. Phys. **22**(4), 505 (1969).

[222]   F. Hoyle and M. Schwarzschild, Astrophys. J., Suppl. **2**(13), (1955).

[223]   F. Hoyle, *Frontiers of Astronomy*, New York, 1960.

[224]   F. Hoyle and W. A. Fowler, Nature **213**, 373 (1967).

[225]   E. Hubble, *The Realm of the Nebulae*, Yale University Press, New Haven, 1937.

[226] C. Hunter, Monthly Notices Roy. Astron. Soc. **126**(4), 299 (1963).

[227] C. Hunter, Monthly Notices Roy. Astron. Soc. **129**(3–4), 321 (1965).

[228] C. Hunter, Stud. Appl. Math. **48**(1), 55 (1969).

[229] C. Hunter, Astrophys. J. **157**(1), 183 (1969).

[230] C. Hunter and A. Toomre, Astrophys. J. **155**(3), 747 (1969).

[231] C. Hunter, Astrophys. J. **162**(1), 97 (1970).

[232] C. Hunter, in *Dynamics of Stellar Systems*, Hayli, ed., D. Reidel, Dordrecht and Boston, 1970.

[233] C. Hunter, Ann. Rev. Fluid Mech., **4**, 219 (1972).

[234] C. Hunter, Monthly Notices Roy. Astron. Soc. **166**, 633 (1974).

[235] C. Hunter, Astron. J. **80**(10), 783 (1975).

[236] G. M. Idlis, Astron. Zn. **3**, 860 (1959).

[237] K. A. Innanen, J. Roy. Astron. Soc. Can. **63**(5), 260 (1969).

[238] J. R. Ipser and K. S. Thorne, preprint, OAP-121 California Inst. Technol., Pasadena, 1968.

[239] J. R. Ipser and K. S. Thorne, Astrophys. J., **154**(1), 251 (1968).

[240] J. D. Jackson, Plasma Phys. **1**, 171 (1960).

[241] J. H. Jeans, Monthly Notices Roy. Astron. Soc. **76**(7), 767 (1916).

[242] J. Jeans, *Astronomy and Cosmology*, Cambridge University Press, Cambridge, 1929.

[243] H. M. Johnson. Astrophys. J. **115**, 124 (1952).

[244] W. H. Julian, Astrophys. J. **155**(1), 117 (1969).

[245] W. H. Julian and A. Toomre, Astrophys. J. **146**(3), 810 (1966).

[246] B. B. Kadomtzev and O. P. Pogutze, Phys. Rev. Lett. **25**(17), 1155 (1970).

[247] F. D. Kahn and L. Woltjer, Astrophys. J. **130**, 705 (1959).

[248] F. D. Kahn and J. E. Dyson, Ann. Rev. Astron. Astrophys. **3**, 47 (1965).

[249] A. J. Kalnajs, Ph.D. thesis, Harvard University, 1965.

[250] A. J. Kalnajs, in *IAU Symposium No. 38*, D. Reidel, Dordrecht, 1970.

[251] A. J. Kalnajs, Astrophys. J. **166**(2), 275 (1971).

[252] A. J. Kalnajs, Astrophys. J. **175**(1), 63 (1972).

[253] A. J. Kalnajs, Astrophys. J. **180**, 1023 (1973).

[254] A. J. Kalnajs and G. E. Athanassoula, Monthly Notices Roy. Astron. Soc. **168**, 287 (1974).

[255] S. Kato, Publ. Astron. Soc. Jpn. **23**, 467 (1971).

[256] S. Kato, Publ. Astron. Soc. Jpn. **25**, 231 (1973).

[257] F. J. Kerr, Monthly Notices Roy. Astron. Soc. **123**, 327 (1962).

[258] F. J. Kerr and G. Westerhout, in *Stars and Stellar Systems*, Chicago University Press, Chicago and London, 1965.

[259] F. J. Kerr, Austr. J. Phys. Astrophys. Suppl., No. 9, (1969).

[260] I. R. King, Astron. J. **70**(5), 376 (1965).

[261] N. Krall and M. Rosenbluth, Phys. Fluids **6**, 254 (1963).

[262] G. P. Kuiper, *Astrophysics*, J. A. Hynek, ed., New York, 1951.

[263] R. M. Kulsrud, J. W.-K. Mark, and A. Caruso, Astrophys. Space Sci. **14**(1), 52 (1971).

[264] R. M. Kulsrud and J. W.-K. Mark, Astrophys. J. **160**, 471 (1970).

[265] P. S. Laplace, Mem. Acad. Sci. (Mécanique Celeste, k. 3, p. VI), 1789 (1787).

[265a] M. J. Lighthill, J. Inst. Math. Appl. **1**, 269 (1965).

[266] E. P. Lee, Astrophys. J. **148**, 185 (1967).

[267] C. C. Lin and F. H. Shu, Astrophys. J. **140**(2), 646 (1964).

[268] C. C. Lin, L. Mestel, and F. Shu, Astrophys. J. **142**(4), 1431 (1965).

[269] C. C. Lin, SIAM J. Appl. Math. **14**(4), 876 (1966).

[270] C. C. Lin and F. H. Shu, Proc. Nat. Acad. Sci. USA **55**(2), 229 (1966).

[271] C. C. Lin, C. Yuan, and F. H. Shu, Astrophys. J. **155**(3), 721 (1969).

[272] C. C. Lin, in *IAU Symposium, No.. 38*, D. Reidel, Dordrecht, 1970.

[273] B. Lindblad, Stockholm Obs. Ann. **20**(6), (1958).

[274] B. Lindblad, Stockholm. Obs. Ann. **22**, 3 (1963).

[275] P. O. Lindblad, Popular Arstok Tidschr. **41**, 132 (1960).

[276] P. O. Lindblad, Stockholm Obs. Ann. **21**, 3 (1960).

[277] P. O. Lindblad, in *Interstellar Matter in Galaxies*, L. Woltjer, ed., New York, 1962.

[278] C. Lundquist, Phys. Rev. **83**, 307 (1951).

[279] D. Lynden-Bell, Monthly Notices Roy. Astron. Soc. **120**(3), 204 (1960).

[280] D. Lynden-Bell, Monthly Notices Roy. Astron. Soc. **123**, 447 (1962).

[281] D. Lynden-Bell, Monthly Notices Roy-Astron. Soc. **124**, 279 (1962).

[282] D. Lynden-Bell, Astrophys. J. **139**, 1195 (1964).

[283] D. Lynden-Bell, Monthly Notices Roy. Astron. Soc. **129**, 299 (1965).

[284] D. Lynden-Bell, *The Theory of Orbits in a Solar System and in Stellar Systems*, 1966.

[285] D. Lynden-Bell, Lect. Appl. Math. **9**, 131 (1967).

[286] D. Lynden-Bell, Monthly Notices Roy. Astron. Soc. **136**, 101 (1967).

[287] D. Lynden-Bell and J. P. Ostriker, Monthly Notices Roy. Astron. Soc. **136**(3), 293 (1967).

[288] D. Lynden-Bell and N. Sanitt, Monthly Notices Roy. Astron. Soc. **143**(2), 176 (1969).

[289] D. Lynden-Bell and A. J. Kalnajs, Monthly Notices Roy. Astron. Soc. **157**, 1 (1972).

[289a] J. W.-K. Mark, Astrophys. J. **169**, 455 (1971).

[290] J. W.-K. Mark, Proc. Nat. Acad. Sci. USA **68**(9), 2095 (1971).

[290a] J. W.-K. Mark, Astrophys. J. **193**, 539 (1974).

[291] J. C. Maxwell, *The Scientific Papers*, Cambridge University Press, Cambridge, 1859, Vol. 1, p. 287.

[292] L. Mestel, Monthly Notices Roy. Astron. Soc. **126**(5–6), 553 (1963).

[293] R. W. Michie, Monthly Notices Roy. Astron. Soc. **125**(2), 127 (1963).

[294]   R. W. Miller, K. H. Prendergast, and W. J. Quirk, Astrophys. J. **161**(3), 903 (1970).

[295]   G. Münch, Publ. Astron. Soc. Pacific, **71**, 101 (1959).

[295a] T. O'Neil, Phys. Fluids **8**, 2255 (1965).

[296]   J. H. Oort, Bull. Astron. Netherl., **6**, 249 (1932).

[297]   J. H. Oort, Scientific Am. **195**, 101 (1956).

[298]   J. H. Oort, F. J. Kerr, and G. Westerhout, Monthly Notices Roy. Astron. Soc. **118**, 319 (1958).

[299]   J. H. Oort, in *Interstellar Matter in Galaxies*, L. Woltjer, ed., W. A. Benjamin, New York, 1962.

[300]   J. P. Ostriker and P. Bodenhiemer, Astrophys. J. **180**, 171 (1973).

[301]   J. P. Ostriker and P. J. E. Peebles, Astrophys. J. **186**(2), 467 (1973).

[302]   P. J. Peebles and R. H. Dicke, Astrophys. J. **154**, 898 (1968).

[303]   J. H. Piddington, Monthly Notices Roy. Astron. Soc. **162**, 73 (1973).

[304]   J. H. Piddington, Astrophys. J. **179**, 755 (1973).

[305]   H. C. Plummer, Monthly Notices Roy. Astron. Soc. **71**, 460 (1911).

[306]   K. H. Prendergast, Astron. J. **69**, 147 (1964).

[307]   K. H. Prendergast and E. Tomer, Astron. J. **75**, 674 (1970).

[308]   W. J. Quirk, Astrophys. J. **167**(1), 7 (1971).

[309]   R.-G. Rohm, Ph.D. thesis, MIT, Cambridge, Mass., 1965.

[310]   P. H. Roberts and K. Stewartson, Astrophys. J. **137**(3), 777 (1963).

[311]   W. W. Roberts, Astrophys. J. **158**, 123 (1969).

[312]   W. W. Roberts, M. S. Roberts, and F. H. Shu, Astrophys. J. **196**, 381 (1975).

[313]   M. N. Rosenbluth, N. Krall, and N. Rostocker., Nucl. Fusion, Suppl. **2**, 143 (1962).

[314]   G. W. Rougoor, Bull. Astron. Inst. Netherl. **17**, 318 (1964).

[315]   V. C. Rubin and W. K. Ford, Astrophys. J. **159**(2), 379 (1970).

[316]   H. N. Russel, *Astronomy, Part 1*, 1926.

[317]   A. Sandage, *The Hubble Atlas of Galaxies*, Carnegie Inst., Washington, 1961.

[318]   A. Sandage, K. C. Freeman, and N. R. Stokes, Astrophys. J. **160**, 831 (1970).

[319]   M. Schmidt, in *Galactic Structure*, A. Blaauw and M. Schmidt, eds., University of Chicago Press, Chicago, 1965.

[320]   W. W. Shane and G. P. Bieger-Smith, Bull. Astron. Netherl. **18**, 263 (1966).

[321]   H. Shapley and H. B. Sawyer, Harv. Obs. Bull. No. 852 (1927).

[322]   F. H. Shu, Astron. J. **73**(10), 201 (1968).

[323]   F. H. Shu, Ph.D. thesis, Harvard University Press, Cambridge, Mass., 1968.

[324]   F. H. Shu, Astrophys. J. **160**(1), 89 (1970).

[325]   F. H. Shu, Astrophys. J. **160**(1), 99 (1970).

[326]   F. H. Shu, R. V. Stachnic, and J. C. Yost, Astrophys. J. **166**(3), 465 (1971).

[327]   E. A. Spiegel, *Symp. Origine Syst. Solaire, Nice, 1972*, Paris, 1972.

[328]   P. Strömgren, in *Proc. IAU Symp. No. 31*, Noordwick, 1966.

[329]   P. Strömgren, in IAU Symp. *No. 31*, D. Reidel, Dordrecht, 1967.

[330]   P. Sweet, Monthly Notices Roy. Astron. Soc. **125**, 285 (1963).

[331]   A. Toomre, Lectures in Geophysical Fluid Dynamics at the Woods Hole Oceanographic Institution, 1966.

[332]   A. Toomre, Astrophys. J. **138**, 385 (1963).

[333]   A. Toomre, Astrophys. J. **139**(4), 1217 (1964).

[334]   A. Toomre, Astrophys. J. **158**, 899 (1969).

[335]   S. D. Tremaine, preprint, California Inst. Technol., Pasadena, 1976.

[336]   P. O. Vandervoort, Astrophys. J. **147**(1), 91 (1967).

[337]   P. O. Vandervoort, Mem. Soc. Roy. Sci. Liege **15**, 209 (1967).

[338]   P. O. Vandervoort, Astrophys. J. **161**, 67, 87 (1970).

[339]   G. de Vaucouleurs, Mem. Mt. Stromlo Obs. **111**(3), (1956).

[340]   G. de Vaucouleurs, Astrophys. J. Suppl. **8**(76), 31 (1963).

[341]   G. de Vaucouleurs, Rev. Popular Astron. **57**(520), 6 (1963).

[342]   G. de Vaucouleurs, Astrophys. J. Suppl. **8**(74), 31 (1964).

[343]   G. de Vaucouleurs, A. de Vaucouleurs, and K. C. Freeman, Monthly Notices Roy. Astron. Soc. **139**(4), 425 (1968).

[344]   G. de Vaucouleurs and K. C. Freeman, Vistas Astron. **14**, 163 (1973).

[345]   L. Volders, Bull. Astron. Netherl. **14**, 323 (1959).

[346]   H. Weaver, in *IAU Symp. No. 38*, D. Reidel, Dordrecht, 1970.

[347]   C. F. Von Weizsäcker, Z. Astrophys. **22**, 319 (1944).

[348]   C. F. Von Weizsäcker, Naturwiss. **33**, 8 (1946).

[349]   G. Westerhout, Bull. Astron. Inst. Netherl. **14**, 215 (1958).

[350]   R. Wielen, Astron. Rechen-Inst., Heidelberg Mitt. Ser. A, No. 47, (1971).

[351]   C. P. Wilson, Astron. J. **80**, 175 (1975).

[352]   R. van der Wooley, Monthly Notices Roy. Astron. Soc. **116**(3), 296 (1956).

[353]   R. van der Wooley, Observatory, **81**(924), 161 (1961).

[354]   C.-S. Wu, Phys. Fluids **11**(3), 545 (1968).

[355]   A. B. Wyse and N. U. Mayall, Astrophys. J. **95**, 24 (1942).

[356]   S. Yabushita, Monthly Notices Roy. Astron. Soc. **143**(3), (1969).

[357]   S. Yabushita, Monthly Notices Roy. Astron. Soc. **133**(3), 247 (1966).

[358]   S. Yabushita, Monthly Notices Roy. Astron. Soc. **142**(2), 201 (1969).

[359]   C. Yuan, Astrophys. J. **158**(3), 871 (1969).

[360]   C. Yuan, Astrophys. J. **158**(3), 889 (1969).

Additional References

[1] L. M. Al'tshul', Dep. No. 50295, VINITI, 1972.

[2] N. N. Bogolyubov and Yu. A. Mitropol'skiy, *Asymptotic Methods in the Theory of Nonlinear Oscillations*, M., Nauka, Moscow, 1974.

[3] N. P. Buslenko, *Statistical Test Method (the Monte-Carlo Method)*, SMB, M., Fizmatgiz, Moscow 1962.

[4] Yu.-I. K. Veltmann, Trudy Astrofiz. Inst. AN Kaz. SSR **5**, 57 (1965).

[5] Yu.-I. K. Veltmann, Publications of the Tartusk. Astr. Observ. **34**, 101 (1964); **35**, 344, 356 (1966).

[6] B. I. Davydov, Dokl. AN SSSR **69**, 165 (1949).

[7] B. P. Demidovich, I. A. Maron and E. Z. Shuvalova, *Numerical Analysis Methods*. GIFML, M., 1963.

[8] V. I. Dokuchayev and L. M. Ozernoy, Preprint FIAN im. P. N. Lebedev, No. 133, S.; ZhETF, **73**, 1587 (1977); Letters to Astron. Zh. **3**, 391 (1977).

[9] S. M. Yermakov and G. A. Mikhaylov, *Course of Statistical Modeling*, M., Nauka, Moscow 1976.

[10] V. Ye. Zakharov, ZhETF **60**, 1713 (1971).

[11] V. Ye. Zakharov, Izvestiya vyzov. Radiofizika **17**, 431.

[12] V. Ye. Zakharov, PMTF No. 2, 86 (1968).

[13] V. Ye. Zakharov, ZhETF **62**, 1945 (1972).

[14] G. M. Idlis, Astron. Zh. **33** (1), 53 (1956).

[15] B. B. Kadomtsev, *Collective Phenomena in Plasma*, M., Nauka, Moscow, 1976.

[16] B. P. Kondrat'yev and L. M. Ozernoy, Letters to Astron. Zh. **5**, 67 (1979).

[17]   V. I. Korchagin and L. S. Marochnik, Astron. Zh. **52**(4), 700 (1975).

[18]   G. G. Kuzmin and Yu.-I. K. Veltmann, Publ. Tartusk. Astr. Observ. **36**, 3, 470 (1968).

[19]   G. G. Kuzmin and Yu.-I. K. Veltmann, Coll: "*Dynamics of Galaxies and Stellar Clusters*," 1973, Alma-Ata, Nauka, Moscow, p. 82.

[20]   A. B. Mikhaylovskiy, V. I. Petviashvili and A. M. Fridman, Letters to ZhETF **26**, 341 (1977).

[21]   L. S. Marochnik, Astrofizika **5**, 487 (1969).

[22]   A. G. Morozov and I. G. Shukhman, Letters to Astron. Zh. **6**, 87 (1980).

[23]   A. G. Morozov, Letters to Astron. Zh. **3**, 195 (1977).

[24]   A. G. Morozov and A. M. Fridman, *Report at the All-Union Conference "Latent Mass in the Universe*," Tallin, January, 1975.

[25]   A. G. Morozov, Astron. Zh **56**, 498 (1979).

[26]   S. N. Nuritdinov, Author's abstract of Thesis, Leningrad, 1975, Astrofizika **11**, 135 (1975).

[27]   L. N. Osipkov, Letters to Astron. Zh. **5**, 77 (1979).

[28]   V. L. Polyachenko, Letters to Astron. Zh. **3**, 99 (1977).

[29]   V. L. Polyachenko and A. M. Fridman, Letters to Astron. Zh. **7**, 136 (1981).

[30]   V. L. Polyachenko and I. G. Shukhman, Letters to Astron. Zh. **3**, 199 (1977).

[31]   V. L. Polyachenko, S. M. Churilov and I. G. Shukhman, Preprint SibIZMIR SO AN SSSR, No. 1–79, Irkutsk, 1979; Astron. Zh. **57**, 197 (1980).

[32]   V. L. Polyachenko and I. G. Shukhman, Astron. Zh. **56**(5), 957 (1979).

[33]   V. L. Polyachenko and I. G. Shukhman, Letters to Astron. Zh. **3**(6), 254 (1977).

[34]   V. L. Polyachenko and I. G. Shukhman, Preprint SibIZMIR SO AN SSSR, No. 31–78, Irkutsk, 1978.

[35]   V. L. Polyachenko and I. G. Shukhman, Astron. Zh. **57**(2), 268 (1980).

[36]   V. L. Polyachenko, Letters to Astron. Zh. (1983), to appear.

[37]   V. L. Polyachenko and I. G. Shukhman, Astron. Zh. **56**(4), 724 (1979).

[38]   V. L. Polyachenko, Astron. Zh. **56**, 1158 (1979).

[39]   V. L. Polyachenko and A. M. Fridman, Astron. Zh., (1983), to appear.

[40]   V. L. Polyachenko and I. G. Shukhman, Preprint SibIZMIR SO AN SSSR, No. 1–78, Irkutsk, 1978.

[41]   V. L. Polyachenko, Letters to Astron. Zh. (1983), to appear.

[42]   V. L. Polyachenko, Letters to Astron. Zh. (1983), to appear.

[43]   V. L. Polyachenko, Letters to Astron. Zh. **7**(3), 142 (1981).

[44]   V. L. Polyachenko and I. G. Shukhman, Astron. Zh. **58**, 933 (1981).

[45]   Yu. M. Rozenraukh, Thesis, IGU, Irkutsk, 1977.

[46]   R. Z. Sagdeev, Vopr. Teor. Plazmy (Plasma Theory Problems) Ed. M. A. Leontovich, No. 4, M., Atomizdat, 1963.

[47]   M. A. Smirnov and B. V. Komberg, Letters to Astron. Zh. **4**, 245 (1978).

[48]   I. M. Sobol', *Numerical Monte-Carlo Methods*, M., Nauka, Moscow, 1973.

[49]   A. M. Fridman, Uspekhi fiz. nauk **125**, 352 (1978).

[50]   A. M. Fridman, Letters to Astron. Zh. **4**, 243 (1978).

[51]   A. M. Fridman, Letters to Astron. Zh. **4**, 207 (1978).

[52]   A. M. Fridman, Letters to Astron. Zh. **5**, 325 (1979).

[53]   S. M. Churilov and I. G. Shukhman, Astron. Zh. **58**, 260 (1981); **59**, 1093 (1982).

[54]   V. D. Shapiro and V. I. Shevchenko, ZhETF **45**, 1612 (1963).

[55]   J. N. Bahcall, Astrophys. J. **209**, 214 (1976).

[56]   J. M. Bardeen, in *IAU Symposium No. 69*, D. Reidel, Dordrecht, 1975.

[57]   F. Bertola and M. Capaccioli, Astrophys. J. **219**, 404 (1978).

[58]   J. Binney, Monthly Notices Roy. Astron. Soc. **177**, 19 (1976).

[59]   M. Clutton-Brock, Astrophys. Space Sci. **16**, 101 (1972).

[60]   G. Contopoulos, Astron. Astrophys. **64**, 323 (1978).

[61]   M. J. Dunkan and J. C. Wheeler, preprint, Astrophys. J. Lett. (1980), Dept. Astron., Univ. Texas, Austin, 1979.

[62]   J. Frank and M. J. Rees, Monthly Notices Roy. Astron. Soc. **176**, 633 (1976).

[63]   A. M. Fridman, preprint, Inst. Zemn, Magn. Ionosf. Rasprostr. Radiovoln Sibir. Otd. Akad. Nauk SSSR, 6–78, Irkutsk, 1978.

[64]   A. M. Fridman, Y. Palous and I. I. Pasha, Monthly Notices Roy. Astron. Soc. **194**, 705 (1981).

[64a] A. M. Fridman and V. L. Polyachenko, Zh ETP **81**, 13 (1981).

[65]   P. B. Globa-Mikhailenko, J. Math. (7 serie) **II**, 1 (1916).

[66]   P. Goldreigh and S. Tremaine, Astrophys. J. **222**, 850 (1978).

[67]   F. Hohl, in *IAU Symposium No. 69*, D. Reidel, Dordrecht, 1975.

[68]   C. Hunter, Astrophys. J. **181**, 685 (1973).

[69]   C. Hunter, Astron. J. **82**, 271 (1977).

[70]   S. Ikeuchi, Progr. Theor. Phys. **57**, 1239 (1977).

[71]   G. Illingworth, Astrophys. J. Lett. **218**, L43 (1977).

[72]   A. J. Kalnajs, Proc. Astron. Soc. Austr. **2**, 174 (1973).

[73]   A. J. Kalnajs, Astrophys. J. **205**, 745, 751 (1976).

[74]   A. J. Kalnajs, Astrophys. J. **212**, 637 (1977).

[75]   J. Katz, Monthly Notices Roy. Astron. Soc. **183**, 765 (1978).

[76]   I. R. King, Astron. J. **71**, 64 (1966).

[77]   I. R. King, in *IAU Symposium No. 69*, D. Reidel, Dordrecht, 1975.

[78]   A. P. Lightman and S. L. Shapiro, Astrophys. J. **211**, 244 (1977).

[79]   D. Lynden-Bell, *C.N.R.S. International Colloquium No. 241*, Centre Nat. de la Rech. Sci., Paris, 1975.

[80]   R. H. Miller, J. Comput. Phys. **21**, 400 (1976).

[81]   R. H. Miller, Astrophys. J. **223**, 122 (1978).

[82]   L. M. Ozernoy and B. P. Kondrat'ev, Astron. Astrophys. **79**, 35 (1979).

[83]   P. J. E. Peebles, Astron. J. **75**, 13 (1970).

[84]   P. J. E. Peebles, Gen. Relativity Gravity, **3**, 63 (1972).

[85]   C. J. Peterson, Astrophys. J. **222**, 84 (1978).

[86]   W. L. W. Sargent, P. J. Young, A. Boksenberg, K. Shortrigge, C. R. Lynds, and F. D. A. Hartwick, Astrophys. J. **221**, 731 (1978).

[87]   L. Schipper and I. R. King, Astrophys. J. **220**, 798 (1978).

[88]   A. Toomre, Annu. Rev. Astron. Astrophys. **15**, 437 (1977).

[89]   P. J. Young, W. L. W. Sargent, A. Boksenberg, C. R. Lynds, and F. D. A. Hartwick, Astrophys. J. **222**, 450 (1978).

[90]   P. J. Young, J. A. Westphal, J. Kristian, C. P. Wilson, and F. P. Landauer, Astrophys. J. **221**, 721 (1978).

[91]   T. A. Zang, Ph.D. thesis, M.I.T., Cambridge, Mass., 1976.

[92]   M. Nishida and T. Ishizawa, 1976, preprint, Kyoto Univ.

[93]   M. Abramowitz and I. A. Stegun, eds., *Handbook of Mathematical Function*, 1964.

[94]   P. L. Schechter and J. E. Gunn, Astrophys. J. **229**, 472 (1979).

[95]   J. R. Cott, Astrophys. J. **201**, 2961 (1975).

[96]   R. B. Larson, Monthly Notices Roy. Astron. Soc. **173**, 671 (1975).

[97]   G. Illingworth, Astrophys. J. **43**, 218 (1977).

[98]   J. J. Binney, Monthly Notices Roy. Astron. Soc. **183**, 501 (1978).

[99]   J. J. Binney, Monthly Notices Roy. Astron. Soc. **190**, 421 (1980).

[100]   T. C. Chamberlin, Astrophys. J. **14**, 17 (1901).

[101]   E. Holmberg, Astrophys. J. **94**, 385 (1941).

[102]   J. Pfleiderer and H. Siedentopf, Z. Astrophys. **51**, 201 (1961).

[103]   J. Pfleiderer, Z. Astrophys. **58**, 12 (1963).

[104]   N. Tashpulatov, Astron. Zh. **46**, 1236 (1969).

[105]   A. Toomre, in *IAU Symposium No. 79*, D. Reidel, Dordrecht, 1977.

[106]   S. Yabushita, Monthly Notices Roy. Astron. Soc. **153**, 97 (1971).

[107]   F. Zwicky, Naturwissenschaften, **26**, 334 (1956).

[108]   N. N. Kozlov, T. M. Eneev, and R. A. Syunyaev, Dokl. Akad. Nauk SSSR **204**, 579 (1972).

[109]   T. M. Eneev, N. N. Kozlov, and R. A. Syunyaev, Astron. Astrophys. **22**, 41 (1973).

[110]   A. Toomre, *IAU Symposium No. 38*, D. Reidel, Dordrecht, 1970.

[111]   A. M. Fridman, *IAU Symposium No. 79*, D. Reidel, Dordrecht, 1977.

[112]   S. M. Churilov and I. G. Shukhman, Astron. Tsirk., No. 1157 (1981).

[113]   L. W. Esposito, J. P. Dilley, and J. W. Fountain, J. Geophys. Res. **85** (A11), 5948 (1980).

[114]   P. Goldreich and S. D. Tremaine, Icarus **34**, 227 (1978).

[115]   S. S. Kumar, Publ. Astron. Soc. Japan **12**, 552 (1960).

[116]   R. H. Sanders and G. T. Wrikon, Astron. Astrophys., **26**, 365 (1973).

[117]   G. Chew, M. Goldberger and F. Low, Proc. Roy. Soc. A, **236**, 112 (1956).

[118]   A. G. Doroshkevich, A. A. Klypin, preprint IPM, No. 2 (1980).

[119] A. G. Morozov, Astron. Zh. **57**, 681 (1980).

[120] A. Lane *et al.*, Science **215**, No. 4532 (1982).

[121] N. C. Lin and P. Bodenheimer, Astrophys. J. Letters **248**, L83 (1981).

[122] G. W. Null, E. L. Lau, and E. D. Biller, Astron. J. **86**, 456 (1981).

[123] V. L. Polyachenko, and A. M. Fridman, Astron. Tsirk., No. 1204 (1981).

[124] Voyager Bulletin, Mission status report, No. 57, Nov. 7, NASA (1980).

[125] W. R. Ward, Geophys. Res. Lett. **8**, No. 6 (1981).

[126] B. B. Kadomtsev, Letters to ZhETP **33**(7), 361 (1981).

[127] I. V. Igumenshev, Thesis, ChGU, Chelyabinsk (1982).

[128] V. A. Ambartsumian, The Scientific Papers, v. 1, Erevan, 1960.

[129] V. A. Ambartsumian, Astron. Zh. **14**, 207 (1937).

[130] M. G. Abrahamian, Astrophysics **14**, 579 (1978).

Index

The page numbers which appear in upright figures refer to Volume I and those which appear in italic refer to Volume II.